AF371690

Global Calculus

Global Calculus

S. Ramanan

Graduate Studies
in Mathematics

Volume 65

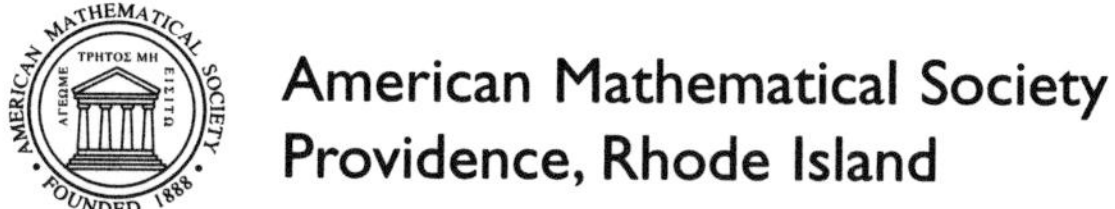

American Mathematical Society
Providence, Rhode Island

2000 *Mathematics Subject Classification.* Primary 14-01, 32-01, 53-01;
Secondary 32Lxx, 32Qxx, 32Wxx, 53Cxx.

For additional information and updates on this book, visit
www.ams.org/bookpages/gsm-65

Library of Congress Cataloging-in-Publication Data
Ramanan, S.
    Global calculus / S. Ramanan.
        p. cm. — (Graduate studies in mathematics, ISSN 1065-7339 ; v. 65)
    Includes bibliographical references and index.
    ISBN 0-8218-3702-8 (alk. paper)
    1. Geometry, Algebraic.   2. Differential operators.   3. Analytic spaces.   4. Differential geom-
etry.   I. Title.   II. Series.

QA564.R34   2004
515′.94—dc22

                                                                        2004057074

Contents

Preface

This book is intended for postgraduate and ambitious senior undergraduate students. The key word is Differential Operators. I have attempted to develop the calculus of such operators in an uncompromisingly global setup.

I have tried to make the book as self-contained as possible. I do make heavy demands on the algebraic side, but at least recall the required results in a detailed way in an appendix.

This book has had an unusually long gestation period. It is incredible for me to realise that the general tone of the book is very much the same as that of a course which I gave way back in 1970 at the Tata Institute, which I (hopefully) improved upon a couple of years later, and expanded into a two quarter graduate course at the University of California, Los Angeles, in 1979-80. The encouragement that the audience gave me, especially Nagisetty Venkateswara Rao (now at the University of Ohio, Toledo), my first student the late Annamalai Ramanathan, and Jost at UCLA, was quite overwhelming. The decision to write it all up was triggered by a strong suggestion of Stefan Mueller-Stach (now at Mainz University). He drove me from Bayreuth to Trieste in Italy, and I used the occasion to clear some of his doubts in mathematics. Apparently happy at my effort, he suggested that I had a knack for exposition and should write books. I took this for more than ordinary politeness, and embarked on this project nearly ten years later.

Chanchal Kumar, Amit Hogadi and Chaitanya Guttikar enthusiastically read portions of the book and made many constructive suggestions. Madhavan of the Chennai Mathematical Institute and Nandagopal of the Tata Institute of Fundamental Research helped with the figures. Ms. Natalya

Pluzhnikov of the American Mathematical Society took extraordinary care in finalising the copy. All through, and particularly in the last phase, my wife Anu's support, physical and psychological, was invaluable.

I utilised the hospitality of various institutions during the course of writing. Besides the Tata Institute of Fundamental Research, my Alma Mater, I would like to mention specially the Institute of Mathematical Sciences at Chennai, the International Centre for Theoretical Physics, Trieste, and at the final stage, Consejo Superior de Investigaciones Científicas in Madrid.

I wish to thank all the individuals and institutions for the help received.

I collaborated for long years with M. S. Narasimhan. I wish to take this opportunity to acknowledge the exciting time that I spent doing mathematical research with him. Specifically, the idea of what I call the composition formula in the last chapter of this book, arose in our discussions. Narasimhan also suggested appropriate references for Chapter 8.

I learnt modern Differential Geometry from Jean-Louis Koszul. His lectures at the Tata Institute, of which I took notes, were an epitome of clarity. I wish to thank him for choosing to spend time in his youth to educate students in what must have seemed at that time the outback.

There is a short summary at the start of each chapter explaining the contents. Here I would like to draw attention to what I think are the new features in my treatment.

In Chapter 1 sheaves make their appearance before differential manifolds. Everyone knows that this is the 'correct' definition but I know of few books that have adopted this point of view. The reason is that generally sheaves are somehow perceived to be more difficult to swallow at the outset. I believe otherwise. In my experience, if sufficient motivation is provided and many illustrative examples given, students take concepts in their stride, and will in fact be all the better equipped in their mathematical life for an early start.

In Chapter 2 I have introduced the notion of the Connection Algebra and believe most computations can, and ought to be, made in this algebra.

In Chapter 3 the treatment of densities and orientation are, I believe, nonconventional. In particular, the change-of-variable formula is not used but is in fact a simple consequence of this approach.

Chapter 4 is a fairly straightforward account of sheaf cohomology.

Connections are treated as tools to lift symbols to differential operators; various tensors connected with connections and linear connections have a natural interpretation from this point of view. These are worked out in Chapter 5. The existence of torsion free linear connections compatible with various structures lead to natural integrability conditions on them. This treatment, which I believe is new, is given in Chapter 6.

The additional structures are themselves studied in Chapter 7 with some emphasis on naturally occurring differential operators.

Chapter 8 is an account of the local theory of elliptic operators, while Chapter 9 contains the composition formula I mentioned earlier. A general vanishing theorem for harmonic sections of an elliptic complex is proved here under a suitable curvature hypothesis. The Bochner, Lichnerowicz and Kodaira vanishing theorems are derived as special cases, followed by a short account of how Kodaira's vanishing theorem leads to the imbedding theorem.

Mistakes, particularly relating to signs and constants, are a professional hazard. I would be grateful to any reader who takes trouble to inform me of such and other errors, misleading remarks, obscurities, etc.

Chapter 1

Sheaves and Differential Manifolds: Definitions and Examples

In Geometry as well as in Physics, one has often to use the tools of differential and integral calculus on topological spaces which are locally like open subsets of the Euclidean space $\mathbb{R}^n$, but do not admit coordinates valid everywhere. For example, the sphere $\{(x, y, z) \in \mathbb{R}^3 : x^2 + y^2 + z^2 = 1\}$, or more generally, the space $\{(x_1, \ldots, x_n) \in \mathbb{R}^n : \sum x_i^2 = 1\}$ are clearly of geometric interest. On the other hand, constrained motion has to do with dynamics on surfaces in $\mathbb{R}^3$. In general relativity, one studies 'space-time' which combines the space on which motion takes place and the time parameter in one abstract space and allows reformulation of problems of Physics in terms of a 4-dimensional object. All these necessitate a framework in which one can work with the tools of analysis, like differentiation, integration, differential equations and the like, on fairly abstract objects. This would enable one to study Differential Geometry in its appropriate setting on the one hand, and to state mathematically the equations of Physics in the required generality, on the other. The basic objects which accomplish this are called *differential manifolds*. These are geometric objects which are locally like domains in the Euclidean space, so that the classical machinery of calculus, available in $\mathbb{R}^n$, can be transferred, first, to small open sets and then patched together. The main tool in the patching up is the notion of a *sheaf*. We will first define sheaves and discuss basic notions related to them, before taking up differential manifolds.

1. Sheaves and Presheaves

At the outset, it is clear that functions of interest to us belong to a class such as continuous functions, infinitely differentiable (or C^∞) functions, real analytic functions, holomorphic functions of complex variables, and so on. Some properties common to all these classes of functions are that

   i) they are all continuous, and

   ii) the condition for a continuous function to belong to the class is of a *local* nature.

By this we mean that for a continuous function to be differentiable, real analytic or holomorphic, it is necessary and sufficient for it to be so in the neighbourhood of every point in its domain of definition. We will start with an axiomatisation of the local nature of the classes of functions we seek to study.

1.1. Definition. Let X be a topological space. An assignment to every open subset U of X, of a set $\mathcal{F}(U)$ and to every pair of open sets U, V with $V \subset U$, of a map (to be called *restriction map*) $\mathrm{res}_{UV} : \mathcal{F}(U) \to \mathcal{F}(V)$ satisfying

$$\mathrm{res}_{VW} \circ \mathrm{res}_{UV} = \mathrm{res}_{UW}$$

for every triple $W \subset V \subset U$ of open sets, is called a *presheaf* of sets. If $\mathcal{F}(U)$ are all abelian groups, rings, vector spaces, ... and the restriction maps are homomorphisms of the respective structures, then we say that $\mathcal{F}$ is a presheaf of abelian groups, rings, vector spaces, ...

1.2. Definition. A presheaf is said to be a *sheaf* if it satisfies the following additional conditions. Let $U = \bigcup_{i \in I} U_i$ be any open covering of an open set U. Then

$\mathbf{S_1}$: Two elements $s, t \in \mathcal{F}(U)$ are equal if $\mathrm{res}_{UU_i} s = \mathrm{res}_{UU_i} t$ for all $i \in I$.

$\mathbf{S_2}$: If $s_i \in \mathcal{F}(U_i)$ satisfy $\mathrm{res}_{U_i U_i \cap U_j} s_i = \mathrm{res}_{U_j U_i \cap U_j} s_j$ for all $i, j \in I$, then there exists an element $s \in \mathcal{F}(U)$ with $\mathrm{res}_{UU_i} s = s_i$ for all i. We will also assume that $\mathcal{F}(\emptyset)$ consists of a single point.

1.3. Examples.

1) As indicated above, the concepts of differentiability, real analyticity, ... are all local in nature, so that it is no surprise that if X is an open subspace of $\mathbb{R}^n$, then the assignment to every open subset U of X, of the set $\mathcal{A}(U)$ of differentiable functions with the natural restriction of functions as restriction maps, gives rise to a sheaf. This sheaf will be called the *sheaf of differentiable functions* on X. Obviously this sheaf is not merely a sheaf of sets, but a sheaf of $\mathbb{R}$-algebras.

2) The assignment of the set of bounded functions to every open set U of X, and the natural restriction maps define a presheaf on X. However, if we are given as in $\mathbf{S}_2$ any compatible set of bounded functions s_i, then while such a data does define a unique function s on U, there is no guarantee that it will be bounded. Thus this presheaf satisfies $\mathbf{S}_1$ but not $\mathbf{S}_2$ and so is not a sheaf. (Can one modify this in order to obtain a sheaf?)

3) Consider the assignment of a fixed abelian group to every nonempty open set U, all restriction maps being identity. We will also assign the trivial group to the empty set. This defines a presheaf, but is not a sheaf in general. (Why?)

4) The *standard n-simplex* Δ_n is defined to be $\{(x_0, x_1, \ldots, x_n) \in \mathbb{R}^{n+1} : \sum x_i = 1, x_i \geq 0\}$ with the induced topology. In algebraic topology, one way of studying the topology of a space X is to look at continuous maps of the standard simplices into X and studying their geometry. We will give the basic definitions here. A *singular n-simplex* in a topological space X is a continuous map of the standard n-simplex Δ_n into X. If A is a fixed abelian group, then an A-valued *singular cochain* in X is an assignment of an element of A to every singular simplex. Now, to every open subset U of X, associate the set $\mathcal{S}^n(U)$ of all singular cochains in U. If V is an open subset of U, then we have an obvious inclusion of the set of singular simplices in V into that in U. Consequently there is also a restriction map $\mathcal{S}^n(U) \to \mathcal{S}^n(V)$. This makes $\mathcal{S}^n$ a presheaf, which we may call the *presheaf of singular cochains in X*. It is obvious that this does *not* satisfy the axiom $\mathbf{S}_1$ for sheaves. For if $X = \bigcup U_i$ is a nontrivial open covering, then we can define a nonconstant cochain which is zero on all simplices whose images are contained in some U_i. On the other hand, if a cochain is defined on simplices with images in some U_i, then one can extend this cochain to all singular simplices by defining the cochain to be zero on simplices whose images are not contained in any of the U_i. Thus this presheaf satisfies $\mathbf{S}_2$.

5) The most characteristic example of a sheaf from our point of view is the one which associates to any open set U the algebra of continuous functions on U. This can be generalised as follows. Let Y be any fixed topological space and consider the assignment to any open set U, of the set of continuous maps from U to Y with the obvious restriction maps. This defines a sheaf of sets, the *sheaf of continuous maps into Y*.

6) We will generalise this example a little further. Let E be a topological space and $\pi : E \to X$ a continuous surjective map. Then associate to each open subset U of X the set of continuous *sections* of π over U (namely, continuous maps $\sigma : U \to E$ such that $\pi \circ \sigma = \mathrm{Id}_U$). This, together with the obvious restriction maps, is a sheaf called the *sheaf of sections of π*. Example 5) is obtained as a particular case on taking $E = Y \times X$ and π to be the second projection.

We will see below that every sheaf arises in this way, that is to say, to every sheaf $\mathcal{F}$ on X, one can associate a space $E = E(\mathcal{F})$ and a map $\pi : E \to X$ as above such that the sheaf of sections of π may be identified with $\mathcal{F}$.

1.4. Sheaf associated to a presheaf.

In the theory of holomorphic functions, one talks of a *germ* of a function at a point, when one wishes to study its properties in an (unspecified) neighbourhood of a point. This means the following. Consider pairs (U, f) consisting of open sets U containing the given point x and holomorphic functions f defined on U. Introduce an equivalence relation in this set by declaring two such pairs (U, f), (V, g) to be *equivalent* if f and g coincide in some neighbourhood of x which is contained in $U \cap V$. An equivalence class is called a *germ*. This procedure can be imitated in the case of a presheaf and this leads to the concept of a *stalk* of a presheaf at a point.

1.5. Definition. Let $\mathcal{F}$ be a presheaf on a topological space X. Then the *stalk $\mathcal{F}_x$* of $\mathcal{F}$ at a point $x \in X$ is the quotient set of the set consisting of all pairs (U, s) where U is an open neighbourhood of x and s is an element of $\mathcal{F}(U)$ under the equivalence relation:

(U, s) is equivalent to (V, t) if and only if there exists an open neighbourhood W of x contained in $U \cap V$ such that the restrictions of s and t to W are the same.

If $s \in \mathcal{F}(X)$ then, for any $x \in X$, the pair (X, s) has an image in the stalk $\mathcal{F}_x$, namely the equivalence class containing it. It is called the *germ* of s at x. We will denote it by s_x.

Let $E = E(\mathcal{F})$ be the germs of all elements at all points of X, that is to say, the disjoint union of all the stalks $\mathcal{F}_x$, $x \in X$. What we intend to do now is to provide the set E with a topology such that

a) the map $\pi : E \to X$ which maps all of $\mathcal{F}_x$ to x, is continuous;

b) if $s \in \mathcal{F}(U)$ then the section $\tilde{s}$ of E over U, defined by setting $\tilde{s}(x) = s_x$, is continuous.

We achieve this by associating to each pair (U, s), where U is an open subset of X and $s \in \mathcal{F}(U)$, the set $\tilde{s}(U)$, and defining a topology on E whose open sets are generated by sets of the form $\tilde{s}(U)$. In order to check that with this topology, the maps $\tilde{s}$ are continuous, we have only to show that $\tilde{s}^{-1}(\tilde{t}(V))$ is open for every open subset V of X and $t \in \mathcal{F}(V)$. This is equivalent to the following

1.6. Lemma. *If $s \in \mathcal{F}(U)$ and $t \in \mathcal{F}(V)$, then the set of points $x \in U \cap V$ such that $s_x = t_x$ is open in X.*

Proof. From the definition of the equivalence relation used to define $\mathcal{F}_x$, we deduce that if $s_a = t_a$ for some point $a \in X$, then the restrictions of s and t to some open neighbourhood N of a are the same. Hence we must have $s_x = t_x$ for every point x of N as well. Thus N is contained in $\{x \in U : s_x = t_x\}$, proving the lemma.

Finally if U is an open set in X, then $\pi^{-1}(U) = \bigcup_{x \in U} \mathcal{F}_x = \bigcup \tilde{s}(V)$ where the latter union is over all open subsets $V \subset U$ and all $s \in \mathcal{F}(V)$. This shows that $\pi^{-1}(U)$ is open in E and hence that π is continuous.

1.7. Definition. The *étale space* associated to the presheaf $\mathcal{F}$ is the set $E(\mathcal{F}) = \bigcup_{x \in X} \mathcal{F}_x$ provided with the topology generated by the sets $\tilde{s}(U)$, where U is any open set in X and $s \in \mathcal{F}(U)$. Moreover, if π is the natural map $E(\mathcal{F}) \to X$ which has $\mathcal{F}_x$ as fibre over x for all $x \in X$, then the sheaf of sections of π is called the *sheaf associated to the presheaf $\mathcal{F}$*.

1.8. Remarks.

1) If we look at the fibre of π over $x \in X$, namely $\mathcal{F}_x$, we see that the topology induced on it is nothing serious, since $\tilde{s}(U)$ (being a section of π), intersects $\mathcal{F}_x$ only in one point, namely s_x. In other words, the induced topology on the fibres is discrete.

2) Consider the constant presheaf $\mathcal{A}$ defined by the abelian group A. Clearly the stalk at any point $x \in X$ is again A so that E can in this case be identified with $A \times X$. Any $a \in A$ gives rise to an element of $\mathcal{A}(U)$ for every open subset U of X. The corresponding section $\tilde{a}$ over U of $\pi : E = A \times X \to X$ is given simply by $x \mapsto (a, x)$ over U. Hence the image, namely $\{a\} \times U$ is open in E. From this one easily concludes that the topology on $E = A \times X$ is the product of the discrete topology on A and the given topology on X. Hence its sections over any open set U are continuous maps of U into the discrete space A. This is the same as locally constant maps on U with values in A.

3) Notice that in the construction of the étale space we did not make full use of the data of a presheaf. It is enough if we are given $\mathcal{F}(U)$ for open sets U running through a base of open sets.

1.9. Definition. If $\mathcal{F}_1, \mathcal{F}_2$ are presheaves on a topological space X, then a *homomorphism* $f : \mathcal{F}_1 \to \mathcal{F}_2$ is an association to each open subset U of X of a homomorphism $f(U) : \mathcal{F}_1(U) \to \mathcal{F}_2(U)$ such that whenever U, V are open sets with $V \subset U$, we have a commutative diagram

$$
\begin{array}{ccc}
\mathcal{F}_1(U) & \xrightarrow{\ f(U)\ } & \mathcal{F}_2(U) \\
\downarrow{\scriptstyle \mathrm{res}_{UV}} & & \downarrow{\scriptstyle \mathrm{res}_{UV}} \\
\mathcal{F}_1(V) & \xrightarrow{\ f(V)\ } & \mathcal{F}_2(V)
\end{array}
$$

When we consider presheaves of abelian groups, rings, ... , and talk of homomorphisms of such sheaves, we require that all the maps $f(U)$ be homomorphisms of these structures.

1.10. Examples.

1) The inclusion of the set of differentiable functions (on a domain in $\mathbb{R}^n$) in the set of continuous functions is clearly a homomorphism of the sheaf of differentiable functions into that of continuous functions.

2) Consider the sheaf of differentiable functions in a domain of $\mathbb{R}^n$. The map $f \mapsto \frac{\partial f}{\partial x_i}$ induces a sheaf homomorphism of the sheaf of differentiable functions into itself, the homomorphism being one of sheaf of abelian groups but *not* of rings.

We will rephrase this as follows. Let V be a (finite-dimensional) vector space over $\mathbb{R}$. Then one has a natural sheaf of differentiable functions on V. One can for example choose a linear isomorphism of V with $\mathbb{R}^n$ and consider functions of V as functions of the coordinate variables $x_1, x_2, \ldots, x_n$. Then differentiability makes sense, independent of the isomorphism chosen. For let $y_1, y_2, \ldots, y_n$ is a set of variables obtained by some other isomorphism of V with $\mathbb{R}^n$, that is to say,

$$
y_i = a_{i1}x_1 + a_{i2}x_2 + \cdots + a_{in}x_n
$$

where (a_{ij}) is an invertible matrix. Then f is differentiable with respect to (x_i) if and only if it is so with respect to (y_i). Indeed differentiation can be defined intrinsically as follows. For any $v \in V$, define $\partial_v f(x) = \lim_{t \to 0} \frac{f(x+tv) - f(x)}{t}$. Then $f \mapsto \partial_v f$ gives rise to a sheaf homomorphism of abelian groups.

3) The inclusion of constant functions in differentiable functions gives a homomorphism of the constant sheaf $\mathbb{R}$ in the sheaf of differentiable functions.

If $\mathcal{F}$ is a presheaf and $\widetilde{\mathcal{F}}$ the sheaf of sections of $E(\mathcal{F})$, then for every open set U in X, we get a natural homomorphism of $\mathcal{F}(U)$ into $\widetilde{\mathcal{F}}(U)$, which maps any s to the section $\tilde{s}$. If $V \subset U$, then for every $x \in V$, the element of $\mathcal{F}_x$ given by (U, s) is the same as that given by $(V, \mathrm{res}_{UV}\, s)$ by the very definition of $\mathcal{F}_x$. This implies that the diagram

$$
\begin{array}{ccc}
\mathcal{F}(U) & \longrightarrow & \widetilde{\mathcal{F}}(U) \\
\downarrow{\scriptstyle \mathrm{res}_{UV}} & & \downarrow{\scriptstyle \mathrm{res}_{UV}} \\
\mathcal{F}(V) & \longrightarrow & \widetilde{\mathcal{F}}(V)
\end{array}
$$

is commutative, proving that the natural homomorphisms $\mathcal{F}(U) \to \widetilde{\mathcal{F}}(U)$ define a homomorphism of presheaves.

1.11. Proposition. *A presheaf $\mathcal{F}$ satisfies Axiom $\mathbf{S}_1$ if and only if the induced map $\mathcal{F}(U) \to \widetilde{\mathcal{F}}(U)$ is injective, for every open set U of X.*

Proof. To say that $\mathcal{F}(U) \to \widetilde{\mathcal{F}}(U)$ is injective is equivalent to saying that if $s_1, s_2 \in \mathcal{F}(U)$ with $(s_1)_x = (s_2)_x$ for all $x \in U$, then $s_1 = s_2$. But then the assumption assures us that $\mathrm{res}_{UN_x}\, s_1 = \mathrm{res}_{UN_x}\, s_2$ for some open neighbourhood $N_x \subset U$ of x. Now Axiom $\mathbf{S}_1$, applied to the covering $U = \bigcup N_x$ says precisely that $s_1 = s_2$. Conversely, if $U = \bigcup U_i$ and s, t are elements of $\mathcal{F}(U)$ satisfying $\mathrm{res}_{UU_i}(s) = \mathrm{res}_{UU_i}(t)$ for all i, then the same is true of $\tilde{s}$ and $\tilde{t}$. But since $\widetilde{\mathcal{F}}$ satisfies $\mathbf{S}_1$, it follows that $\tilde{s} = \tilde{t}$. Now if we assume that $\mathcal{F}(U) \to \widetilde{\mathcal{F}}(U)$ is injective, it follows that $s = t$, so that we conclude that $\mathcal{F}$ satisfies $\mathbf{S}_1$.

1.12. Proposition. *A presheaf $\mathcal{F}$ is a sheaf if and only if the natural maps $\mathcal{F}(U) \to \widetilde{\mathcal{F}}(U)$ are all isomorphisms.*

Proof. In fact, if all these maps are isomorphisms, then $\mathcal{F}$ is isomorphic to the sheaf $\widetilde{\mathcal{F}}$ and it follows that $\mathcal{F}$ is itself a sheaf. On the other hand, if $\mathcal{F}$ is a sheaf, then we conclude from the above proposition that the maps $\mathcal{F}(U) \to \widetilde{\mathcal{F}}(U)$ are injective and we need only to verify that they are surjective. An element $\sigma \in \widetilde{\mathcal{F}}(U)$ is a section over U of the étale space of $\mathcal{F}$. Hence, for every $x \in U$, there is a neighbourhood N_x and an element $s^{(x)} \in \mathcal{F}(N_x)$ which represents the equivalence class $\sigma(x) \in \mathcal{F}_x$. The section $\widetilde{s^{(x)}}$ over N_x of the étale space given rise to by $s^{(x)}$, and the section σ, coincide at x. It follows that the two sections coincide in a neighbourhood N'_x of x, contained in N_x. This means that there exist an open covering $U = \bigcup_x N'_x$ and $s^{(x)} \in \mathcal{F}(N'_x)$ such that $(s^{(x)})_a = \sigma(a)$ for all $a \in N'_x$. In particular, $s^{(x)}$

and $s^{(y)}$ give rise to the same section of $\widetilde{\mathcal{F}}$ over $N'_x \cap N'_y$. But, thanks to the injectivity of the natural map $\mathcal{F}(N'_x \cap N'_y) \to \widetilde{\mathcal{F}}(N'_x \cap N'_y)$ we conclude that the restrictions of $s^{(x)}$ and $s^{(y)}$ to $N'_x \cap N'_y$ are the same. Since $\mathcal{F}$ is actually a sheaf, this implies that there exists $s \in \mathcal{F}(U)$ whose restriction to N'_x is $s^{(x)}$ for all $x \in U$. Thus we have $s_x = \sigma(x)$ for all $x \in U$, or, what is the same, $\tilde{s} = \sigma$, proving that $\mathcal{F}(U) \to \widetilde{\mathcal{F}}(U)$ is surjective.

Somewhat subtler is the relationship between Axiom $\mathbf{S}_2$ and the surjectivity of $\mathcal{F}(U) \to \widetilde{\mathcal{F}}(U)$. If $\mathcal{F}$ satisfies $\mathbf{S}_2$, then any section of $\widetilde{\mathcal{F}}$ gives rise, as above, to an open covering $\{N_x\}$ and elements $s^{(x)}$ of $\mathcal{F}(N_x)$. In order to piece all these elements together and obtain an element of $\mathcal{F}(U)$ we need to check that the restrictions of $s^{(x)}$ and $s^{(y)}$ to $N_x \cap N_y$ coincide, at least after passing to a smaller covering. We have the following set-topological lemma.

1.13. Lemma. *Let $\{U_i\}_{i \in I}$ be a locally finite open covering of a topological space U and $\{V_i\}_{i \in I}$ be a shrinking. Then for every $x \in U$, there exists an open neighbourhood M_x such that $I_x = \{i \in I : M_x \cap V_i \neq \emptyset\}$ is finite, and if $i \in I_x$ then x belongs to $\overline{V}_i$ and M_x is a subset of U_i. If M_x and M_y intersect, then there exists $i \in I$ such that $M_x \cup M_y \subset U_i$.*

Proof. Since $\{U_i\}$ is locally finite, so is the shrinking, and the existence of M_x such that the corresponding I_x is finite, is trivial. We will now cut down this neighbourhood further in order to satisfy the other conditions. We intersect M_x with $U \setminus \overline{V}_i$ for all $i \in I_x$ for which $x \notin \overline{V}_i$. We thus obtain an open neighbourhood of x, and the closures of all $V_i, i \in I_x$ then contain x. It can be further intersected with $\bigcap_{i \in I_x} U_i$, and the resulting neighbourhood satisfies the first assertion of the lemma. Now if $M_x \cap M_y \neq \emptyset$, then for any $z \in M_x \cap M_y$, choose $i \in I$ such that $z \in V_i$. Then M_x intersects V_i and hence $M_x \subset U_i$. Similarly M_y is also contained in U_i proving the second assertion.

This lemma can be used to deduce that under a mild topological hypothesis, the natural maps $\mathcal{F}(U) \to \widetilde{\mathcal{F}}(U)$ are surjective for all open sets U, if the presheaf $\mathcal{F}$ satisfies Axiom $\mathbf{S}_2$.

1.14. Proposition. *If every open subset U of X is paracompact, and the presheaf $\mathcal{F}$ satisfies $\mathbf{S}_2$, then the map $\mathcal{F}(U) \to \widetilde{\mathcal{F}}(U)$ is surjective for all U.*

Proof. Firstly, given an element σ of $\widetilde{\mathcal{F}}(U)$, there is a locally finite open covering $\{U_i\}$ of U, and elements $s_i \in \mathcal{F}(U_i)$ for all i, with the property that for all $x \in U_i$, the elements s_i have the image $\sigma(x)$ in $\widetilde{\mathcal{F}}_x$. Let $\{V_i\}$ be a shrinking of $\{U_i\}$. For every $x \in U$, choose M_x as in Lemma 1.13. We may also assume that the restrictions to M_x of any of the s_i for which $i \in I_x$, is the same, say $s^{(x)}$. It follows that the restrictions of $s^{(x)}$ and $s^{(y)}$ to $M_x \cap M_y$

are the same as the direct restriction of some s_j to $M_x \cap M_y$, proving in view of Axiom $\mathbf{S}_2$, that there exists $s \in \mathcal{F}(U)$ whose restriction to M_x is $s^{(x)}$ for all $x \in U$. This proves that $\mathcal{F}(U) \to \widetilde{\mathcal{F}}(U)$ is surjective.

1.15. Exercises.

1) Let X be a topological space which is the disjoint union of two proper open sets U_1 and U_2. Define $\mathcal{F}(U)$ to be (0) whenever U is an open subset of either U_1 or U_2. For all other open sets U, define $\mathcal{F}(U) = A$, where A is a nontrivial abelian group. If $U \subset V$ and $\mathcal{F}(U) = A$, then define the restriction map to be the identity homomorphism. All other restriction maps are zero. Show that $\mathcal{F}$ is a presheaf such that $\widetilde{\mathcal{F}} = (0)$.

2) In the above, does $\mathcal{F}$ satisfy Axiom $\mathbf{S}_2$?

Subsheaves.

1.16. Definition. A sheaf $\mathcal{G}$ is said to be *a subsheaf* of a sheaf $\mathcal{F}$ if we are given a homomorphism $\mathcal{G} \to \mathcal{F}$ satisfying either of the following equivalent conditions.

1) $\mathcal{G}_x \to \mathcal{F}_x$ is injective for all $x \in X$.

2) For any open subset U of X, $\mathcal{G}(U) \to \mathcal{F}(U)$ is injective.

To see that the above conditions are equivalent, note that 1) implies that $E(\mathcal{G}) \to E(\mathcal{F})$ is injective and hence the set of sections of $\mathcal{G}$ over any set is also mapped injectively into the set of sections of $\mathcal{F}$. Conversely, assume 2), and let $a, b \in \mathcal{G}_x$ have the same image in $\mathcal{F}_x$. Then there exist a neighbourhood U of x and elements $s, t \in \mathcal{G}(U)$ such that $s_x = a, t_x = b$. Moreover, by replacing U with a smaller neighbourhood we may also assume that the images of s and t are the same in $\mathcal{F}(U)$. This implies by our assumption that $s = t$ as elements of $\mathcal{G}(U)$, as well. Hence $s_x = t_x$ in $\mathcal{G}_x$.

2. Basic Constructions

When $\mathcal{F}$ is a sheaf, it is legitimate to call elements of $\mathcal{F}(U)$ *sections* of $\mathcal{F}$ over an open set U, since they can be identified with sections of the associated étale space. Continuous sections of the étale space make sense, on the other hand, over any subspace of X.

2.1. Proposition. *If K is a closed subspace of a paracompact topological space X, then any section over K of a sheaf $\mathcal{F}$ on X is the restriction to K of a section of $\mathcal{F}$ over a neighbourhood of K.*

Proof. In fact, any section s over K is defined by open sets U_i of X that cover K and elements $s_i \in \mathcal{F}(U_i)$. Since K is also paracompact, we may assume (by passing to a refinement, if necessary) that the covering is locally finite. Let $U = \bigcup U_i$ and $\{V_i\}$ be a shrinking of this covering. Choose for every $x \in U$, an open neighbourhood M_x of x as in Lemma 1.13. Denote by I_x the set of all $i \in I$ such that $x \in \overline{V}_i$. Then our choice of M_x is equivalent to saying that i belongs to I_x if and only if M_x intersects V_i. Consider the subset $W = \{x \in U : (s_i)_x$ is independent of i for all $i \in I_x\}$. For any $y \in M_x$, we have $I_y \subset I_x$. For, if $i \in I_y$, then $y \in \overline{V}_i$ and hence M_x which is a neighbourhood of y has nonempty intersection with V_i. This means that $i \in I_x$. Clearly, the set $\{y \in M_x : (s_i)_y$ is independent of $i \in I_x\}$ is an open set containing x and contained in W. Hence W is open. On the other hand we have the inclusion $K \subset W$. It is now clear that the s_i actually give a section of $\mathcal{F}$ over W.

2.2. Remark. Let us consider the sheaf of continuous functions on $\mathbb{R}^n$ for example. A section of the corresponding étale space over a closed set K is the same as a continuous function in a neighbourhood of K with the understanding that two such continuous functions are to be considered equivalent if they coincide in a neighbourhood of K. Any such 'germ' gives, on restriction to K, a continuous function on K. On the other hand any continuous function on K can be extended to some neighbourhood of K. Thus we have a surjection of the set of sections over K of the sheaf of continuous functions into the set of continuous functions on K. But this is not injective, even when K consists of a single point. For in this case, a section is simply an element of the stalk at the point, which consists of germs of continuous functions at the point.

2.3. Inverse images.

We used the construction of $E(\mathcal{F})$ as a means to pass from a presheaf to a sheaf. However, even when one starts with a sheaf the construction of the étale space is useful in some applications. One such is the notion of the *inverse image* of a sheaf. Let $f : X \to Y$ be a continuous map of topological spaces. If $\mathcal{F}$ is a sheaf on Y, we seek to define a sheaf $f^{-1}\mathcal{F}$ on X. Let $E(\mathcal{F})$ be the étale space of $\mathcal{F}$. Then form the *fibre product* of the map $f : X \to Y$ and the map $\pi : E(\mathcal{F}) \to Y$. (It is the subspace of the topological space $X \times E(\mathcal{F})$ consisting of points (x, a) such that $f(x) = \pi(a)$.) This space comes with a natural continuous map into X. The sheaf of continuous sections of this space is called *the inverse image* of $\mathcal{F}$ by the map f.

In the particular case when $X \subset Y$ is a subspace, the inverse image is also called the *restriction* of $\mathcal{F}$ to X. We sometimes use the notation $\mathcal{F}|X$ for this restriction.

2.4. Glueing together.

Let (U_i) be an open covering of a space Y and for each i, let $\mathcal{F}_i$ be sheaves on U_i. Then we wish *to glue all these together* and obtain a sheaf on the whole of Y. For this we need some *glueing data*. If the $\mathcal{F}_i$ are all restrictions to U_i of the same sheaf $\mathcal{F}$ on Y, then the restrictions of $\mathcal{F}_i$ and $\mathcal{F}_j$ to $U_i \cap U_j$ are the same as direct restrictions of $\mathcal{F}$ to $U_i \cap U_j$. So, we assume as data, isomorphisms $m_{ij} : \mathcal{F}_j|U_i \cap U_j \to \mathcal{F}_i|U_i \cap U_j$. Actually we need more, namely *compatibility* of these isomorphisms. Consider the set $U_i \cap U_j \cap U_k$. We have restrictions to it of the sheaves $\mathcal{F}_i$, $\mathcal{F}_j$ and $\mathcal{F}_k$. Besides, the isomorphisms m_{ij}, m_{jk} and m_{ik} restrict to isomorphisms (two by two) of these three sheaves on $U_i \cap U_j \cap U_k$ as well. We will use the same notation for these restrictions. Then the compatibility condition that we have in mind is that they should satisfy

$$m_{ij} \circ m_{jk} = m_{ik}.$$

One can easily verify that given such data as above, one can glue the sheaves $\mathcal{F}_i$ together and obtain a sheaf $\mathcal{F}$ on the whole of Y with natural isomorphisms of $\mathcal{F}|U_i$ with $\mathcal{F}_i$.

Actually, in a certain sense, the construction involved in glueing is the converse of the construction of the inverse image. If X is the topological union of the spaces U_i, then it is clear that the $\mathcal{F}_i$ build a sheaf $\overline{\mathcal{F}}$ on it, and what is needed is a sheaf on Y whose inverse image under the natural map $X \to Y$ is $\overline{\mathcal{F}}$.

2.5. Remark. Regarding our glueing construction above, we wish to make the following remark. Let $\mathcal{F}$ be a sheaf on X. Let (U_i) be an open covering and let $\mathcal{G}_i$ be subsheaves of $\mathcal{F}|U_i$, for all i. Then, in order to glue the $\mathcal{G}_i$ together, we only need to check that $\mathcal{G}_i|U_i \cap U_j$ is *the same* subsheaf of $\mathcal{F}|U_i \cap U_j$ as $\mathcal{G}_j|U_i \cap U_j$. Then we can, not only glue them together, but also get a homomorphism of $\mathcal{G}$ into $\mathcal{F}$ which makes it a subsheaf.

2.6. Definition. Let $\mathcal{F}$ be a sheaf on a space X. Let $f : X \to Y$ be a continuous map. Then one can define a sheaf on Y called *the direct image of $\mathcal{F}$ by f* as follows. To any open set U in Y, associate $\mathcal{F}(f^{-1}(U))$. It is easy to check that this defines a sheaf. We will denote this by $f_*(\mathcal{F})$.

This is related to the inverse image very closely. In fact, let $\mathcal{F}$ be a sheaf on X in the above situation, and $\mathcal{G}$ a sheaf on Y. We may consider on the one hand, the direct image $f_*(\mathcal{F})$ and on the other, the inverse image

$f^{-1}(\mathcal{G})$. Then one can see easily that there is a natural bijection between homomorphisms $\mathcal{G} \to f_*(\mathcal{F})$ and $f^{-1}(\mathcal{G}) \to \mathcal{F}$. Indeed, let us start with a homomorphism $T : \mathcal{G} \to f_*(\mathcal{F})$. This gives, for every open subset $U \subset Y$ a homomorphism $T(U) : \mathcal{G}(U) \to (f_*\mathcal{F})(U) = \mathcal{F}(f^{-1}(U))$. If $x \in f^{-1}(U)$, then we have a natural map $\mathcal{F}(f^{-1}(U)) \to \mathcal{F}_x$. Composing with $T(U)$, we get a homomorphism $\mathcal{G}(U) \to \mathcal{F}_x$. It is obvious that as U varies over neighbourhoods of $f(x)$, this is compatible with restrictions, and so induces a map $\mathcal{G}_{f(x)} = f^{-1}(\mathcal{G})_x \to \mathcal{F}_x$. This is easily checked to be continuous on the étale space and hence gives a homomorphism of $f^{-1}(\mathcal{G})$ into $\mathcal{F}$. It is this association that gives the bijection, as claimed.

2.7. Modules over a sheaf of algebras.

Before we leave this preliminary account of sheaves and move on to differential manifolds, we would like to introduce one more notion which is very useful in our context. As we have observed, the sheaf of continuous functions on a topological space, that of differentiable functions in $\mathbb{R}^n$, that of holomorphic functions in a domain in $\mathbb{C}^n$, etc. are all sheaves of algebras. Let then $\mathcal{A}$ be a sheaf of algebras over a topological space X. On the other hand, let $\mathcal{M}$ be a sheaf of abelian groups. Then we say that $\mathcal{M}$ *is a sheaf of modules over* $\mathcal{A}$, or simply an $\mathcal{A}$-*module* if for every U, open in X, the abelian group $\mathcal{M}(U)$ comes provided with a structure of an $\mathcal{A}(U)$-module in such a way that the restriction maps res_{UV} respect the module structures in the obvious sense, namely

$$\mathrm{res}_{UV}(fs) = \mathrm{res}_{UV}(f)\,\mathrm{res}_{UV}(s)$$

for all $f \in \mathcal{A}(U)$ and $s \in \mathcal{M}(U)$.

3. Differential Manifolds

After these preliminaries, we are now ready to define the concept of differential manifolds. These objects provide the proper setting for developing differential and integral calculus. In Physics, these are called *configuration spaces* and may be thought of as the set of all possible states of the system of which one wishes to study the dynamics.

3.1. Definition. *A differential manifold M (of dimension n) consists of*

a) a topological space which is Hausdorff and admits a countable base for open sets, and

b) a sheaf $\mathcal{A}^M = \mathcal{A}$ of subalgebras of the sheaf of continuous functions on M.

These are required to satisfy the following local condition. For any $x \in M$, there is an open neighbourhood U of x and a homeomorphism of U with an open set V in $\mathbb{R}^n$ such that the restriction of $\mathcal{A}$ to U is the inverse image of the sheaf of differentiable functions on V.

The homeomorphisms referred to above, defined in a neighbourhood of any point, are called *coordinate charts*. This is because composing with this homeomorphism the coordinate functions in $\mathbb{R}^n$, one obtains functions $x_1, \ldots, x_n$. We think of M as a global object on which action takes place, and it is locally described by using coordinates.

3.2. Examples.

1) Since the concept of differential manifolds is based on the notion of differentiability in $\mathbb{R}^n$, it is clear that $\mathbb{R}^n$ together with the sheaf of differentiable functions is a differential manifold. Slightly more abstractly, any finite-dimensional vector space over $\mathbb{R}$ is a differential manifold.

2) If $(M, \mathcal{A})$ is a differential manifold, and U an open subset of M, then the subspace U, together with the restriction $\mathcal{A}|U$ of $\mathcal{A}$ is a differential manifold as well. This will be referred to as *an open submanifold* of M.

3) Combining these two examples, we see that any open subspace of $\mathbb{R}^n$ (or any finite-dimensional vector space) is a differential manifold. In particular, the space $GL(n, \mathbb{R})$ of invertible (n, n)-matrices, which is actually the open set in the vector space of all (n, n)-matrices given by the nonvanishing of the determinant, is a manifold of dimension n^2. Incidentally, it is also a group under matrix multiplication and is called the *General Linear group.*

4) Let f be a differentiable function in $\mathbb{R}^n$. Consider the closed subspace (*the zero locus of f*) of $\mathbb{R}^n$ given by

$$Z_f = \{x \in \mathbb{R}^n : f(x) = 0\}.$$

Then Z_f has a natural structure of differential manifold, if at every $x \in Z_f$, at least one of the partial derivatives of f does not vanish. In fact, consider the association to any open set $U \subset Z_f$, of the set of functions on U, which can be extended to a differentiable function on a neighbourhood of U in $\mathbb{R}^n$. This gives a presheaf on Z_f. (Actually it will turn out that it is a sheaf, but we do not need it here.) Let $\mathcal{A}^{Z_f}$ be the associated sheaf. Then by our assumption, for any $x \in Z_f$, there exists a neighbourhood N of x in $\mathbb{R}^n$ such that, one of the partial derivatives, say $\frac{\partial f}{\partial x_n}$, is nonzero. By the implicit function theorem, the projection to $\mathbb{R}^{n-1}$ taking $(x_1, \ldots, x_n)$ to $(x_1, \ldots, x_{n-1})$ is a differentiable isomorphism of $N \cap Z_f$ with an open set V in $\mathbb{R}^{n-1}$

(that is to say, a differentiable bijective map from N onto V whose inverse is also differentiable). It is now clear that this isomorphism gives the local requirement of the sheaf $\mathcal{A}^{Z_f}$. If we further specialise the function to be $f(x) = \sum x_i^2 - 1$, then Z_f is the set of vectors of unit length in $\mathbb{R}^n$ which we call the *unit sphere S^{n-1}*.

On the contrary, if f is taken to be the function xy defined in $\mathbb{R}^2$ with coordinates x, y, then clearly it does not satisfy the criterion given above. In fact, both the partial derivatives, $\frac{\partial f}{\partial x}$ and $\frac{\partial f}{\partial y}$ vanish at $(0,0)$. So we cannot conclude that Z_f is a differential manifold in this case. In fact, the topological space Z_f cannot have a structure of a differential manifold, since it is easy to see that no neighbourhood of 0 in Z_f is homeomorphic to an open interval in $\mathbb{R}$.

5) The above example can be generalised further, by taking, instead of one function, finitely many functions. So, let $f = (f_i), 1 \leq i \leq r$, be finitely many differentiable functions in $\mathbb{R}^n$. Then the closed subspace

$$Z_f = \{x \in \mathbb{R}^n : f_i(x) = 0 \text{ for all } i\}$$

has a natural structure of a differential manifold, if f satisfies the following condition. For every $x \in Z_f$, the rank of the (r, n) matrix $(\frac{\partial f_i}{\partial x_j})$ is r. One may think of f as a function into $\mathbb{R}^r$.

We will now take for f the function on (n, n)-matrices with values also in $\mathbb{R}^{n^2}$ given by $A \mapsto AA' - I_n$, where A' denotes the transpose of A. Then one can check that the above criterion is fulfilled and hence the topological space $\{A \in M_n(\mathbb{R}) : AA' = I_n\}$ is a differential manifold. This is actually a subgroup of $GL(n, \mathbb{R})$ as well, called *the orthogonal group* and is usually denoted $O(n, \mathbb{R})$. Similarly take the map $f : M(n, \mathbb{C})$ into itself given by $A \mapsto A\overline{A}' - I_n$, and get the set $f = 0$ as a subgroup of $GL(n, \mathbb{C})$. This group is called the *unitary group* and denoted $U(n)$. (Are these manifolds connected?)

6) Consider the *real projective space* $\mathbb{RP}^n$ defined to be the quotient of the unit sphere S^n by the identification of antipodal points x and $-x$. We may define a sheaf on $\mathbb{RP}^n$ by associating to any open set $U \subset \mathbb{RP}^n$, the algebra of differentiable functions on its inverse image in S^n which are invariant under the antipodal map. For any point $a \in S^n$, the open neighbourhood of a consisting of those $y \in S^n$ whose distance from a is less than 1 is mapped homeomorphically on an open set in $\mathbb{RP}^n$, and it is easy to see that this takes the sheaf of differentiable functions on S^n isomorphically onto the sheaf defined above. Thus we see that $\mathbb{RP}^n$ is a differential manifold in a natural way.

3.3. Exercises.

1) Show that the topological subspace of the space of all (n, n)-matrices, consisting of those matrices whose determinants are 1, is a differential manifold. This is also a subgroup of the group $GL(n, \mathbb{R})$ mentioned above and is denoted $SL(n, \mathbb{R})$. It is called the *Special Linear group*.

2) Is the same true of matrices with determinant 0?

3.4. Glueing up differential manifolds.

Often, a structure of a differential manifold is given on a Hausdorff topological space M with a countable base for open sets, by the following procedure. Suppose $\{U_i\}$ is an open covering, and that each U_i is provided with a subsheaf $\mathcal{A}_i$ of the sheaf of continuous functions making $(U_i, \mathcal{A}_i)$ a differential manifold. If we can glue all these sheaves together to get a sheaf of algebras on M, then it is clear that it would make M a differential manifold. We have already seen (2.5) how we can glue them together. What we need is simply that $\mathcal{A}_i|U_i \cap U_j$ is the same as $\mathcal{A}_j|U_j \cap U_i$. In other words, the open submanifold $U_i \cap U_j$ of U_i is the same as the open submanifold $U_i \cap U_j$ of U_j.

In particular, if $(V_i, \mathcal{A}_i)$ are open submanifolds of $\mathbb{R}^n$, then the glueing data may also be formulated as follows. The space M is covered by open sets U_i. For each i, one is given a homeomorphism c_i of U_i with the open subset V_i of $\mathbb{R}^n$. If U_i and U_j intersect, then $U_i \cap U_j$ has as images in V_i and V_j, two open sets which we may call V_{ij} and V_{ji}. Then $c_j \circ c_i^{-1}$ gives a homeomorphism $V_{ij} \to V_{ji}$. One may use the homeomorphism c_i to transport the sheaf of differentiable functions on V_i to a sheaf of algebras on U_i. But in order to glue these together, we need to know that its restriction to $U_i \cap U_j$ is the same as the restriction of the transported sheaf on U_j. This can be achieved if and only if the above homeomorphism $V_{ij} \to V_{ji}$ is differentiable for every i, j. (Note that the inverse is also differentiable, by reversing the roles of i and j.) This is in fact the traditional definition of a differential manifold.

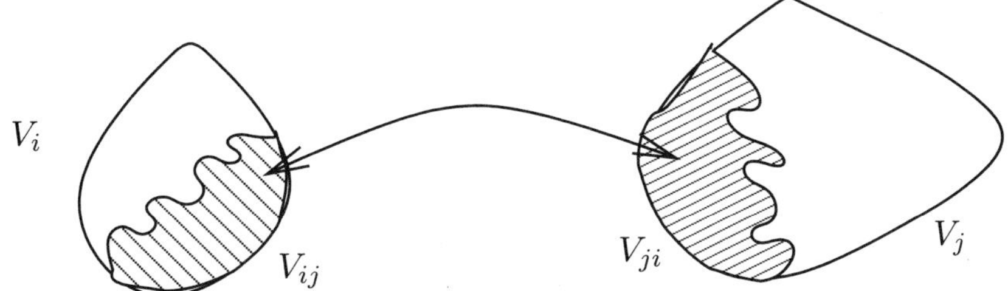

For example, consider the sphere in $\mathbb{R}^n$. The map

$$(x_1, \ldots, x_n) \mapsto (x_1, \ldots, \hat{x}_i, \ldots, x_n)$$

is a homeomorphism of the open set consisting of points of the sphere for which $x_i > 0$ onto the open unit ball in $\mathbb{R}^{n-1}$. We can therefore provide these open sets with the differential manifold structure of the unit ball. We could do the same with the open sets of the sphere in which $x_i < 0$. It is clear that all these open sets, as i varies, cover S^{n-1}. To glue these up, we have only to check that the map

$$(y_1, \ldots, y_{n-1}) \mapsto (y_1, \ldots, \sqrt{1 - |y|^2}, \ldots, \hat{y}_j, \ldots, y_{n-1}),$$

where the term $\sqrt{1 - |y|^2}$ occurs at the ith place, is a differentiable isomorphism of the open set of the unit ball onto the open ball.

Another example is provided by the complex projective space. Notice first that the real projective space $\mathbb{RP}^n$ may also be defined as the quotient of $\mathbb{R}^{n+1} \setminus \{0\}$ by the equivalence relation: $(x_0, x_1, \ldots, x_n) \sim (y_0, y_1, \ldots, y_n)$ if and only if there exists a nonzero real number a such that $y_i = ax_i$, for all i. An analogous definition makes sense over complex numbers as well. In other words, define the *complex projective space* $\mathbb{CP}^n$ to be the quotient of $\mathbb{C}^{n+1} \setminus \{0\}$ by the equivalence relation:

$$(z_0, \ldots, z_n) \sim (z'_0, \ldots, z'_n)$$

if and only if there exists $a \in \mathbb{C}^\times$ such that $z'_i = az_i$ for all i. Since for any such $z = (z_0, \ldots, z_n)$ at least one z_i is nonzero, $\mathbb{CP}^n$ is covered by open sets which are images of sets $U_i, i = 0, \ldots, n$ under the natural map $\mathbb{C}^{n+1} \setminus \{0\} \to \mathbb{P}^n$, where $U_i = \{z = (z_0, \ldots, z_n) : z_i \neq 0\}$. It is clear that the subspace $z_i = 1$ of U_i is mapped homeomorphically onto the image of U_i in $\mathbb{CP}^n$. On the other hand, it is homeomorphic to $\mathbb{C}^n$ under the projection which omits the ith coordinate. This can be used as above to put a structure of a differential manifold on it. To check the condition for glueing up, we only have to check that the map

$$\{(z_0, \ldots, z_n), z_i = 1, z_j \neq 0\} \mapsto \{(z_0, \ldots, z_n), z_i \neq 0, z_j = 1\}$$

given by $(z_0, \ldots, z_n) \mapsto (z_0/z_j, \ldots, z_n/z_j)$ is a differentiable isomorphism. This is of course obvious.

A little more abstractly, we could have replaced $\mathbb{C}^{n+1}$ by any complex vector space V of dimension $n + 1$. Then one considers the open set $V \setminus \{0\}$ and introduces the equivalence relation: $v \sim v'$ if and only if there exists $a \in \mathbb{C}^\times$ such that $av = v'$. The quotient is defined to be the *projective space $P(V)$ associated to V*. If f is any nonzero linear form on V, the set $\{v \in V : f(v) = 1\}$ is mapped homeomorphically onto an open set X_f of $P(V)$. What we did above amounts to using this homeomorphism to define a differential structure on X_f. We may glue all these structures together to obtain a differential structure on the whole of $P(V)$.

One may also think of the points of $P(V)$ as one-dimensional subspaces of V. Then one may generalise this by considering all r-dimensional vector subspaces of V for any fixed $r < \dim(V)$. The set thus formed is called the *Grassmannian of r-dimensional subspaces* of V. We will indicate two ways in which one can provide this set with the structure of a differential manifold. Firstly, let W be any such subspace. Consider the image of the one-dimensional space $\Lambda^r(W)$ in $\Lambda^r(V)$. Thus to every element of the Grassmannian we have associated an element of the *projective space* $P(\Lambda^r(V))$. Then one checks the following assertion.

3.5. Lemma. *If ω is any nonzero element of $\Lambda^r(V)$, then the linear map $V \to \Lambda^{r+1}(V)$ given by $v \mapsto v \wedge \omega$ has kernel of dimension $\leq r$. Moreover, the kernel is of dimension r if and only if $\omega \in \Lambda^r(W)$ for some r-dimensional subspace W of V.*

Proof. In fact, it is obvious that if ω belongs to $\Lambda^r(W)$, then every element of W is in the kernel of the above map. Let $\{e_i\}, 1 \leq i \leq n$, be a basis of V such that the first r of these generate W. In other words, $e_1 \wedge \cdots \wedge e_r$ can be taken to be ω. Let $v = \sum a_i e_i$; then $v \wedge \omega = 0$ if and only if $a_i = 0$ for all $i > r$. Thus the kernel of the map $v \mapsto v \wedge \omega$ is precisely W.

Conversely, if the kernel contains an r-dimensional subspace W, again take a basis like the one above. Writing out ω in terms of a basis as $\sum_{i_1 < \cdots < i_r} a_{i_1,\ldots,i_r} e_{i_1} \wedge \cdots \wedge e_{i_r}$, we deduce that if $e_j \wedge \omega = 0$, then $a_{i_1,\ldots,i_r} = 0$ whenever j does not belong to the set $\{i_1, \ldots, i_r\}$. Since we have assumed that $e_j \wedge \omega = 0$ for all $j \leq r$, we see that the only nonzero coefficient in the expression for ω is $a_{1,\ldots,r}$, that is to say, $\omega \in \Lambda^r(W)$. This also shows in particular that the dimension of the kernel is $\leq r$.

The above lemma asserts in fact that the Grassmannian is imbedded in $P(\Lambda^r V)$ as a closed subset. Using this description, one can check that the Grassmannian is actually a closed submanifold. This imbedding is called the *Plücker imbedding*.

Another way of introducing the structure of differential manifold on the Grassmannian is the following. In order to introduce a coordinate system in a neighbourhood of any r-dimensional subspace U_0 of V, we proceed as follows. Fix an $(n - r)$-dimensional subspace W of V supplementary to U_0. Consider the set of all r-dimensional subspaces U of V which are supplementary to W. This set evidently contains U_0. The projection of V onto W corresponding to the direct sum decomposition $V = U \oplus W$ restricted to U_0 gives a linear map A_U of U_0 into W. When $U = U_0$ this map is the zero map. In general the linear map determines U as the image of $\eta - A_U$, where η is the inclusion of U_0 in V. This sets up a bijection between

$\mathrm{Hom}(U_0, W)$ and the above set, thereby coordinatising the Grassmannian in a neighbourhood of U_0.

3.6. Definition. If $(M, \mathcal{A})$ is a differential manifold, then sections of $\mathcal{A}$ over an open set U of M are called *differentiable functions* on U.

Since $\mathcal{A}$ is a sheaf, the notion of a differentiable function on M is of a local nature, so that the following properties are obvious.

i) If f is a nowhere vanishing differentiable function, then so is $1/f$.

ii) If (U_i) is a locally finite covering and (f_i) is a family of differentiable functions with support in U_i, then $\sum f_i$ is also a differentiable function.

iii) If φ is a differentiable function on $\mathbb{R}^m$ and $f_1, \ldots, f_m$ are differentiable functions on M, then $\varphi(f_1, \ldots, f_m)$ is also a differentiable function.

3.7. Definition. A continuous map f of a differential manifold M into another differential manifold N is said to be *differentiable* if for any $x \in M$, and for every differentiable function φ in a neighbourhood U of $f(x)$ in N, the composite $\varphi \circ f$ is a differentiable function on $f^{-1}(U)$.

From the definition of a differentiable map it follows that there is a homomorphism of the sheaf $\mathcal{A}^N$ into $f_*(\mathcal{A}^M)$ and therefore also a homomorphism of $\mathcal{A}^M$ into $f^{-1}(\mathcal{A}^N)$. This is called the *structure homomorphism* associated to f.

3.8. Remark. It is clear from the definition above that if M, N, P are differential manifolds, and $f : M \to N$, $g : N \to P$ are differentiable maps, then the composite $g \circ f : M \to P$ is also differentiable. However, if $f : M \to N$ is a differentiable map which is bijective, then one *cannot* conclude that the inverse map is also differentiable. The hackneyed counterexample is the function $x \mapsto x^3$ of $\mathbb{R}$ into $\mathbb{R}$, which is differentiable and bijective, but whose inverse $x \mapsto x^{\frac{1}{3}}$ is not differentiable at 0 .

3.9. Definition. A differentiable map $f : M \to N$ of differential manifolds is said to be *a diffeomorphism* if there is a differentiable inverse.

A fact, basic to the study of differentiable functions, is that there are lots of them. We will now formulate this precisely.

Differentiable partition of unity.

3.10. Proposition. *The following are equivalent.*

i) *Given any locally finite open covering $(U_i)_{i \in I}$ of M, there exist differentiable functions φ_i on M with values in the closed interval $[0,1]$ such that the support of φ_i is contained in U_i and $\sum \varphi_i = 1$.*

ii) *Given open sets U, V with $\overline{V} \subset U$, there exists a positive differentiable function whose support is contained in U and which does not vanish anywhere in V.*

Proof. i) implies ii). Consider the open covering $(M \setminus \overline{V}, U)$. By i), there exist $[0,1]$-valued functions φ, ψ with supports respectively in $M \setminus \overline{V}$ and U such that $\varphi + \psi = 1$ everywhere. Then ψ satisfies the requirement in ii).

ii) implies i). Let (V_i) be a shrinking of (U_i), i.e. an open covering with $\overline{V_i} \subset U_i$ for all i. By Assumption ii), we see that there exist positive differentiable functions φ_i such that supp $\varphi_i \subset U_i$ and φ_i are nonzero everywhere on V_i. Now since the covering is locally finite, the sum $\varphi = \sum \varphi_i$ makes sense, is differentiable and is nonzero everywhere. The family of functions $\{\psi_i = \varphi_i/\varphi\}$, satisfies i).

Of course the point of the above proposition is that the equivalent properties stated there are actually true. We will now prove this fact. Firstly, in order to prove ii), it is enough to do it locally in the following sense. For every $m \in M$ there exists a neighbourhood N_x such that there is a function as in ii) with V replaced by $V \cap N_x$ and U by $U \cap N_x$. Then one replaces the sets N_x by a locally finite refinement and notes that the sum of the corresponding functions fulfils the requirement. Taking N_x to be a coordinate neighbourhood of x, we therefore reduce the problem to proving the following.

3.11. Proposition. *Let S_1, S_2 be concentric spheres in $\mathbb{R}^n$ centered at 0, with $S_1 \subset S_2$. Then there exists a differentiable function which is nonzero everywhere inside S_1 and has support contained in the interior of S_2.*

Proof. Clearly it is enough to construct a differentiable function on $\mathbb{R}^n$ which is nonzero everywhere inside the unit ball and zero in the complement. The function $x \mapsto \exp(\frac{1}{\sum x_i^2 - 1})$ for all x inside the unit ball and 0 in the complement, is such a function.

3.12. Definition. Let (U_i) be a locally finite open covering of a differential manifold M. A family (φ_i) of differentiable functions on M with values in $[0,1]$ is said to be a *partition of unity with respect to the covering* (U_i), if the support of φ_i is contained in U_i for every i and $\sum \varphi_i = 1$.

We will now derive a simple consequence of the existence of a partition of unity.

3.13. Proposition. *Any section of $\mathcal{A}$ over a closed set can be extended to a differentiable function on M. In other words, given a differentiable function φ in a neighbourhood of a closed set K, there exists a differentiable function $\tilde{\varphi}$ on M which coincides with φ in a neighbourhood of K.*

Proof. In fact, if $K \subset U$, and f is a differentiable function on U, consider the partition of unity with respect to the covering $U, M \setminus \overline{V}$, where V is a neighbourhood of K with $\overline{V} \subset U$. Thus there is a differentiable function φ on M which is 1 on V and with support in U. The function $f\varphi$ on U has support contained in the support of φ. Hence the function $f\varphi$ on U and the constant function 0 on $M \setminus (\operatorname{supp} f)$ coincide on the intersection, thereby giving rise to a differentiable function on the whole of M as required.

3.14. Proposition. *Let M, N be differential manifolds. If $f : M \to N$ is a continuous map such that for every differentiable function φ on N the composite $\varphi \circ f$ is differentiable, then f is differentiable.*

Proof. In fact, for any $x \in M$ and any differentiable function φ in a neighbourhood of $f(x)$, we have to show that $\varphi \circ f$ is differentiable in a neighbourhood of x. Let $\tilde{\varphi}$ be a differentiable function on N coinciding with φ in a neighbourhood of $f(x)$; then we are given that $\tilde{\varphi} \circ f$ is differentiable. But then $\tilde{\varphi} \circ f$ and $\varphi \circ f$ coincide in a neighbourhood of x, proving our assertion.

3.15. Product manifolds.

Let M and N be differential manifolds. Then one can provide the topological space $M \times N$ with the structure of a differential manifold in the following way. Cover M and N by coordinate charts $c_i : U_i \to V_i$ and $c_j' : U_k' \to V_k'$. Then we may cover $M \times N$ by $U_i \times U_k'$. On each open set $U_i \times U_k'$ one may define a coordinate chart $c_i \times c_k'$ onto an open set in $\mathbb{R}^m \times \mathbb{R}^n$. The compatibility condition that needs to be verified, namely the differentiability of $(c_j \times c_l') \circ (c_i \times c_k')^{-1}$, follows obviously from those of $c_j \circ c_i^{-1}$ and $c_l' \circ c_k'^{-1}$. Thus we have provided the space $M \times N$ with the structure of a differential manifold. Now it is easy to check from this definition, that a mapping of any differential manifold L into $M \times N$ is differentiable if and only if its composites with the projections to M and N are both differentiable. This manifold is (therefore) called *the product of M and N.*

3.16. Submanifolds.

We have already defined the notion of an open submanifold. One may define a *closed submanifold* as follows. Let M be a closed subset of a differential manifold N with the property that for every $m \in M$, there exists a coordinate chart (U, c) around m in N such that $M \cap U$ is the set of common zeros of some of the coordinates of the chart c.

The rationale of the definition is quite clear. If U is an open submanifold of $\mathbb{R}^n$, then the set of points (x) in U satisfying $x_1 = x_2 = \cdots = x_r = 0$, should clearly be defined to be a closed submanifold of U. Note that this set is itself a manifold with the remaining coordinates serving as a coordinate chart.

Combining the notions of an open submanifold and a closed submanifold, we may define a *locally closed submanifold*, to be a closed submanifold of an open submanifold.

3.17. Immersed manifolds.

There is a more general notion of a subset of a manifold that has the structure of a manifold, which is formally similar to the above notion, but subtler. Suppose M and N are differential manifolds. First of all, assume that we have an injective differentiable map $M \to N$. Secondly, for any point $m \in M$, we require that there is a coordinate chart (U, c) in N containing the image of m and a neighbourhood U' of m *in* M such that U' maps into U and its image is the set of common zeros of some of the coordinates in the chart c. Then we say that M is a *manifold immersed* in N.

The notion of an immersed manifold is somewhat delicate for the following reason. Let us identify M with its image in the following discussion. Notice that we have not required that for every point m of M, there is a coordinate chart (U, c) of N such that $U \cap M$ is defined by the vanishing of some of the coordinates of the chart c. The difference is not slight! Indeed, the topology of the immersed manifold M is *not* necessarily that of its image induced from that of N.

Let us consider an example. We know that the real line $\mathbb{R}$ and *the torus* $S^1 \times S^1$ are both differential manifolds. Identify S^1 with the submanifold of $\mathbb{C}$ consisting of complex numbers of absolute value 1. Let $\alpha \in \mathbb{R}$. Consider the map $f : \mathbb{R} \to S^1 \times S^1$ given by $f(x) = (\exp(2\pi i x), \exp(2\pi \alpha i x))$. It is clearly differentiable. It is also injective if α is irrational. For, if $f(x) = f(y)$, then $x - y \in \mathbb{Z}$ on the one hand and also, $\alpha(x - y) \in \mathbb{Z}$ on the other. Consider the maps $g : \mathbb{R} \to \mathbb{R} \times \mathbb{R}$ given by $x \mapsto (x, \alpha x)$ and $h : \mathbb{R} \times \mathbb{R} \to S^1 \times S^1$ given by $(x, y) \mapsto (\exp(2\pi i x), \exp(2\pi i y))$. Clearly we have $f = h \circ g$. It is obvious that g imbeds $\mathbb{R}$ as a closed submanifold of $\mathbb{R} \times \mathbb{R}$. On the other hand, h is a local isomorphism of differential manifolds and we conclude that f makes

$\mathbb{R}$ an immersed manifold in $S^1 \times S^1$ in our sense. It is easy to see that the image with the induced topology is not even locally connected. Indeed this shows that the image is not a subspace at any point.

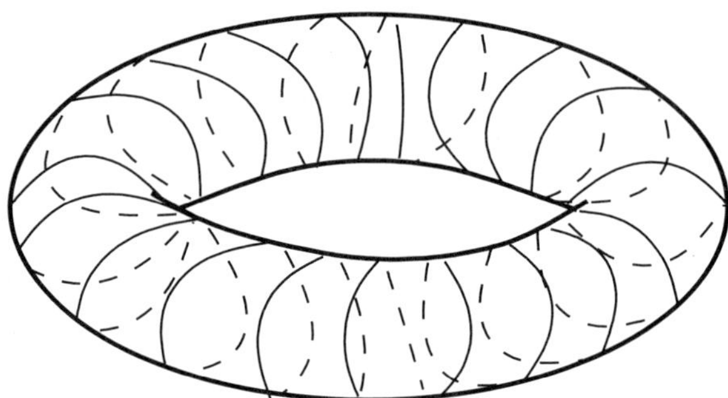

If α is irrational, it goes round and round infinitely. If it is rational, it rewinds at a finite stage.

We will indicate many simpler examples as well. A figure like 6 (open at the top end) can be realized as a submanifold of $\mathbb{R}^2$ by mapping $\mathbb{R}$ differentiably like

Clearly this figure with the topology induced from that of $\mathbb{R}^2$ is not a manifold at the nodal point.

Again a figure like 8 can be realised as an immersed manifold of $\mathbb{R}^2$ in two different ways, namely by parametrising it as follows.

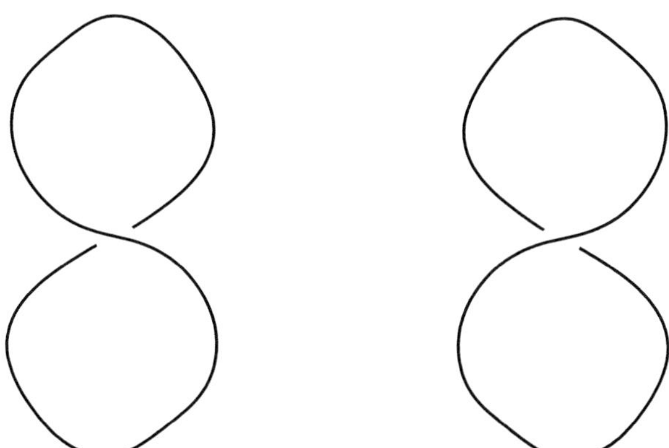

This example also drives home the point that *a closed submanifold is not just an immersed manifold whose point set is closed.* It also needs to have the induced topology.

4. Lie Groups; Action on a Manifold

4.1. Definition. *A Lie group G consists* of two structures on the same set G, namely it is a differential manifold and has also a group structure. The two structures are interrelated by the assumptions that the group law $G \times G \to G$ and the group inverse $G \to G$ are both differentiable.

4.2. Examples.

1) Any countable group with the discrete topology is a Lie group in our sense. (Countability is required because in our definition, manifolds are supposed to have a countable base of open sets.)

2) The real line $\mathbb{R}$ is a Lie group under addition since the maps $\mathbb{R} \times \mathbb{R} \to \mathbb{R}$ given by $(x, y) \mapsto x + y$ and $\mathbb{R} \to \mathbb{R}$ given by $x \mapsto -x$ are differentiable.

3) It is also clear that $\mathbb{R}^n$ is a Lie group under addition.

4) The multiplicative group $\mathbb{C}^\times$ consisting of nonzero complex numbers is an open submanifold of $\mathbb{C}$ and is actually a Lie group under multiplication.

5) The groups $GL(n, \mathbb{R})$ or $GL(n, \mathbb{C})$, which are open submanifolds of $\mathbb{R}^{n^2}$ and $\mathbb{C}^{n^2}$, are Lie groups. Indeed, the group composition is the restriction of a polynomial map $\mathbb{R}^{n^2} \times \mathbb{R}^{n^2} \to \mathbb{R}^{n^2}$. If V is any vector space of finite-dimension over $\mathbb{R}$ or $\mathbb{C}$, then the group $GL(V)$ of linear automorphisms is a Lie group.

6) The orthogonal group $O(n, \mathbb{R})$ (resp. the unitary group $U(n)$) is a Lie group under matrix multiplication.

4.3. Exercise. The quotient of $GL(V)$ by its centre, namely nonzero scalar matrices (or automorphisms), is called the *projective linear group $PGL(V)$*. Show that it is a Lie group.

4.4. Definition. A *homomorphism $G \to H$* of Lie groups is a group homomorphism which is also differentiable. A homomorphism of G into $GL(V)$ is said to be a *representation* of G in the vector space V.

It is clear that the composite of a homomorphism $G_1 \to G_2$ of Lie groups and another from G_2 to G_3 is a homomorphism from G_1 to G_3.

4.5. Definition. A *Lie subgroup H* of a Lie group G is a Lie group H with an injective homomorphism of H into G.

4.6. Remark. For any $g \in G$, the (right) translation map $G \to G$ given by $x \mapsto xg$ is of course differentiable, being the composite of the inclusion $G \to G \times G$ given by $g \mapsto (x, g)$ and the group operation. Its inverse is translation by g^{-1}. Hence right (and similarly left) translations are diffeomorphisms.

4.7. Definition. Let G be a Lie group and M a differential manifold. An *action* of G on M is a differentiable map $G \times M \to M$ denoted $(g, m) \mapsto gm$ such that $g_1(g_2 m) = (g_1 g_2)(m)$ for all $g_1, g_2 \in G$ and $m \in M$ and $1.m = m$ for all $m \in M$.

In Physics, the role of the Lie group is that of *the symmetries* of the system. Often the most important physical insight turns out to be the intuition for the appropriate group of symmetries.

4.8. Examples.

1) The proper orthogonal group $SO(3)$ acting on $\mathbb{R}^3$ and the group generated by it and translations (called the *Euclidean motion group*) are the group of symmetries in the study of motion of rigid bodies.

2) The group of all linear transformations of $\mathbb{R}^4$ which leave the symmetric bilinear form

$$((x_1, x_2, x_3, x_4), (y_1, y_2, y_3, y_4)) \mapsto -x_1 y_1 + x_2 y_2 + x_3 y_3 + x_4 y_4$$

   invariant, is called the *homogeneous Lorentz group*. If we consider the group generated by this group and translations, then it is called the *inhomogeneous Lorentz group*. Both these groups act on the differential manifold $\mathbb{R}^4$. This is the symmetry group for the theory of special relativity.

In quantum physics, one is often interested in representations of G in the projective unitary group (namely the group of unitary operators, modulo scalars) of a Hilbert space, but we will not deal with them in this book.

Exercises

1) Which of the following are sheaves on $\mathbb{R}^n$:
   a) For every open set define $\mathcal{F}(U)$ to be the space of square integrable functions on U.
   b) $\mathcal{F}(U)$ is the set of Lebesgue measurable functions on U.
   c) $\mathcal{F}(U)$ consists of continuous functions on U which are restrictions of continuous functions on $\mathbb{R}^n$.

2) Show that the stalk at 0 of the sheaf of differentiable functions on $\mathbb{R}^n$, $n \geq 1$, is an infinite-dimensional vector space over $\mathbb{R}$.

3) Show that the étale space associated to the sheaf of differentiable functions on $\mathbb{R}$ is not Hausdorff.

4) Show that any section of the sheaf $\mathcal{A}$ of continuous functions on a closed set of a normal topological space X can be extended to a section over the whole of X.

5) Determine for what values of a_i and c is the intersection of the hyperplane $\sum a_i x_i = c$ with the sphere $\sum x_i^2 = 1$, a closed submanifold of $\mathbb{R}^n$.

6) Let M be a differential manifold and f a differentiable function on it. Realise the open submanifold of M given by $f \neq 0$ as a closed submanifold of $M \times \mathbb{R}$.

7) Consider the map $x \mapsto x^2$ of $GL(2, \mathbb{R})$ (resp. $GL(2, \mathbb{C})$) into itself and find its image. Is the image a submanifold?

8) Show that the space of nonzero nilpotent $(2, 2)$ matrices is a closed submanifold of the space of nonzero matrices.

9) If G_1, G_2 are Lie groups, show that the product manifold $G_1 \times G_2$ with the direct product structure is also a Lie group.

10) Interpret the Jordan canonical form for matrices, as describing the orbits under the action of $GL(2, \mathbb{C})$ on the space $M(2, \mathbb{C})$ of all matrices given by $g.A = gAg^{-1}$. Determine which orbits are closed submanifolds.

Chapter 2

Differential Operators

We will now proceed to develop the formal machinery necessary to carry the notions of differential calculus in the Euclidean space over to arbitrary manifolds. The first step in this programme is to define differential operators on manifolds. We deal only with linear differential operators even if we do not say so explicitly each time. We will start with first order operators.

1. First Order Differential Operators

Let M be a differential manifold. We first define *homogeneous* first order operators on the algebra of differentiable functions, taking as characteristic example, an operator like

$$Df = \sum \varphi_i \frac{\partial f}{\partial x_i}$$

where φ_i are some differentiable functions on $\mathbb{R}^n$ or on an open subset U of $\mathbb{R}^n$. One of the basic properties of such an operator (sometimes called the *Leibniz property*) is

1.1. $$D(fg) = (Df)g + f(Dg)$$

for any two differentiable functions f, g on U. A k-linear homomorphism of a k-algebra into itself satisfying the Leibniz property, is for this reason called a *derivation*. Notice that a consequence of the definition is that $D(1) = 0$, for $D(1) = D(1.1) = D(1).1 + 1.D(1) = 2D(1)$. It follows that $D(\lambda) = 0$ for all $\lambda \in k$.

We propose to take the purely algebraic property 1.1 as the definition of a linear homogeneous differential operator of order 1 on an arbitrary manifold. We will now provide the justification for doing so.

Firstly, the algebraic condition implies that the operator D is local in the following sense.

1.2. Proposition. *If D is a linear operator on the $\mathbb{R}$-algebra of differentiable functions satisfying the Leibniz property, then the value of Df in any open set V depends only on the restriction of f to V.*

Proof. In fact, if $f = g$ in an open neighbourhood N of a point x in V, consider a differentiable function φ which is 1 in a smaller neighbourhood of x and vanishes outside N. Then we clearly have $\varphi(f - g) = 0$. Applying D and using 1.1 we get $(D\varphi)(f - g) + \varphi D(f - g) = 0$. In particular, we see that $\varphi D(f - g) = 0$ on N, proving that $D(f - g) = Df - Dg = 0$ in a neighbourhood of x. Since x is an arbitrary point of V, our assertion is proved.

Consequently, any map $\mathcal{A}(M) \to \mathcal{A}(M)$ which satisfies the Leibniz property (in particular, an operator of the form $\sum \varphi_i \frac{\partial}{\partial x_i}$) induces a sheaf homomorphism of $\mathcal{A}$ into itself, the homomorphism being one of $\mathbb{R}$-vector spaces.

1.3. Proposition. *Let U be an open submanifold of $\mathbb{R}^n$. If $D : \mathcal{A}^U \to \mathcal{A}^U$ is a sheaf homomorphism of $\mathbb{R}$-vector spaces, satisfying the Leibniz rule, then D is an operator of the form $f \mapsto \sum \varphi_i \frac{\partial f}{\partial x_i}$ for some differentiable functions φ_i on U.*

Proof. Clearly, if D is to be of the form $\sum \varphi_i \frac{\partial}{\partial x_i}$, then applying D to the functions x_i, we see that φ_i ought to be Dx_i. Replacing D by $D - \sum (Dx_i) \frac{\partial}{\partial x_i}$ we deduce that it is enough to prove the following. If D is a sheaf derivation such that $Dx_i = 0$ for all i, then D is the zero homomorphism. Let then $a = (a_1, \ldots, a_n)$ be a point of U and f any differentiable function in a neighbourhood of a. Then f can be written as the sum of the constant function $f(a)$ and $\sum (x_i - a_i) g_i$ in some neighbourhood of a. Let us assume this for the moment. Now $Df = \sum (x_i - a_i) Dg_i$ in view of our assumption. But $(x_i - a_i)$ vanishes at a, implying that $(Df)(a) = 0$. Since a is any point in U, it follows that $Df = 0$, as was to be proved.

It remains to prove our assertion about the decomposition of f. Indeed we have

$$
\begin{aligned}
f(x) - f(a) &= \int_0^1 \frac{d}{dt}\{f(tx + (1-t)a)\}dt \\
&= \int_0^1 \sum \frac{\partial}{\partial x_i} f(tx + (1-t)a) \cdot \frac{d}{dt}(tx_i + (1-t)a_i)dt \\
&= \sum (x_i - a_i) \int_0^1 \frac{\partial}{\partial x_i} f(tx + (1-t)a)dt.
\end{aligned}
$$

The above considerations motivate the following definition.

1.4. Definition. A *homogeneous first order differential operator* on a differential manifold M is a linear operator on $\mathcal{A}(M)$ satisfying the Leibniz rule. Equivalently, it is a sheaf homomorphism $\mathcal{A} \to \mathcal{A}$ which satisfies the Leibniz rule. A *linear differential operator of order at most one* on functions is of the form $f \mapsto Df + \varphi.f$ where D is a homogeneous operator as above.

From the geometric point of view, a homogeneous first order operator is called a *vector field* or an *infinitesimal transformation*. We will presently explain this alternate terminology.

2. Locally Free Sheaves and Vector Bundles

The set $\mathcal{T}(M)$ of all homogeneous first order operators on M has some nice structure. Firstly it forms a vector space over $\mathbb{R}$ in an obvious way. In fact, if $D_1, D_2 \in \mathcal{T}(M)$ then $D_1 + D_2$ is defined by

$$(D_1 + D_2)(\varphi) = D_1\varphi + D_2\varphi$$

for all $\varphi \in \mathcal{A}(M)$. It is obvious that it is also a derivation. Moreover if $f \in \mathcal{A}(M)$ then fD defined by

$$(fD)(\varphi) = fD\varphi$$

is also a derivation, thus making $\mathcal{T}(M)$ an $\mathcal{A}(M)$-module. Thirdly, if D_1, D_2 are two such operators, then the operator $[D_1, D_2]$ defined by

$$[D_1, D_2](\varphi) = D_1(D_2(\varphi)) - D_2(D_1(\varphi))$$

is also one such. For, if $\varphi, \psi \in \mathcal{A}(M)$, then we have

$$\begin{aligned}
[D_1, D_2](\varphi.\psi) &= D_1(D_2(\varphi.\psi)) - D_2(D_1(\varphi.\psi)) \\
&= D_1((D_2\varphi)\psi + \varphi(D_2\psi)) \\
&\quad - D_2((D_1\varphi)\psi + \varphi(D_1\psi)) \\
&= (D_1(D_2\varphi))\psi + (D_2\varphi)(D_1\psi) \\
&\quad + (D_1\varphi)(D_2\psi) + \varphi(D_1(D_2\psi)) \\
&\quad - (D_2(D_1\varphi))\psi - (D_1\varphi)(D_2\psi) \\
&\quad - (D_2\varphi)(D_1\psi) - \varphi(D_2(D_1\psi)) \\
&= ([D_1, D_2]\varphi)\psi + \varphi([D_1, D_2]\psi).
\end{aligned}$$

We will refer to the map $(D_1, D_2) \mapsto [D_1, D_2]$ as the *bracket operation*. Finally, since any $D \in \mathcal{T}(M)$ may also be described as a sheaf homomorphism of $\mathcal{A}$ into itself satisfying the Leibniz rule, it is clear that if $V \subset U$ are open subsets of M, a restriction map $\mathcal{T}(U) \to \mathcal{T}(V)$ may also be defined, making the assignment to any U of $\mathcal{T}(U)$ a sheaf as well. This sheaf will be denoted $\mathcal{T}$.

2.1. Definition. A *Lie algebra* over a commutative ring k is a module V over k with a k-bilinear operation

$$(X, Y) \mapsto [X, Y]$$

which satisfies

  i) $[X, X] = 0$;

  ii) (Jacobi's identity) $[X, [Y, Z]] + [Y, [Z, X]] + [Z, [X, Y]] = 0$,

for all $X, Y, Z \in V$. A *homomorphism* of a Lie algebra V into another Lie algebra W is a k-linear map $f : V \to W$ which satisfies

$$f([X, Y]) = [f(X), f(Y)]$$

for all $X, Y \in V$.

2.2. Examples.

1) Consider the n^2-dimensional vector space $M(n, k)$ of (n, n) matrices with coefficients in a commutative ring k. Define the bilinear map $(X, Y) \mapsto XY - YX$ to be the bracket operation (using matrix multiplication). Then it is a Lie algebra over k. More generally, any (associative) algebra can be regarded as a Lie algebra by defining $[x, y] = xy - yx$ where on the right side we use the algebra multiplication. If V is a vector space of dimension n, the algebra $\mathrm{End}(V)$ can be regarded as a Lie algebra and this is the abstract version of the matrix Lie algebra.

2) We can use various subspaces of $\mathrm{End}(V)$ or $M(n)$ which are closed under the bracket operation and obtain Lie algebras. For example, if b is a bilinear form on V, we may consider $\{X \in \mathrm{End}(V) : b(Xv, w) + b(v, Xw) = 0$ for every $v, w \in V\}$. It is easy to check that if X and Y satisfy this condition, $XY - YX$ also does.

3) Any vector space with the bracket operation defined by $[v, w] = 0$ gets a Lie algebra structure. A Lie algebra in which all the bracket operations are 0 is called an *abelian* Lie algebra.

4) Let A be any associative algebra over k. Then the space of k-derivations of A form a Lie algebra over k under the operation $[D_1.D_2] = D_1 D_2 - D_2 D_1$, for any two derivations D_1, D_2.

The relevance of this definition here is of course that the bracket operation we defined above on the space of homogeneous first order operators, endows it with the structure of a Lie algebra over $\mathbb{R}$. Although $\mathcal{T}(M)$ is a Lie algebra over $\mathbb{R}$ and is a module over $\mathcal{A}(M)$, it is *not* a Lie algebra over

$\mathcal{A}(M)$ since the bracket operation we defined above is not $\mathcal{A}(M)$-bilinear. In fact, we have

$$
\begin{aligned}
[D_1, fD_2](\varphi) &= D_1(f(D_2\varphi)) - (fD_2)(D_1(\varphi)) \\
&= D_1f.D_2\varphi + fD_1(D_2\varphi) - fD_2(D_1\varphi) \\
&= (D_1f.D_2 + f[D_1, D_2])(\varphi).
\end{aligned}
$$

It is obvious that the restriction maps $\mathcal{T}(U) \to \mathcal{T}(V)$ are all Lie algebra homomorphisms so that the sheaf $\mathcal{T}$ is a *sheaf of Lie algebras* over $\mathbb{R}$. Besides, we now have the following situation. On the one hand, $\mathcal{A}$ and $\mathcal{T}$ are sheaves on M and on the other $\mathcal{T}(U)$ is an $\mathcal{A}(U)$-module for every open subset U of M. The module structure is compatible with the restriction maps in an obvious sense. We recall the definition of a sheaf of $\mathcal{A}$-modules, already indicated in [Ch. 1, 2.7].

2.3. Definition. i) Let $\mathcal{A}$ be a sheaf of rings over a topological space X. Let $\mathcal{M}$ be a sheaf of abelian groups over X with a structure of $\mathcal{A}(U)$-module on $\mathcal{M}(U)$ for every open subset U of X. Then we say that $\mathcal{M}$ is a sheaf of $\mathcal{A}$-modules or that it is an *$\mathcal{A}$-module*, if

$$
\text{res}_{UV}(am) = (\text{res}_{UV}\, a)(\text{res}_{UV}(m))
$$

for all $a \in \mathcal{A}(U)$ and $m \in \mathcal{M}(U)$ and open sets $V \subset U$.

ii) An *$\mathcal{A}$-homomorphism* of an $\mathcal{A}$-module $\mathcal{M}_1$ into another $\mathcal{A}$-module $\mathcal{M}_2$ is a homomorphism f of sheaves of abelian groups such that the homomorphisms $f(U) : \mathcal{M}_1(U) \to \mathcal{M}_2(U)$ are $\mathcal{A}(U)$-linear, for all open sets U in X.

2.4. Examples.

1) With this definition, we see that if $(M, \mathcal{A})$ is a differential manifold, then $\mathcal{T}$ is an $\mathcal{A}$-module, besides being a sheaf of Lie algebras over $\mathbb{R}$. This sheaf will be called the *tangent sheaf.*

2) The direct sum $\mathcal{A}^r$ of $\mathcal{A}$ with itself r times (where $\mathcal{A}$ is a sheaf of rings) is obviously an $\mathcal{A}$-module.

3) If $(M, \mathcal{A})$ is a differential manifold and $Z \subset M$ is a closed set, one may consider for each open subset U of M, the set of all elements of $\mathcal{A}(U)$ which vanish on $Z \cap U$. This gives a sheaf $\mathcal{I}_Z$ which is clearly an $\mathcal{A}$-module. This is called the *ideal sheaf* of Z in M.

The local structure of the sheaf $\mathcal{T}$ as an $\mathcal{A}$-module is quite simple. We have seen that if (U, x) is a coordinate system, the sheaf $\mathcal{T}|U$ is actually isomorphic to $\mathcal{A}^n$, the map $(a_i) \mapsto \sum a_i \frac{\partial}{\partial x_i}$ providing such an isomorphism.

2.5. Definition. Let $\mathcal{A}$ be a sheaf of rings on a topological space X. An $\mathcal{A}$-module $\mathcal{M}$ is said to be *locally free of rank r* if every $x \in X$ has a neighbourhood U such that the restriction of $\mathcal{M}$ to U is isomorphic to $\mathcal{A}^r|U$ as an $\mathcal{A}|U$-module.

With this definition, our observation above implies that the tangent sheaf of a differential manifold is locally free as an $\mathcal{A}$-module, where $\mathcal{A}$ is the sheaf of differentiable functions.

2.6. Example. Let N be a closed submanifold of a differential manifold M. We defined in 2.4, Example 3) the sheaf of ideals $\mathcal{I}$ by the prescription $\mathcal{I}(U) = \{f \in \mathcal{A}(U) : f(x) = 0 \text{ for all } x \in U \cap N\}$. If N has dimension $n - 1$, then this sheaf is a locally free $\mathcal{A}$-module of rank 1. In order to see this, first observe that this being a local statement, we may assume that M is the unit ball in $\mathbb{R}^n$ and that N is the closed submanifold given by $x_n = 0$. If f is a function vanishing on N, then it can be written as $x_n.g$. In fact, we have

$$f(x_1, \ldots, x_n) = \int_0^1 \frac{\partial}{\partial t} f(x_1, \ldots, tx_n) dt = x_n \int_0^1 \frac{\partial f}{\partial x_n}(x_1, \ldots, tx_n) dt.$$

In other words, the ideal of functions vanishing on N is a principal ideal generated by x_n.

Any function f which generates the ideal of functions vanishing on N, can be written as $x_n.g$ with g nonvanishing. This implies that at any point of N, the function $\frac{\partial f}{\partial x_n} = g + x_n \frac{\partial g}{\partial x_n}$ does not vanish at any point of N. In other words, a function f is a generator of the ideal at 0 if and only if it vanishes on N and at least one of the partial derivatives of f is nonzero at 0.

2.7. Exercise. Show that the $\mathbb{R}$-linear map of the maximal ideal $\mathcal{M}$ of functions vanishing at 0 into $\mathbb{R}^r$ taking f to $((\frac{\partial f}{\partial x_1})_0, \ldots, (\frac{\partial f}{\partial x_r})_0)$ induces an isomorphism of $\mathcal{M}/\mathcal{M}^2$ onto $\mathbb{R}^r$. Conclude that if $r > 1$, then the ideal sheaf $\mathcal{M}$ of 0 is not locally free.

2.8. Definition. Let A be a constant sheaf of rings. Then any locally free sheaf of A-modules is called a *local system*.

Let $(M, \mathcal{A})$ be a differential manifold. Since sections of $\mathcal{A}$ are simply differentiable functions on U, it is natural to call sections of the $\mathcal{A}$-module $\mathcal{A}^r = \bigoplus^r \mathcal{A}$, *systems of functions* or *vector-valued functions*. In the general case of a locally free sheaf $\mathcal{E}$, let us discuss whether we can think of sections as some sort of functions. Locally it is indeed possible since we have assumed that $\mathcal{E}$ is locally isomorphic to $\mathcal{A}^r$. If $x \in M$, and $\mathcal{M}_x$ is the ideal of $\mathcal{A}_x$ consisting of germs of functions at x which vanish at x, one may consider for any $f \in \mathcal{A}_x$ its image in $\mathcal{A}_x/\mathcal{M}_x$. The natural evaluation map $\mathcal{A}_x \to \mathbb{R}$ taking any f to $f(x)$, gives an isomorphism of $\mathcal{A}_x/\mathcal{M}_x$ with $\mathbb{R}$. Therefore one may think of the evaluation map as taking the image of f in $\mathcal{A}_x/\mathcal{M}_x$. Guided by this fact, we may take the $\mathbb{R}$-vector space $\mathcal{E}_x/\mathcal{M}_x\mathcal{E}_x$ to be the space in which the sections of $\mathcal{E}$ take values at the point $x \in M$. The (important) difference between the case of $\mathcal{A}$ and the general case of a locally free sheaf, is that the vector space associated to $x \in M$, namely $E_x = \mathcal{E}_x/\mathcal{M}_x\mathcal{E}_x$ *depends* on the point $x \in M$. In other words, there is no natural isomorphism of these vector spaces at two different points. Consider the set union E of all these sets, namely $\bigcup_{x \in M} E_x$. This comes with a natural map π into M, with $\pi^{-1}(x) = E_x$. For any $x \in M$ there exists an open neighbourhood U such that $\mathcal{E}|U$ is isomorphic to $\mathcal{A}^r$, so that $\pi^{-1}(U)$ may be identified with $U \times \mathbb{R}^r$ and the map $\pi^{-1}(U) \to U$ with the projection $U \times \mathbb{R}^r \to U$. In particular, the set $\pi^{-1}(U)$ can be provided with the differential structure of the product. This is obviously independent of the isomorphism $\mathcal{E}|U \to \mathcal{A}^r$ chosen and hence may be patched together to yield a differential structure on E. It is easy to see that the topology we have introduced is Hausdorff and that it admits a countable base for open sets. Now we have the following structures on E.

a) E is a differential manifold.

b) $\pi : E \to M$ is a differentiable map.

c) For any $x \in M$, there exist an open neighbourhood U of x and a diffeomorphism $\pi^{-1}(U) \to U \times \mathbb{R}^r$ such that π becomes the projection $U \times \mathbb{R}^r \to U$ following this isomorphism.

d) There is a structure of a vector space on each fibre $\pi^{-1}(x)$ for every $x \in M$, which is compatible with the isomorphism above. This means that if we identify $\pi^{-1}(x)$ with $\mathbb{R}^r$ using the local isomorphism in c) above, then the identification is a linear isomorphism.

2.9. Definition. A differentiable map $\pi : E \to M$ satisfying a) – d) above, is called a *differentiable vector bundle of rank r*. A *homomorphism* of a vector bundle E into F is a differentiable map $E \to F$ which makes the diagram

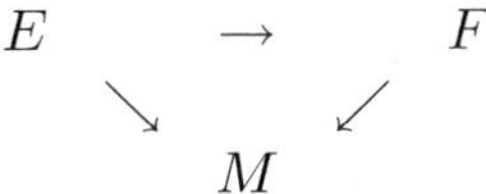

commutative such that the induced maps on fibres are all linear.

The isomorphism of the type described in c) above is referred to as a *local trivialisation of the bundle E*.

We have associated, to any locally free sheaf $\mathcal{E}$ of $\mathcal{A}$-modules, a differentiable vector bundle $E \to M$. Conversely, if $\pi : E \to M$ is a differentiable vector bundle, then the sheaf of differentiable sections of π is a locally free sheaf of $\mathcal{A}$-modules. Moreover, if $\mathcal{F}$ is another locally free sheaf of $\mathcal{A}$-modules with associated vector bundle F, then there is a natural bijection between the set of $\mathcal{A}$-linear sheaf homomorphisms $\mathcal{E} \to \mathcal{F}$ and the set of vector bundle homomorphisms $E \to F$. Thus one may use the notions of 'locally free sheaf' and 'differentiable vector bundle' interchangeably.

2.10. Remark. One has however to guard against the following possible confusion. A *subbundle* of E is a vector bundle F together with a homomorphism $F \to E$ which is injective on *all* fibres. It gives rise to an injective $\mathcal{A}$-homomorphism of $\mathcal{F}$ into $\mathcal{E}$, namely an $\mathcal{A}$-subsheaf of $\mathcal{E}$. Conversely, the inclusion of an $\mathcal{A}$-module $\mathcal{F}$ in $\mathcal{E}$ gives rise to a homomorphism F into E, but the latter need not be injective on *all* fibres.

2.11. Exercise. Consider the inclusion of the ideal sheaf of any point in $\mathbb{R}$ in the sheaf $\mathcal{A}$. Show that the corresponding homomorphism at the vector bundle level is not injective on all fibres.

If $\varphi : M \to M'$ is a differentiable map and E is a differentiable vector bundle on M', then one can define a differentiable vector bundle $\varphi^* E$ called the *pull-back* of E as follows. Take the subspace of $M \times E$ defined by

$$\varphi^* E = \{(m, x) \in M \times E : \varphi(m) = \pi(x)\}.$$

It is obvious that it is Hausdorff and admits a countable base for open sets. This subspace comes with two natural maps, namely, $\pi' : \varphi^* E \to M$ given by $\pi'(m, x) = m$ and $\tilde{\varphi} : \varphi^* E \to E$ given by $\tilde{\varphi}(m, x) = x$. If U is an open set over which E is trivial and $V = \varphi^{-1}(U)$, then $\pi'^{-1}(V)$ can be identified with $V \times \mathbb{R}^r$ so that $\varphi^* E$ is a differential manifold, with respect to which π' is differentiable. Clearly the fibre $\pi'^{-1}(m)$ over $m \in M$ can be identified with the fibre $\pi^{-1}(\varphi(m))$ and hence has a natural vector space structure. It is now easy to verify that $\pi' : \varphi^* E \to M$ satisfies the conditions a) – d) and hence $\varphi^* E$ is a vector bundle.

2.12. Definition. The vector bundle $\varphi^* E$ defined above is called the *pull-back of the vector bundle E by the map $\varphi : M \to M'$*. For every differentiable section s of E one can define a section of $\varphi^* s$ of $\varphi^* E$ again called the *pull-back of s* in such a way that

$$\tilde{\varphi}(\varphi^* s(m)) = s(\varphi(m))$$

for all $m \in M$.

Suppose $\mathcal{E}$ is a locally free sheaf of $\mathcal{A}$-modules on M' and φ is a differentiable map $M \to M'$. Then the inverse image $\varphi^{-1} \mathcal{A}^{M'}$ as defined in

Chapter 1, 2.3 is a sheaf of algebras over M. The inverse image $\varphi^{-1}(\mathcal{E})$ is *not* a sheaf of $\mathcal{A}^M$-modules but only a locally free sheaf of $\varphi^{-1}(\mathcal{A}^{M'})$-modules. The companion map of φ, namely the map $\underline{\varphi} : \varphi^{-1}(\mathcal{A}^{M'}) \to \mathcal{A}^M$ may be used to *define* $\varphi^*\mathcal{E}$ as $\varphi^{-1}\mathcal{E} \otimes_{\varphi^{-1}(\mathcal{A}^{M'})} \mathcal{A}^M$, where $\mathcal{A}^M$ is considered as a sheaf of $\varphi^{-1}\mathcal{A}^{M'}$-modules via $\underline{\varphi}$. With this definition it is easy to see that the vector bundle associated to $\varphi^*\mathcal{E}$ can be canonically identified with the pull-back φ^*E of the vector bundle E associated to $\mathcal{E}$.

2.13. Remarks.

1) All the constructions which we have made above are also valid for locally free sheaves of $\mathcal{A}^M \otimes \mathbb{C}$-modules. The corresponding vector bundles are bundles of vector spaces over $\mathbb{C}$. If E is a complex vector bundle, one can define $\overline{E}$ by setting $\overline{E} = E$ as a differential manifold but changing the vector space structure on the fibres of $\overline{E}$ by redefining multiplication by $i = \sqrt{-1}$ as multiplication by $-i$ in E. Given any real vector bundle E one can associate to it a complex vector bundle $E_\mathbb{C}$ by tensoring with $\mathbb{C}$. Of course in such a case, we have a canonical isomorphism of $\overline{E}_\mathbb{C}$ with $E_\mathbb{C}$.

2) We refer to complex vector bundles of rank 1 as *line bundles*. Real vector bundles of rank 1 are not that interesting. In fact, if L is one such, then consider the relation given by declaring two points v, v' of L to be equivalent if there exists a *positive* real number a such that $v' = av$. Then the quotient space maps onto M and has precisely two points on each fibre. In view of the local triviality of the line bundle, the quotient is easily seen to be a two-sheeted (étale) covering space M' of M. If M is simply connected, this covering consists therefore of two copies of M. Choosing one of them is equivalent to choosing for all $m \in M$, one of the components of $L_m \setminus \{0\}$, where L_m is the fibre of L at m. We may call elements of that component positive vectors. Local trivialisation of L is equivalent to the data consisting of an open covering $\{U_i\}$ and sections s_i on each U_i which are everywhere nonzero. We may in the above situation, actually choose positive sections s_i, i.e. sections whose values are in the chosen component in the fibre.

We may then take a partition of unity $\{\varphi_i\}$ for the covering and get a global everywhere nonzero section $\sum \varphi_i s_i$ of L. This shows that L is globally trivial. This argument actually shows that even if M were not simply connected, we can pull back L to the two-sheeted covering M' and trivialise it on M'.

3) If M is a closed submanifold of $\mathbb{R}^n$ of dimension $n - 1$, then we have remarked already in (2.6) that its ideal sheaf is a locally free $\mathcal{A}$-module of rank 1. By 2) this sheaf is globally trivial. Therefore there is a section which generates the local ideal everywhere, that is to say, there is a function f which vanishes precisely on M and generates the ideal sheaf locally at all points.

Let us get back to our sheaf $\mathcal{T}$, the tangent sheaf of M. Applying the above considerations to it, we get a vector bundle, which is called the *tangent bundle T*. For any $x \in M$, the fibre over x is called the *tangent space* at x. Elements of this space are called *tangent vectors at x*. A differentiable section of this vector bundle may therefore be called a *vector field* and thus the concept of a homogeneous differential operator of first order and that of a vector field are essentially equivalent.

The tangent space at a point $x \in M$ is thus the quotient space $\mathcal{T}_x/\mathcal{M}_x\mathcal{T}_x$. If X is a germ of a vector field at x then the map $f \mapsto Xf$ induces a derivation of the $\mathbb{R}$-algebra $\mathcal{A}_x$. The map $f \mapsto (Xf)(x)$ gives rise to a map $t : \mathcal{A}_x \to \mathbb{R}$ which is a derivation in the sense that

$$t(fg) = (tf)g(x) + f(x)t(g)$$

for all $f, g \in \mathcal{A}_x$. By this correspondence, one easily verifies that the tangent space at x can be identified with the set of linear maps $\mathcal{A}_x \to \mathbb{R}$ satisfying the above condition.

In the case when $M = \mathbb{R}^n$, or more abstractly a vector space V of dimension n, the tangent bundle is trivial. The tangent space at any point $v \in V$ can be identified with the vector space itself. In fact, we associate to any $x \in V$ the derivation $\partial_v : \mathcal{A}_x \to \mathbb{R}$ given by $f(x) \mapsto \lim_{t\to 0} \frac{f(x+tv)-f(x)}{t}$. If N is a submanifold of V, then the tangent bundle of N is a subbundle of the trivial bundle $N \times V$. Thus the tangent space at any point $P \in N$ is a subspace of V. We visualise this geometrically as the coset space of this subspace in V which contains P. In other words, the geometric tangent space is then the space parallel to the abstract tangent space, passing through P.

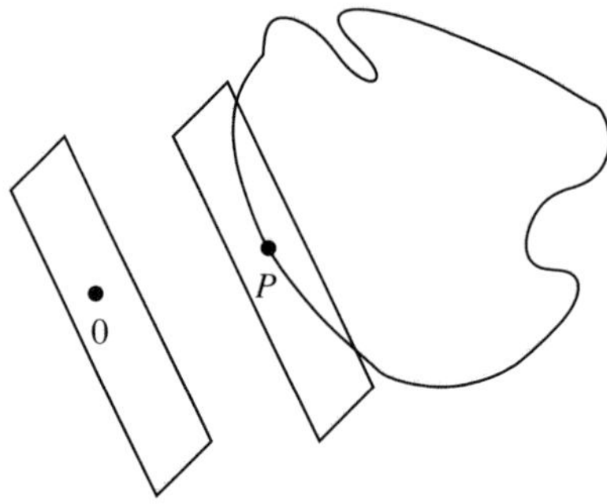

2.14. Exercise. If a submanifold N of $\mathbb{R}^n$ is given by $f = 0$ where f is a differentiable function with the property that at least one of the partial derivatives of f is nonzero at every point of N, write down the equation of the tangent space at a point $n \in N$.

2.15. Differential of a map.

If $\varphi : M \to M'$ is a differentiable map of a manifold M into M' and t is a tangent vector at $m \in M$, then one can define a tangent vector $\varphi(t)$ at $m' = \varphi(m)$, by setting $\varphi(t)(f) = t(f \circ \varphi)$ for all $f \in \mathcal{A}_{m'}$. This defines a linear map $T_m(M) \to T_{m'}(M')$ called the *differential* of φ. The tangent space at m is to be thought of as the linear approximation of M near m and the differential of φ as the linear approximation of the map φ. Globally speaking, the differential gives a vector bundle homomorphism of T_M into $\varphi^*(T_{M'})$, or what is the same, a sheaf homomorphism of $\mathcal{T}_M$ into $\varphi^*(\mathcal{T}_{M'})$. This homomorphism is usually denoted by $d\varphi$. Occasionally, we may write φ for this differential as well.

If (U, x) (resp. (V, y)) is a coordinate system in M (resp. M'), such that $\varphi(x) = y$ and $\varphi(U) \subset V$, then φ is given by functions $\varphi_i(x_1, \ldots, x_n)$. Then its differential takes $\frac{\partial}{\partial x_k}$ to the vector at $\varphi(x)$ which takes the coordinate functions y_j to $\frac{\partial}{\partial x_k}(\varphi_j(x))$. Hence we deduce that the differential of φ maps $\frac{\partial}{\partial x_k}$ to $\sum \frac{\partial}{\partial x_k}(\varphi_j(x))\frac{\partial}{\partial y_j}$. In other words, the matrix of the linear map $T_x \to T_{\varphi(x)}$ with respect to the standard bases, is given by $A_{ij} = \frac{\partial}{\partial x_i}(\varphi_j)$. This matrix is called the *Jacobian* of the map φ.

If f is a differentiable function, namely a differentiable map $M \to \mathbb{R}$, then df is thus a homomorphism $T_M \to f^*(T_{\mathbb{R}})$. But $T_{\mathbb{R}}$ is a trivial bundle since the vector field $\frac{d}{dx}$ forms a basis for the tangent spaces at all points of $\mathbb{R}$. Hence df may be regarded as a homomorphism of T_M into the trivial bundle of rank 1, or what is the same, a section of the dual vector bundle. This may simply be called the *differential* of f.

2.16. Normal bundle.

If $f : N \to M$ is a differentiable map such that the differential of f is injective at all points of N, then the differential of f, namely the homomorphism $T_N \to f^*T_M$, is a subbundle inclusion. The cokernel of this is called the *normal bundle* of N in M. Thus we have the exact sequence of vector bundles:

$$0 \to T_N \to f^*(T_M) \to \mathrm{Nor}(N, M) \to 0.$$

Under this assumption, the implicit function theorem assures us of the existence of a local coordinate system (U, x) at a point $n \in N$ and a local coordinate system (V, y) at $f(n)$ such that $f|U$ is injective and $f(U) \subset V$ is given by the vanishing of some of the coordinates, say $y_1, \ldots, y_r$. The tangent space to N is freely generated by $\frac{\partial}{\partial y_{r+1}}, \ldots, \frac{\partial}{\partial y_n}$.

If in addition f is injective, it follows that N is a manifold immersed in M. In that case, the notion of normal bundle is more geometric. Suppose now that N is a closed submanifold of $\mathbb{R}^n$. Then the tangent space at $P \in N$ is a subspace of $\mathbb{R}^n$, and the normal space is a quotient. But using the Euclidean metric on $\mathbb{R}^n$, we may identify this quotient with a subspace as well. We may visualise it as the coset of this latter space, passing through P.

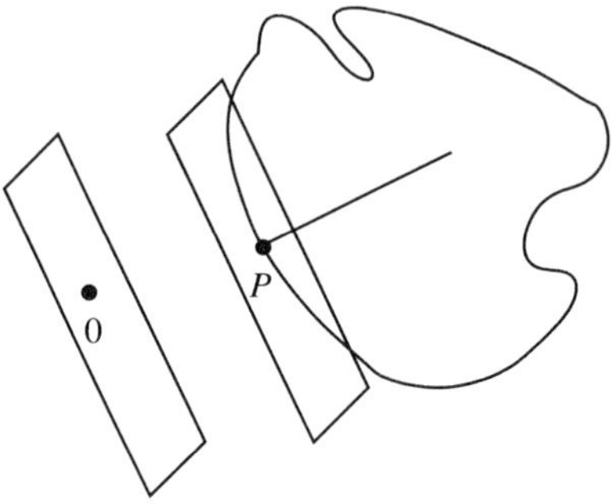

2.17. Exercises.

1) Let N be a submanifold of $\mathbb{R}^n$ given by $f = 0$ as in 2.14. Write down the equation of the normal space at any point $n \in N$.

2) If M is a closed submanifold of $\mathbb{R}^n$ of dimension $n - 1$, show that its normal bundle is trivial.

On the other hand, the differential of f may even be an isomorphism at all points, without the map f being injective.

2.18. Definition. If the differential of $f : N \to M$ is an isomorphism at all points of N, we call it an *étale map*.

2.19. Example. The map $t \mapsto \exp(it)$ of $\mathbb{R} \to S^1$ is an étale map.

If an étale map is also injective, it follows from the definition that it is an inclusion of an open submanifold.

3. Flow of a Vector Field

A geometric way of looking upon differentiation in $\mathbb{R}^n$ is the following. Suppose given a vector $a = (a_1, \ldots, a_n)$. For any $t \in \mathbb{R}$, consider the transformation $\varphi_t : (x_1, \ldots, x_n) \mapsto (x_1 + ta_1, \ldots, x_n + ta_n)$ of $\mathbb{R}^n$ into itself. This is a homomorphism of the additive group $\mathbb{R}$ into the group of diffeomorphisms of $\mathbb{R}^n$. For any differentiable function f on $\mathbb{R}^n$, define

$$(D_a(f))(x) = \lim_{t \to 0} \frac{f(\varphi_t(x)) - f(x)}{t}.$$

This exists and in fact defines the differential operator $\sum a_i \frac{\partial}{\partial x_i}$. We may generalise this idea to an arbitrary differential manifold. Let (φ_t), $t \in \mathbb{R}$, be a *one-parameter group of diffeomorphisms* of M in the sense that

i) the map $\mathbb{R} \times M \to M$ given by $(t, x) \to \varphi_t(x)$ is differentiable.

ii) the map $\varphi_0 : M \to M$ is the identity.

iii) $\varphi_t \circ \varphi_{t'} = \varphi_{t+t'}$ for all $t, t' \in \mathbb{R}$.

Then we may define differentiation of functions f with respect to the above data by setting

$$(X_\varphi f)(x) = \lim_{t \to 0} \frac{f(\varphi_t(x)) - f(x)}{t}.$$

It is easy to see that X_φ is indeed a homogeneous first order operator (i.e. a vector field). In fact, the linearity of X_φ is obvious, while we have for every $f, g \in \mathcal{A}(M)$,

$$\begin{aligned}(X_\varphi(fg))(x) &= \lim_{t \to 0} f(\varphi_t(x)) \frac{g(\varphi_t(x)) - g(x)}{t} \\ &\quad + \lim_{t \to 0} g(x) \frac{f(\varphi_t(x)) - f(x)}{t} \\ &= f(x)(X_\varphi g)(x) + g(x)(X_\varphi f)(x).\end{aligned}$$

3.1. Examples.

1) Take $M = \mathbb{R}$ and $\varphi_t(x) = x + t$ for all $t, x \in \mathbb{R}$.

2) Take $M = \mathbb{R}$ and $\varphi_t(x) = e^t x$, for all $t, x \in \mathbb{R}$.

In these two cases, we see that the associated operators are respectively $\frac{d}{dx}$ and $x \frac{d}{dx}$.

3) Again take $M = \mathbb{R}$ and consider the function $\varphi_t(x) = \frac{x}{1 - tx}$. It is easy to see formally that $\varphi_t \circ \varphi_{t'} = \varphi_{t+t'}$. But, for any given $x \neq 0$, $\varphi_t(x)$ is only defined for $t < 1/x$. We are dealing here with a *local* 1-parameter group of *local* automorphisms. In other words, for any x there exist a neighbourhood U and $\epsilon > 0$, such that $\varphi_t(y)$ is defined for all $|t| < \epsilon$ and $y \in U$. The group condition iii) above is satisfied to the extent it makes sense. But notice that the associated vector field is still meaningful.

3.2. Exercises.

1) Compute the vector field given in Example 3) above.

2) Determine the vector field given by the one-parameter group (φ_t) whose action on $\mathbb{R}^2$ is given by

$$\varphi_t(v, w) = (\cos(t)v + \sin(t)w, -\sin(t)v + \cos(t)w).$$

With this extended notion we have a converse.

3.3. Theorem. *Let X be a vector field on a differential manifold M. Then for every $x \in M$, there exist an open neighbourhood U of x, $\epsilon > 0$, and maps $\varphi_t : U \to M$ for all $|t| < \epsilon$, satisfying:*

i) *the map $(-\epsilon, \epsilon) \times U \to M$ given by $(t, y) \mapsto \varphi_t(y)$ is differentiable.*

ii) *the map $\varphi_0 : U \to M$ is the inclusion.*

iii) *$(\varphi_t \circ \varphi_{t'})(y) = \varphi_{t+t'}(y)$ for all $|t|, |t'| < \epsilon$ and $y \in M$ such that $|t+t'| < \epsilon$ and $y, \varphi_{t'}(y)$ are in U.*

iv) *$X_\varphi = X$.*

Proof. Since X_φ and X are both vector fields, the assertion is purely local and so we may replace M by an open set in $\mathbb{R}^n$ and prove the existence of maps φ_t such that $X_\varphi x_i = X x_i$ for all i. This reduces our task to showing the following. Given differentiable functions a_i, we need to find functions

$$\varphi(t, x) : (-\epsilon, \epsilon) \times U \to \mathbb{R}^n$$

for some neighbourhood U of a given point such that

$$\lim_{t \to 0} \frac{\varphi_i(t, x) - x_i}{t} = a_i(x), \quad \text{for all } i.$$

This implies that

$$
\begin{aligned}
\tfrac{\partial}{\partial t} \varphi_i(t, x) &= \lim_{t' \to 0} \frac{\varphi_i(t' + t, x) - \varphi_i(t, x)}{t'} \\
&= \lim_{t' \to 0} \frac{\varphi_i(t', \varphi(t, x)) - \varphi_i(t, x)}{t'} \\
&= a_i(\varphi(t, x)).
\end{aligned}
$$

We also have the initial condition $\varphi_i(0, x) = x_i$. So we start with this equation and note that it has a unique solution in a neighbourhood of $(0, x)$ in $\mathbb{R} \times \mathbb{R}^n$. To prove that iii) is satisfied, we use the uniqueness of the solution. In fact, both $\varphi(t + t', x)$ and $\varphi(t, \varphi(t', x))$ are solutions of the equation $\frac{d}{dt} \psi_i(t, x) = a_i(\psi(t, x))$ with the initial condition $\psi_i(0, x) = \varphi_i(t', x)$. Finally equation iv) is obvious from the construction.

From this point of view, the term 'infinitesimal transformation' is an appropriate alternative to that of a 'vector field'.

3.4. Definition. The one-parameter group associated to a vector field is called the *flow* of the vector field.

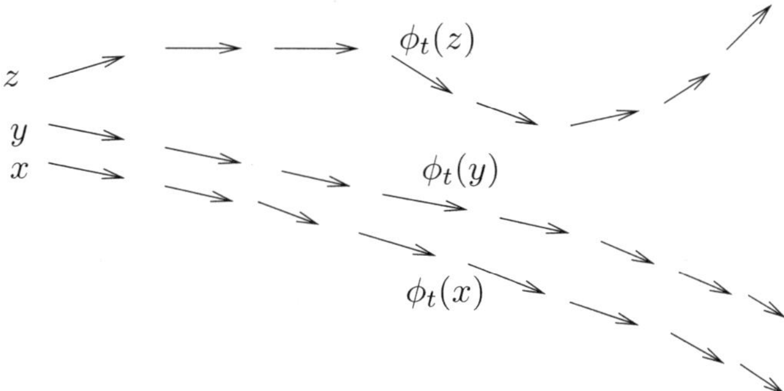

3.5. Remarks.

1) Although a vector field gives rise in general only to a local 1-parameter group, a limited globalization is possible. Indeed, given a compact set $K \subset M$, we can define φ_t in a neighbourhood of K for all $|t| < \epsilon$. For, by Theorem 3.3 this can be done in a neighbourhood of every point of K. Since K can be covered by finitely many of these neighbourhoods, $\varphi_t(y)$ are defined in the same open set $|t| < \epsilon$ for small enough ϵ, and for y in a neighbourhood of K. In particular, if M is itself compact, then φ_t is defined as an automorphism of M for all $|t| < \epsilon$, and hence by iteration, we get in this case, a *global* flow.

2) It is obvious that if X depends differentiably on some parameters s, then the one-parameter group is defined for small values of the parameters and depends differentiably on them.

3.6. Definition. A vector field which gives rise to a global flow is said to be *complete*.

We have seen above that any vector field on a compact manifold is complete. It is easily seen that the vector field $x^2 \frac{d}{dx}$ on $\mathbb{R}$ is not complete.

3.7. Exercise. Let M be a compact manifold and X a vector field. If $m \in M$, determine when the restriction of X to the open submanifold $M \setminus \{m\}$ is complete.

3.8. Definition. If X is a vector field and (φ_t) the flow corresponding to it, the orbit of a point $m \in M$ under φ_t, namely the map $t \mapsto \varphi_t(m)$, is called an *integral curve* for X.

3.9. Remark. The integral curve of a vector field X has the property that the differential of this map at t maps $\frac{d}{dt}$ to $X_{\varphi_t(m)}$. This characterises the curve. In particular, the curve degenerates to a constant map if and only if $X_m = 0$. If X is 0 at a point m, we say that m is a *singularity* of X. Our remark amounts to saying that the one-parameter group φ_t fixes a point m if and only if m is a singularity of X.

Suppose that m is not a singularity. Then by continuity, we see that $X_x \neq 0$ for all x in a neighbourhood of m. Now the integral curve has injective differential at all points near 0 and hence it is an immersed manifold of dimension 1. Actually there is a local coordinate system (U, x) in which X is given by $\frac{\partial}{\partial x_1}$. Indeed, suppose $X = \sum a_i \frac{\partial}{\partial x_i}$, where by our assumption one of the a_i's, say a_1, is nonzero. Consider the coordinate system given by $(y_1, \ldots, y_n)$ where $y_i = \varphi_i(x_1, 0, x_2, \ldots, x_n)$. We now compute the partial derivatives $\frac{\partial y_i}{\partial x_j}$ at m given by $x_i = 0$ for all i:

$$\frac{\partial y_i}{\partial x_1} = a_i(0); \qquad \frac{\partial y_i}{\partial x_j} = \delta_{i,j} \quad \text{for } j \geq 2.$$

This shows that $(y_1, \ldots, y_n)$ is a coordinate system in a neighbourhood of m. It is easy to see that this coordinate system serves the purpose.

Invariant vector fields.

We wish to study now vector fields on a Lie group.

3.10. Definition. Suppose M is differential manifold and a Lie group G acts on it. Then a vector field X on M is said to be *invariant* under the action if the transform of X by any element of G is the same as X, that is to say, for every $m \in M$ and $g \in G$, X_{gm} is the image of the vector X_m under the differential at m of the map $x \mapsto gx$ of M into itself.

From the uniqueness of the flow corresponding to a vector field, we deduce that if the flow of X is φ_t, then the flow corresponding to gX is given by $\psi_t(m) = g\varphi_t(g^{-1}m)$. Therefore, if X is invariant under G, then the flows ψ_t and φ_t are the same, so that φ_t commutes with the action of G for all t.

Notice also that from the definition of Lie brackets of vector fields it follows that if X and Y are G-invariant, then $[X, Y]$ is also invariant. In particular, the vector space of invariant vector fields is actually a Lie algebra. We have already remarked that G acts on itself by left translations and so *left invariant vector fields of a Lie group form a Lie algebra over* $\mathbb{R}$.

3.11. Definition. Let G be a connected Lie group. The Lie algebra of *vector fields* which are left invariant (i.e. invariant under left translation by elements of G) is called the *Lie algebra of G*, and is often denoted by $Lie(G)$ or $\mathfrak{g}$.

3.12. Remark. There is a natural linear map of $\mathfrak{g}$ into the tangent space $T_1(G)$ at 1, given by $X \mapsto X_1$. It is an isomorphism and the inverse associates to any vector $v \in T_1(G)$, the vector field X given by $X_g = L_g(v)$, where L_g denotes the differential at 1 of the left translation by g. So $\mathfrak{g}$ is a Lie algebra of dimension $n = \dim(G)$.

3.13. Examples.

1) The left invariant vector fields on the additive Lie group $\mathbb{R}$ are of the form $a\frac{d}{dt}$, $a \in \mathbb{R}$. Hence the Lie algebra of $\mathbb{R}$ is canonically isomorphic to the abelian Lie algebra $\mathbb{R}$. In the same way, if we take the additive Lie group underlying a vector space V, then its Lie algebra is identified with the abelian Lie algebra V.

2) On the other hand, if we take the multiplicative group $\mathbb{R}^\times$ or its connected component $\mathbb{R}^+$ containing 1, then the invariant vector fields are scalar multiples of $t\frac{d}{dt}$. Thus its Lie algebra is also the abelian Lie algebra $\mathbb{R}$.

3) The Lie algebra of the group $GL(n, \mathbb{R})$ or the connected component $GL(n, \mathbb{R})^+$ containing 1, can be identified with $M(n, \mathbb{R})$. It is clear that the tangent space at 1 of this open submanifold is canonically $M(n)$ as a vector space. It only remains to compute the Lie algebra structure. Denote by x_{ij} the function which associates to any matrix A its (i,j)th coefficient. Let $A \in M(n)$ and X be the left invariant vector field on $GL(n)$ such that $X_1(x_{ij}) = A_{ij}$. For any $s \in GL(n, \mathbb{R})^+$, we have $X_s(x_{ij}) = X_1(x_{ij} \circ L_s)$, where L_s is left translation by s. Hence the function $X x_{ij}$ is given by $s \mapsto \sum_k x_{ik}(s) X_1 x_{kj} = \sum x_{ik} A_{kj}$. If Y is any left invariant vector field with $Y_1(x_{ij}) = B_{ij}$, then we have $(XY - YX)_1(x_{ij}) = X_1(Y x_{ij}) - Y_1(X x_{ij}) = X_1(\sum x_{ik} B_{kj}) - Y_1(\sum x_{ik} A_{kj}) = \sum A_{ik} B_{kj} - \sum B_{ik} A_{kj} = (AB - BA)_{ij}$. In other words, if we identify left invariant vector fields on $GL(n)$ with $M(n)$, then the Lie bracket is given by the bracket associated with the multiplication in the matrix algebra.

3.14. Exercises.

1) Show that the left invariant vector fields on $GL(n)$ are generated by $E_{p,q} = \sum x_{i,p} \frac{\partial}{\partial x_{i,q}}$ and that the right invariant vector fields by $F_{p,q} = \sum x_{q,j} \frac{\partial}{\partial x_{p,j}}$.

2) Deduce that $E_{p,q}$ and $F_{r,s}$ commute. Explain this in terms of their flows.

The integral curve through 1 of a left invariant vector field, namely $t \mapsto \varphi_t(1)$, is actually a homomorphism of the group $\mathbb{R}$ into G. In fact, left invariance of X implies that for every $g \in G$, the automorphism $x \mapsto g\varphi_t(x)$ is the same as $x \mapsto \varphi_t(gx)$. Taking $g = \varphi_{t'}(1)$, we get $\varphi_{t'}(1)\varphi_t(1) = \varphi_t(\varphi_{t'}(1)) = \varphi_{t+t'}(1)$.

3.15. Definition. A differentiable group homomorphism of the Lie group $\mathbb{R}$ into a Lie group G is called a *one-parameter group* .

If $\rho : \mathbb{R} \to G$ is a 1-parameter group, then $\rho(\frac{d}{dt})$ gives a vector at 1 which in turn defines a left invariant vector field X. It is clear that ρ gives an integral curve for X.

3.16. Remarks.

1) If the vector field X is 0, the corresponding one-parameter group is the constant homomorphism $t \mapsto 1$. Even if X is not 0, the map $t \mapsto \varphi_t(1)$ mentioned above, may not be injective. If we take $G = S^1 = \{(x,y) \in \mathbb{R}^2 : x^2 + y^2 = 1\}$, then the left invariant vector fields form a 1-dimensional vector space generated by $X = x\frac{\partial}{\partial y} - y\frac{\partial}{\partial x}$. The image of $\frac{d}{dt}$ under the map $t \mapsto (\cos at, \sin at)$ is easily computed to be aX. Hence it is the one-parameter group of the vector field aX. Its kernel is the subgroup $\frac{2\pi}{a}\mathbb{Z}$ of $\mathbb{R}$.

2) Also the image of a one-parameter group is not in general closed. For example, consider the case when $G = S^1 \times S^1$ and $X = (\frac{d}{dt}, a\frac{d}{dt})$. Then the induced one-parameter group is given by

$$t \mapsto ((\cos t, \sin t), (\cos at, \sin at)).$$

This is a closed submanifold if and only if a is rational. See the figure in [Ch. 1, 3.17].

3) If $g \in G$, then (the differential of) the inner automorphism $x \mapsto gxg^{-1}$ of G takes the vector field X to another left invariant vector field which we may denote gXg^{-1}. If $t \mapsto c(t)$ is the flow of X, then the flow of gXg^{-1} is given by $t \mapsto gc(t)g^{-1}$.

3.17. Definition. The representation of a Lie group G into $GL(\mathfrak{g})$ which associates to $g \in G$ the automorphism $\mathrm{Ad}(g) = X \mapsto gXg^{-1}$ is called the *adjoint representation* of G.

3.18. Remark. The linear automorphism $\mathrm{Ad}(g)$ is actually an automorphism of the *Lie algebra* $\mathfrak{g}$.

3.19. Exercise. Show that if the image of the one-parameter group is closed, then it it is actually a closed submanifold.

If G and H are Lie groups and $T : G \to H$ is a homomorphism of Lie groups, then the differential at 1 is a linear map $T_1(G) \to T_1(H)$. As such it gives a linear map $\mathfrak{t} : \mathfrak{g} \to \mathfrak{h}$ as well, using the isomorphisms $T_1(G) \to \mathfrak{g}$ and $T_1(H) \to \mathfrak{h}$.

3.20. Proposition. *Let X, Y be left invariant vector fields of a Lie group G. If $T : G \to H$ is a homomorphism of Lie groups, then we have*

$$\mathfrak{t}[X, Y] = [\mathfrak{t}(X), \mathfrak{t}(Y)].$$

Proof. Let f be a differentiable function in a neighbourhood of 1 in H. Then for any $Z \in \mathfrak{h}$ and any $x \in G$, we have $(Zf)(Tx) = Z_1(L_{Tx}f)$, since Z is an invariant vector field. Let us take $Z = \mathfrak{t}X$, that is to say, Z is the left invariant vector field whose value in $T_1(H)$ is the image of $X_1 \in T_1(G)$ by the differential at 1 of T. Then $Z_1(L_{Tx}f) = X_1(L_{Tx}f \circ T) = X_1(L_x(f \circ T)) = X_x(f \circ T)$ since X is left invariant. In other words, $(\mathfrak{t}X)(f) \circ T = X(f \circ T)$. Hence if $Y \in \mathfrak{g}$, we may replace f by $(\mathfrak{t}Y)f$ in this equation and obtain $(\mathfrak{t}X)(\mathfrak{t}Y)(f) \circ T = X((\mathfrak{t}Y)f \circ T) = XY(f \circ T)$. Interchanging X and Y and subtracting, we get $[\mathfrak{t}X, \mathfrak{t}Y](f) \circ T = [X, Y](f \circ T)$. Evaluating at 1, we see that the image of $[X, Y]$ under the differential of T at 1 is actually $[\mathfrak{t}(X), \mathfrak{t}(Y)]$. This proves our assertion.

We have remarked that if a vector field depends differentiably on some parameters, then its flow also depends differentiably on the parameters. From this it also follows that the one-parameter group associated to a left invariant vector field depends differentiably on it, meaning that there exists a differentiable map $\rho : T_1(G) \times \mathbb{R} \to G$ such that for any $X \in T_1(G)$ the map $t \mapsto \rho(X, t)$ is the corresponding one-parameter group. The restriction of ρ to $T_1(G) \times \{1\}$ is a differentiable map $\mathfrak{g} \to G$. This is called the *exponential mapping*. Let $v \in T_1(G)$. Then we will evaluate at v, the differential at 0 of the exponential. We might as well restrict ρ to $\mathbb{R}v \times \mathbb{R}$ first in order to compute this differential. In other words, we consider the map $(sv, t) \mapsto c(sv, t)$. But this is the same as $c(v, st)$. Setting $t = 1$, we need to compute the differential of $s \mapsto c(v, s)$ at $s = 0$. By definition it is v. In other words, we have shown

3.21. Proposition. *The differential at 0 of the exponential map from $\mathfrak{g}$ to G is the identity map. In particular, the exponential map gives a diffeomorphism of a neighbourhood of 0 in $\mathfrak{g}$ onto a neighbourhood of 1 in G.*

It follows from the uniqueness assertion regarding one-parameter groups, that if $T : G \to H$ is a differentiable homomorphism of connected Lie groups, the induced map $\mathfrak{t}$ commutes with the exponential map in the sense that $\exp_H \circ \mathfrak{t} = T \circ \exp_G$.

If $T, T' : G \to H$ are two homomorphisms of connected Lie groups such that the induced homomorphisms $\mathfrak{t}$, $\mathfrak{t}'$ are the same, then $T = T'$. For from Proposition 3.21, our assumption implies that T and T' coincide in a neighbourhood of 1 in G. But if G is connected, any neighbourhood U of 1 generates G as a group. Hence T, T' coincide on the whole of G.

Moreover, if a homomorphism $T : G \to H$ of connected Lie groups is injective, then the induced homomorphism $\mathfrak{t}$ is also injective. For if X is in the kernel of $\mathfrak{t}$, then the image of the one-parameter group $t \mapsto T(\exp(tX))$ has tangent 0 at 1 and should be the constant homomorphism. By our assumption, the map $t \mapsto \exp(tX)$ is also constant, and hence $X = 0$.

Conversely, if $\mathfrak{t}$ is injective, the kernel of T cannot intersect the exponential neighbourhood, and is therefore a discrete normal subgroup of G.

Finally, if $\mathfrak{t}$ is an isomorphism of Lie algebras, then T has discrete kernel N and goes down to a homomorphism of the Lie group G/N into H. This induces an isomorphism at the Lie algebra level. Since the differential of T at 1 and hence at any other point is an isomorphism, it is a diffeomorphism onto an open subgroup of H. Since H is connected, the image coincides with H. In other words, $T : G \to H$ is an isomorphism.

3.22. Remark. We have associated to every Lie group, a Lie algebra and to every Lie group homomorphism, a Lie algebra homomorphism of the corresponding Lie algebras. This correspondence helps one to understand an analytical object such as a Lie group, by a purely algebraic object, namely its Lie algebra.

4. Theorem of Frobenius

Suppose M is a differential manifold and we are given a subbundle E of T_M. We seek to find conditions under which one can assert that at every $m \in M$, there exists a local coordinate system (U, x) such that $\frac{\partial}{\partial x_1}, \ldots, \frac{\partial}{\partial x_k}, k = \mathrm{rk}(E)$ generate the subbundle E at all points of U. Note that this implies that all sections of $\mathcal{E}$ over U are of the form $\sum_{i=1}^{k} f_i \frac{\partial}{\partial x_i}$. Hence if we consider two sections of E as vector fields and take their bracket, the resulting vector field is also a section of E. The theorem of Frobenius asserts that this condition is also sufficient.

4.1. Definition. A subbundle E of the tangent bundle of a differential manifold M is said to be *integrable* if for any two differentiable sections X and Y of E, the bracket $[X, Y]$ is also a section of E.

4.2. Theorem. *If E is an integrable subbundle of T_M, then at every point of M, there exists a local coordinate system (U, x) such that $E|U$ is generated (freely) by sections $\frac{\partial}{\partial x_1}, \ldots, \frac{\partial}{\partial x_k}, k = \mathrm{rk}(E)$.*

Proof. We will prove this by induction on the rank of E. We have seen in (3.9) that the theorem is true for $k = 1$. Let $m \in M$. Notice that if $X_1, \ldots, X_k$ are vector fields which freely generate E in a coordinate neighbourhood at m, then all of these vector fields are nonsingular. In particular, X_k is of the form $\frac{\partial}{\partial x_1}$ in a suitable coordinate system (U, x). Writing out the other X_i's in terms of the basis $(\frac{\partial}{\partial x_i})$, we see that by subtracting from each of $X_i, i < k$, a multiple of X_k, one may assume that they are all of the form $\sum_{i \geq 2} f_i \frac{\partial}{\partial x_i}$. Direct expansion of $[X_i, X_j]$, $i \leq j < k$, shows that the subbundle of E generated by $X_i, i < k$ is also integrable. Our induction assumption then implies that there is a coordinate system (V, y) at m such that $X_i = \frac{\partial}{\partial y_i}$ for all $i < k$. In this coordinate system, suppose $X_k = \sum f_i \frac{\partial}{\partial y_i}$. Again by subtracting from X_k the linear combination $\sum_{i<k} f_i \frac{\partial}{\partial y_i}$ of $X_i, i < k$, we may assume that $X_k = \sum_{i \geq k} f_i \frac{\partial}{\partial y_i}$. Moreover, since X_k is nonsingular, at least one of the f_i's, say f_k, is invertible. We may replace X_k by X_k/f_k. Then we have $X_i = \frac{\partial}{\partial y_i}$ for all $i < k$ and $X_k = \frac{\partial}{\partial y_k} + \sum_{i>k} g_i \frac{\partial}{\partial y_i}$. Now $[X_j, X_k] = \sum_{i>k} \frac{\partial g_i}{\partial y_j} \frac{\partial}{\partial y_i}$ on the one hand, and is a linear combination of $X_i, i \leq k$ on the other. This implies that $\frac{\partial g_i}{\partial y_j}$ are all zero. In other words, g_i are independent of y_j, $j < k$. Let us now take the coordinate neighbourhood in the form $V_1 \times V_2$ with V_1 (resp. V_2) a domain in $\mathbb{R}^{k-1}$ (resp. $\mathbb{R}^{n-k+1}$) with coordinates $y_1, \ldots, y_{k-1}$ (resp. y_i, $i \geq k$). Since X_k can be regarded as a nonsingular vector field in V_2, we can find a coordinate system $z_1, \ldots, z_{n-k+1}$ in which X_k has the expression $\frac{\partial}{\partial z_1}$. Now it is clear that $(y_i, 1 \leq i \leq k - 1, z_j, 1 \leq j \leq n - k + 1)$ is a coordinate system with the required property.

4.3. Definition. An immersed manifold $\varphi : N \to M$ is said to be *integral* for a subbundle E if the differential of φ at any point $p \in N$ maps the tangent space $T_p(N)$ isomorphically onto the fibre $E_{\varphi(p)}$.

In view of Theorem 4.2, there do exist integral submanifolds for an integrable subbundle. In fact, take a coordinate system (U, x) as in Theorem 4.2 (where for convenience of notation we will assume that U is an open cube in $\mathbb{R}^n$ in the coordinate system). Then the closed submanifolds $S_{(a)} = \{(x) :$

$x_i = a_i$ for all $i > k\}$, are obviously all integral manifolds. An integral manifold obtained in this way is called *a slice*.

Suppose $\varphi : V \to U$ is any connected immersed manifold which is integral for E. Then the functions $x_i \circ \varphi$ satisfy $\frac{\partial \varphi_i}{\partial x_j} = 0$ for all $i > k$ and $j \leq k$. By our assumption, this implies that $v\varphi_i = 0$ for all $i > k$ and all tangent vectors v at any point of V. Hence φ_i is a constant on V. The manifold V is therefore contained in the slice $S_{(a)}$, and indeed as an open submanifold.

4.4. Remark. If $\varphi : M \to N$ is an integral manifold, then for every $p \in N$, there exists a neighbourhood U such that $\varphi(M) \cap U$ is a countable union of closed submanifolds of U. We may of course also assume that these are connected components of $\varphi(M) \cap U$.

4.5. Exercise. Explain in the above light which of the examples of immersed submanifolds given in [Ch. 1, 3.17] are integral curves and which not, for a suitable line subbundle of the tangent bundle.

We will now show that immersed integral manifolds do not admit pathologies of the kind we pointed out in [Ch. 1, 3.17].

4.6. Proposition. *If $\varphi : N \to M$ is any integral manifold and L is any other manifold with a map $f : L \to N$, then f is differentiable if and only if $\varphi \circ f$ is.*

Proof. If f is differentiable, the composite $\varphi \circ f$ is of course differentiable. In proving the converse, the key point is that the differentiability of $\varphi \circ f$ implies the continuity of f. Let $l \in L$ and (U, x) be an open cube containing $\varphi \circ f(l)$ as in Theorem 4.2. The open submanifold $(\varphi)^{-1}(U)$ of N is a countable union of open connected manifolds, each of which is an open submanifold of a slice. The map $\varphi \circ f$ maps a connected open neighbourhood W of l into the union of these slices. But the image in $\mathbb{R}^{n-k}$ is countable and connected and hence consists of a single point. In other words, the map $\varphi \circ f$ maps W into a locally closed submanifold of M. Hence it is differentiable as a map into the submanifold as well.

4.7. Corollary. *If $\varphi : N \to M$ and $\varphi' : N' \to M$ are two connected integral manifolds for E with the same image, then they are diffeomorphic.*

Proof. Indeed the natural maps $N \to N'$ and $N' \to N$ are both differentiable, by Proposition 4.6.

We can now take the set of all (connected) integral manifolds for a given integrable subbundle E (identifying them with their images) and partially order them by inclusion. Clearly it is an inductive family and therefore there exists a maximal element. These are called *maximal integral manifolds*.

We will now give an application of the Frobenius theorem to Lie groups.

4.8. Corollary. *Let G be a connected Lie group and $\mathfrak{h}$ a Lie subalgebra of the Lie algebra $\mathfrak{g}$ of G. Then there is a Lie subgroup H of G with $\mathfrak{h}$ as its Lie algebra.*

Proof. Consider the subbundle E which is left G invariant such that $E_1 =$ the subspace $\mathfrak{h}$ of $T_1 = \mathfrak{g}$. Its sections are generated over $\mathcal{A}$ by a basis (X_i) of the vector space $\mathfrak{h}$. In other words, all sections are of the form $\sum f_i X_i$. If $X = \sum f_i X_i$ and $Y = \sum g_i X_i$, then $[X, Y] = \sum f_i(X_i g_j)X_j - \sum g_i(X_i f_j)X_j + \sum f_i g_j[X_i, X_j]$ is again a section of E. Hence E is integrable. Let H be a maximal integral submanifold for E, containing 1. For any $h \in H$, the left translation by h^{-1} of H gives another maximal integral submanifold for the same integrable subbundle, since E is invariant under left translations. But since $h \in H$, this translate contains 1 and so coincides with H. Hence $h^{-1} \in H$ and H is closed under multiplication. In order to show that H is a Lie group, we have to verify for example that the map $H \times H \to H$ given by $(h_1, h_2) \mapsto h_1 h_2^{-1}$ is differentiable. By Proposition 4.6, it is enough to check that this map, considered as one from $H \times H$ into G, is differentiable. But this latter map is the composite of the inclusion of $H \times H$ in $G \times G$ and the corresponding group multiplication map of $G \times G$ into G. This completes the proof.

4.9. Remarks.

1) Even in the case of a Lie subgroup H the topology of H may not coincide with that of the image, as the illustration in [Ch. 1, 3.17] shows. However we have the following comforting situation in the case of Lie groups. Since H induces injection on the Lie algebra, it is an integral submanifold of G for the left invariant subbundle of the tangent bundle of G defined by $\mathfrak{h}$. Hence the subgroup H can be provided with the structure of an immersed manifold.

2) A connected Lie subgroup is a locally closed manifold only if it is actually a closed submanifold. For its closure is a connected subgroup in which it is open. But an open subgroup is necessarily closed. In this case, let $\mathfrak{p}$ be a subspace of $\mathfrak{g}$, supplementary to $\mathfrak{h}$. The exponential image of this space in the exponential neighbourhood intersects H only at 1. The coset space G/H is a Hausdorff topological space with a countable base for open sets. Besides there is a neighbourhood of the trivial coset which can be provided with a differentiable structure, via the exponential map. From this we easily conclude that G/H is a differential manifold and that the natural map $G \to G/H$ admits a differentiable section.

3) One can show that if H is a closed subgroup of G then it is automatically a Lie subgroup, and therefore the above considerations do apply.

4) It can also be proved that an arcwise connected subgroup of a Lie group is a Lie subgroup.

Suppose that H and G are connected Lie groups with $\mathfrak{h}$, $\mathfrak{g}$ as their Lie algebras. Given any Lie algebra homomorphism $\mathfrak{t}$ of $\mathfrak{h}$ into $\mathfrak{g}$, consider the *graph* of $\mathfrak{t}$, namely the Lie subalgebra of $\mathfrak{h} \times \mathfrak{g}$ defined by the set of elements $(X, \mathfrak{t}X), X \in \mathfrak{h}$. Let $\tilde{H}$ be the corresponding connected Lie subgroup of $H \times G$. The projection homomorphism of $\tilde{H}$ into H induces an isomorphism of Lie algebras. Hence there exists a discrete normal subgroup N such that $\tilde{H}/N$ is mapped isomorphically on H. Thus although the map $\mathfrak{t}$ may not come from a homomorphism of H into G, there is an étale covering of H from which there is a homomorphism giving rise to $\mathfrak{t}$.

4.10. Remark. We have carried out the correspondence between Lie groups and Lie algebras except for one particular. It is also true that *every* Lie algebra is actually the Lie algebra of a Lie group. This would follow for example, if we can show that there is an injective homomorphism of the Lie algebra into $\mathfrak{gl}(n, \mathbb{C})$, in view of Corollary 4.8. This latter assertion is known as Ado's theorem, and we do not prove nor use it in this book.

5. Tensor Fields; Lie Derivative

The usual algebraic operations that one performs on vector spaces may also be done on vector bundles. Thus if E and F are differentiable vector bundles, then their *tensor product* $E \otimes F$ may be defined as follows. Consider the set $E \otimes F = \bigcup_{x \in M} E_x \otimes F_x$. It can be made into a differential manifold by noting that when x varies on a small enough open set U around any point, this union can be identified with $U \times (\mathbb{R}^k \otimes \mathbb{R}^l)$ by using local trivialisations of E and F. Identifying $\mathbb{R}^k \otimes \mathbb{R}^l$ with $\mathbb{R}^{kl}$, we may introduce a differentiable structure on $\bigcup_{x \in U} E_x \otimes F_x$. It is easy to see that these structures do not depend on the particular local trivialisations one uses, and consequently glue up to provide a differential structure on $E \otimes F$. It is clear how to define a vector bundle structure on this space. The dual E^* as well as the tensor, exterior and symmetric powers of a vector bundle E may also be defined in a similar way. If $\mathcal{E}$ and $\mathcal{F}$ are the locally free sheaves of $\mathcal{A}$-modules corresponding to E and F, then the tensor product $E \otimes F$ is the vector bundle associated to the *presheaf* which assigns to any open set U the $\mathcal{A}(U)$-module $\mathcal{E}(U) \otimes_{\mathcal{A}(U)} \mathcal{F}(U)$. Its stalk at any point $x \in M$ is easily seen to be $\mathcal{E}_x \otimes_{\mathcal{A}_x} \mathcal{F}_x$.

On a differential manifold M, many interesting geometric objects are described by *sections of vector bundles*, usually the tensor powers of the tangent bundle and its dual, the *cotangent bundle*. We will only deal with differentiable sections even if we do not say so explicitly each time.

Sections of the cotangent bundle are called *differential forms of degree* 1 and sections of its pth exterior power $\Lambda^p(T^*)$ are called *differential forms of degree p*. Note that there are no nonzero differential forms of degree greater than the dimension of M. A differential form of degree p may also be regarded as an *alternating $\mathcal{A}(M)$-multilinear form of degree p* on the space of vector fields (i.e. takes the value 0, whenever two of the argument vector fields are equal). With this interpretation, one can see that the exterior product of two differential forms α, β of degree p and q is given by

5.1.
$$(\alpha \wedge \beta)(X_1, \ldots, X_{p+q})$$
$$= \sum \epsilon_\sigma \alpha(X_{\sigma(1)}, \ldots, X_{\sigma(p)}) \beta(X_{\sigma(p+1)}, \ldots, X_{\sigma(p+q)}),$$

where σ runs through the so-called 'shuffle' permutations and ϵ_σ is the signature of σ. (A shuffle is a permutation which preserves the relative orders in the two subsets, that is to say, it is a permutation of $\{1, \ldots, k+l\}$ such that $\sigma(1) < \cdots < \sigma(p)$ and $\sigma(p+1) < \cdots < \sigma(p+q)$.)

In order to develop differential calculus on a differential manifold M, we first need the notion of differentiation of a tensor field. We then seek to define the derivative of a tensor field, namely a section of $\bigotimes^r T \bigotimes^s T^*$, with respect to a vector field X. The most natural definition that one might give, following the geometric definition of a vector field as an infinitesimal transformation given above, is the following.

5.2. Definition. If ω is a tensor field, then *the Lie derivative $L(X)(\omega)$ of ω with respect to X* is

$$\lim_{t \to 0} \frac{(\varphi_t^* \omega - \omega)}{t}$$

where φ_t is the flow determined by X.

It is not difficult to show that the above limit exists if (as we assume) the vector field X and the tensor field in question are differentiable. Note that the automorphisms φ_t give also isomorphisms φ_t^* between tensors at any point $m \in M$ and those at $\varphi_t(m)$. With this understanding, the above definition amounts to saying that the value of the tensor field $L(X)(\omega)$ at any point $m \in M$ is the tensor $\lim_{t \to 0} \frac{(\varphi_t^* \omega_{\varphi_t(m)} - \omega_m)}{t}$. This is the obvious generalisation of the action of X on functions, that is to say, we have $L(X)f = Xf$ for functions f.

Imitating the proof for the product rule for differentiation, one can show the following.

5.3. Proposition. *Assume given an $\mathcal{A}$-bilinear sheaf homomorphism $(\omega,\omega') \mapsto B(\omega,\omega')$ which associates to any two tensor fields ω, ω' of fixed types, a third one. Assume that for any diffeomorphism φ of an open submanifold U with U', we have $B(\varphi^*\omega, \varphi^*\omega') = \varphi^* B(\omega,\omega')$. Then for any vector field X on M, the following identity holds:*

$$L(X)B(\omega,\omega') = B(L(X)\omega, \omega') + B(\omega, L(X)\omega').$$

Proof. In fact, by definition, $(L(X)B(\omega,\omega'))_p$ is

$$= \lim_{t\to 0} \frac{\varphi_t^* B(\omega,\omega')_{\varphi_t(p)} - B(\omega,\omega')_p}{t}$$

$$= \lim_{t\to 0} \frac{B(\varphi_t^*\omega_{\varphi_t(p)}, \varphi_t^*\omega'_{\varphi_t(p)}) - B(\omega_p, \omega'_p)}{t}$$

$$= \lim_{t\to 0} \frac{B(\varphi_t^*\omega_{\varphi_t(p)} - \omega_p, \varphi_t^*\omega'_{\varphi_t(p)}) + B(\omega_p, \varphi_t^*\omega'_{\varphi_t(p)} - \omega'_p)}{t}$$

$$= B\left(\lim_{t\to 0} \frac{\varphi_t^*\omega_{\varphi_t(p)} - \omega_p}{t}, \varphi_t^*\omega'_{\varphi_t(p)}\right) + B\left(\omega_p, \lim_{t\to 0} \frac{\varphi_t^*\omega'_{\varphi_t(p)} - \omega'_p}{t}\right)$$

$$= B(L(X)\omega, \omega')_p + B(\omega, L(X)\omega')_p.$$

5.4. Corollary. *i) If ω, ω' are tensor fields, then*

$$L(X)(\omega \otimes \omega') = L(X)\omega \otimes \omega' + \omega \otimes L(X)\omega'.$$

ii) If α, β are differential forms, then

$$L(X)(\alpha \wedge \beta) = L(X)\alpha \wedge \beta + \alpha \wedge L(X)\beta.$$

iii) If ω is a differential form of degree 1 and X, Y are vector fields, then

$$X(\omega(Y)) = (L(X)\omega)(Y) + \omega(L(X)Y).$$

Proof. We simply have to take in the above proposition the map B to be

i) $(\omega,\omega') \mapsto \omega \otimes \omega'$;

ii) $(\alpha, \beta) \mapsto \alpha \wedge \beta$;

iii) $(\omega, Y) \mapsto \omega(Y)$.

5.5. Remark. From the definition of Lie derivation it is easy to deduce that if $L(X)\omega = 0$, then ω is left invariant under the flow of X, i.e. φ_t takes ω to itself. Hence we give the following definition.

5.6. Definition. Let X be a vector field and ω a tensor field. Then we say that ω is *invariant* under X if $L(X)\omega = 0$.

5.7. Computation of $L(X)$.

In Proposition 5.3, we assumed that B is $\mathcal{A}$-bilinear, but used mainly that it was $\mathbb{R}$-bilinear. The only point at which we used $\mathcal{A}$-linearity was when we took the limit inside the argument in B. Notice that if X, ω are differentiable (as we always assume), then the limit as t tends to zero, of $\frac{(\varphi_t)^*\omega - \omega}{t}$, namely $L(X)(\omega)$, exists even uniformly on compact sets. Indeed this exists even uniformly for the partial derivatives. We do not wish to go into this question extensively because it is irrelevant to the present discussion. We mention it only to point out that the conclusion of Proposition 5.3 is valid even when B is not $\mathcal{A}$-bilinear but satisfies a suitable continuity axiom. We will apply this to the following examples in which in any case the identity can be directly checked.

5.8. Examples.

1) $(Y, f) \mapsto Yf$ where Y is a vector field and f is a function. The corresponding identity gives $X(Yf) = L(X)(Y) + YXf$, which computes the Lie derivatives on vector fields to be

$$L(X)(Y) = XY - YX.$$

2) $(Y, Z) \mapsto [Y, Z]$, where Y, Z are vector fields. Then we get, using the above computation, that

$$[X, [Y, Z]] = [[X, Y], Z] + [Y, [X, Z]]$$

which is just the Jacobi identity.

3) The above identity can also be rewritten as

$$L([X, Y]) = L(X)L(Y) - L(Y)L(X)$$

on vector fields. It is easy to see that this is valid even for derivatives of other tensor fields.

5.9. Remark. From Remark 5.5, we conclude that if X and Y are commuting vector fields, that is to say, $[X, Y] = 0$, then the flow φ_t of X leaves Y fixed. If ψ_t is the flow of Y, then it follows that the group $\varphi_{t'} \circ \psi_t \circ (\varphi_{t'})^{-1}$ gives rise to the vector field $\varphi(Y) = Y$. The uniqueness of the flow therefore implies that $\varphi_{t'}$ and ψ_t commute for all t, t'.

Corollary 5.4 also computes $L(X)$ on differential forms. Firstly, on forms of degree 1, we have, by 5.4, iii),

$$(L(X)\omega)(Y) = X\omega(Y) - \omega([X, Y]),$$

while 5.4, ii) gives the computation on differential forms of higher degree. Indeed, we get

$$(L(X)\alpha)(X_1,\ldots,X_p) = X\alpha(X_1,\ldots,X_p) - \sum_{i=1}^{p}\alpha(X_1,\ldots,[X,X_i],\ldots,X_p).$$

While the Lie derivative is a good notion of differentiation of a tensor field with respect to a vector field, it *does not* lead to a notion of differentiation of a tensor field with respect to a tangent vector at a point. The reason for this is that the value of the Lie derivative of a tensor field at a point depends on the value of the vector field not only at that point, but in a *neighbourhood*. It is easy to see that from the algebraic point of view this is due to the fact that the map $X \mapsto L(X)$ is itself a differential operator and *not* an $\mathcal{A}$-linear map. For instance, if $f \in \mathcal{A}(M)$, and α is a 1-form, we have

$$
\begin{aligned}
(L(fX)\alpha)(Y) &= (fX)\alpha(Y) - \alpha([fX,Y]) \\
&= f(X\alpha(Y)) - \alpha(f[X,Y] - (Yf)X) \\
&= fX(\alpha(Y)) - f\alpha([X,Y]) + (Yf)\alpha(X) \\
&= f(L(X)\alpha)(Y) + (Yf)\alpha(X).
\end{aligned}
$$

Thus we have

5.10. $$L(fX)\alpha = fL(X)\alpha + \alpha(X)df.$$

In order to get a good notion of differentiation on tensor fields, we need some additional structure on the manifold. We will deal with this in some detail in Chapter 5.

6. The Exterior Derivative; de Rham Complex

6.1. Definition. Let α be a differential form of degree p. Then we define *the exterior derivative $d\alpha$ of α* to be the differential form of degree $p+1$ given by the formula

$$(d\alpha)(X_1,\ldots,X_{p+1}) = \sum_{i=1}^{p+1}(-1)^{i+1}X_i\alpha(X_1,\ldots,\hat{X}_i,\ldots,X_{p+1})$$

$$+ \sum_{1\leq i<j\leq p+1}(-1)^{i+j}\alpha([X_i,X_j],X_1,\ldots,\hat{X}_i,\ldots,\hat{X}_j,\ldots,X_{p+1})$$

where a hat over a symbol means that the symbol does not occur.

6.2. Remark. Firstly, one easily checks that $d\alpha$ is $\mathcal{A}$-linear. To show that it is alternating, notice that if $X_k = X_l = X$, $k < l$, the terms of the first type on the right side of the definition reduce to two terms, namely those corresponding to $i = k$ and $i = l$, and it is clear that these cancel out. On the other hand, in the second type of terms, we have to consider the summation over terms with $i = k$, $j \neq l$ and $i \neq k$, $j = l$, while the term corresponding to $i = k$, $j = l$ is zero as it involves $[X_k, X_l]$ in one of the arguments. It is easy to see that the above two terms also cancel out, proving that $d\alpha$ is indeed alternating.

6.3. Example. If α is a form of degree 0, namely a function f, then the definition gives $df(X) = Xf$. If α is of degree 1, then $d\alpha(X, Y) = X\alpha(Y) - Y\alpha(X) - \alpha([X, Y])$.

6.4. Definition. If X is a vector field and α is a differential form of degree p, we define the *inner product* $\iota_X \alpha$ of α with respect to X to be a form of degree $p - 1$ given by

$$(\iota_X \alpha)(X_1, \ldots, X_{p-1}) = \alpha(X, X_1, \ldots, X_{p-1}).$$

6.5. Lemma. *For forms α, β of degree p, q we have*

$$\iota_X(\alpha \wedge \beta) = \iota_X \alpha \wedge \beta + (-1)^p \alpha \wedge \iota_X(\beta).$$

Proof. This is a simple consequence of Definitions 6.1 and 6.4.

6.6. Remarks.

1) The space of all differential forms constitutes a graded associative algebra. It has also a $\mathbb{Z}/2$-gradation coming from its gradation. If a linear endomorphism preserves the parity of forms, we say that it is $\mathbb{Z}/2$-homogeneous of even type. Similarly if an endomorphism takes odd forms to even ones, and even to odd, we say it is homogeneous of odd type. A linear endomorphism is said to be a *derivation of even type* if it is homogeneous of even degree and is a derivation in the usual sense. But if an endomorphism D is homogeneous of odd degree and satisfies $D(\alpha \wedge \beta) = D\alpha \wedge \beta + (-1)^{\deg \alpha} \alpha \wedge D\beta$, we say that it is a *derivation of odd type*. For example, by Lemma 6.5, the inner product ι_X is a derivation of odd type while $L(X)$ is one of even type. If both of them are of even type, we have seen that $D_1 D_2 - D_2 D_1$ is also a derivation of even type. If one of them is of odd type and the other even, then $D_1 D_2 - D_2 D_1$ is still a derivation, but of odd type.

Finally if both of them are odd derivations, then $D_1 D_2 + D_2 D_1$ is a derivation of even type.

2) In terms of the linear map ι_X, formula 5.10 can be rewritten as

$$L(fX)(\alpha) = fL(X)(\alpha) + df \wedge \iota_X \alpha$$

for a 1-form α. Using the fact that $L(fX), fL(X)$ and the map $\alpha \mapsto df \wedge \iota_X \alpha$ are all derivations, one sees that this formula is valid for forms of arbitrary degree. In particular, if α is of degree n, we have $0 = \iota_X(df \wedge \alpha) = \iota_X(df) \wedge \alpha - df \wedge \iota_X \alpha = (Xf)\alpha - df \wedge \iota_X \alpha$, so that $L(fX)(\alpha) = fL(X)\alpha + (Xf)(\alpha)$.

6.7. Proposition. *The exterior derivative d has the following properties.*

i) *It is additive, namely, $d(\alpha + \beta) = d\alpha + d\beta$.*

ii) *For any vector field X, we have $d\iota_X + \iota_X d = L(X)$ on differential forms.*

iii) *If α and β are differential forms of degree p and q, respectively, then*

$$d(\alpha \wedge \beta) = d\alpha \wedge \beta + (-1)^p \alpha \wedge d\beta.$$

In other words, d is a derivation of odd type.

iv) $d \circ d = 0.$

Proof. i) is obvious from the definition of d.

ii) In fact, we have
$$(d\iota_X + \iota_X d)(\alpha)(X_1, \ldots, X_p) = \sum (-1)^{i+1} X_i (\iota_X \alpha)(X_1, \ldots, \hat{X}_i, \ldots, X_p)$$
$$+ \sum (-1)^{i+j} (\iota_X \alpha)([X_i, X_j], \ldots, \hat{X}_i, \ldots, \hat{X}_j, \ldots X_p) + (d\alpha)(X, X_1, \ldots, X_p)$$
$$= X\alpha(X_1, \ldots, X_p) + \sum (-1)^i \alpha([X, X_i], X_1, \ldots, \hat{X}_i, \ldots, X_p)$$
$$= (L(X)\alpha)(X_1, \ldots, X_p)$$
for any form α of degree p and vector fields $X_1, \ldots, X_p$.

iii) is proved by induction on $p + q$. Our assertion is obvious from the definition if one of p, q is 0. So let us assume that $p + q > 0$. In order to prove the equality of the differential forms on the two sides, it is enough to show that they are the same after applying ι_X for any arbitrary X. Now we may use ii) and the induction assumption to complete the proof.

iv) Since d is a derivation of odd type, we have, by Remark 6.6, 1), that $d \circ d + d \circ d = 2d^2$ is an even derivation. In order to show that it is 0, it is enough to check this on the generators of the $\mathbb{R}$-algebra of differential forms, namely functions and 1-forms. Firstly, $(d^2 f)(X, Y) = X(df)(Y) - Y(df)(X) - df([X, Y]) = XY(f) - YX(f) - [X, Y](f) = 0$. Using the definition we may check similarly that it is 0 also on 1-forms. Alternatively, we note that it is enough to prove that it is 0 on small open sets. In a domain in $\mathbb{R}^n$, however, it is clear that any 1-form is a linear combination of forms of the type $\alpha = f(x)dx$. But then $d\alpha = df \wedge dx$. It follows therefore that d^2 vanishes on α.

6.8. Local computation.

We have seen that on a domain U in $\mathbb{R}^n$, the operators $\frac{\partial}{\partial x_1}, \ldots, \frac{\partial}{\partial x_n}$ generate the space of all vector fields as an $\mathcal{A}(U)$-module. The dual basis for differential forms of degree 1 is $(dx_1, \ldots, dx_n)$ where $(x_1, \ldots, x_n)$ are the coordinate functions and dx_i are their exterior derivatives. For, if f is a function and X a vector field, we have by definition, $df(X) = Xf$. In particular, $(dx_i)(\frac{\partial}{\partial x_j}) = \frac{\partial x_j}{\partial x_i} = \delta_{ij}$. Hence every differential 1-form can be uniquely expressed as $\alpha = \sum f_i dx_i$. If f is any function, we have by definition, $(df)(\frac{\partial}{\partial x_i}) = \frac{\partial f}{\partial x_i}$. Hence $df = \sum \frac{\partial f}{\partial x_i} dx_i$. More generally, any r-differential form ω can be expressed as $\omega = \sum_{i_1 < \cdots < i_r} a_{i_1,\ldots,i_r} dx_{i_1} \wedge \cdots \wedge dx_{i_r}$, where $a_{i_1,\ldots,i_r}$ are functions. Using 6.7 we obtain

$$d\omega = da_{i_1,\ldots,i_r} \wedge dx_{i_1} \wedge \cdots \wedge dx_{i_r}$$

and since $da_{i_1,\ldots,i_r}$ has been computed above, this completes the local computation of the exterior derivative. Moreover, this also gives the computation of $L(X)$ for any vector field in local coordinates. One notes that these local expressions are first order *differential operators*, in the sense that the effect of d on α involves the derivatives of the coefficient functions.

6.9. A digression on complexes.

We recall the notions of complexes, morphisms of complexes, homotopy between morphisms, ... of A-modules, where A is any ring. Analogous notions are available for sheaves and will be dealt with in Chapter 4, but we need the basic definitions here. A *complex* of A-modules is a sequence M^i of modules and homomorphisms $M^i \to M^{i+1}$ for all $i \in \mathbb{Z}$ such that the successive composites $M^{i-1} \to M^i \to M^{i+1}$ are all 0. The homomorphisms are called the *differentials* of the complex, and usually all of them are denoted by d. The complex itself is often denoted by M°.

6.10. Examples.

1) Recall that a singular r-cochain of a topological space is simply an assignment to every singular r-simplex, of an element of a fixed abelian group A. For every $0 \leq i \leq r$, consider the restriction ∂_i to the standard $r + 1$-simplex, of the linear map $\mathbb{R}^{r+1} \to \mathbb{R}^{r+2}$ defined on the elements of the standard basis as follows: $\partial_i(e_j) = e_j$ for $0 \leq j \leq i$ and $\partial_i(e_j) = e_{j+1}$ for $i+1 \leq j \leq r+1$. This gives a map $\Delta_r \to \Delta_{r+1}$. The ith *face* of a singular $r + 1$-simplex $s : \Delta_{r+1} \to X$ is defined to be the composite $s \circ \partial_i$. Then we define the differential of a singular i-cochain c by setting $dc(s) = \sum (-1)^i c(s \circ \partial_i)$. Then one can check

that $d^2 = 0$ and the resulting complex is called the *singular complex* of X, with coefficients in A.

2) We have checked above that the exterior derivative on differential forms gives rise to a complex called the *de Rham complex*.

If M° is a complex, one defines its *cohomology groups H^i* to be the groups given by $\frac{Z^i}{B^i}$ where Z^i is the kernel of the map $M^i \to M^{i+1}$ and B^i is the image of the map $M^{i-1} \to M^i$. Notice that since M° is a complex, we have the inclusion $B^i \subset Z^i$. Thus H^i can be taken as a measure of deviation of B^i from being Z^i. Another way of saying it is that $H^i = 0$ if and only if $M^{i-1} \to M^i \to M^{i+1}$ is *exact*, that is to say, the kernel of $M^i \to M^{i+1}$ is the same as the image of $M^{i-1} \to M^i$.

6.11. Definition. The cohomology of the singular complex with coefficients in an abelian group A is called the *singular cohomology* of the space with coefficients in A.

A *morphism f°* of a complex M° into another complex N° is a sequence of homomorphisms $f^i : M^i \to N^i$ for all i such that the diagrams

$$
\begin{array}{ccc}
M^{i-1} & \to & M^i \\
\downarrow{\scriptstyle f^{i-1}} & & \downarrow{\scriptstyle f^i} \\
N^{i-1} & \to & N^i
\end{array}
$$

are all commutative. The maps (f^i) induce in an obvious way homomorphisms $Z^i(M^\circ) \to Z^i(N^\circ)$ and $B^i(M^\circ) \to B^i(N^\circ)$ and consequently a homomorphism $H^i(M^\circ) \to H^i(N^\circ)$ of the cohomology groups.

6.12. Remark. A continuous map $f : X \to Y$ of topological spaces gives in an obvious way a morphism of the singular complex of Y into that of X. In fact, any singular r-simplex $s : \Delta \to X$ in X gives rise to one in Y, namely $f \circ s$. In particular, it gives rise to a homomorphism of the singular cohomology group of Y into that of X.

Two morphisms f°, g° of complexes from M° into N° are said to be *homotopic* if there exist homomorphisms $k^i : M^i \to N^{i-1}$ for all i such that

$$dk^i + k^{i+1}d = f^i - g^i.$$

If $f^\circ, g^\circ : M^\circ \to N^\circ$ are two morphisms of complexes which are homotopic, then the induced homomorphisms on the cohomology groups are the same. This is trivial to verify from the definition of homotopy. In particular, if M° and N° are homotopically equivalent in the sense that there are morphisms in both directions such that both composites are homotopic, then their cohomology groups are isomorphic.

6.13. Definition. The complex

$$0 \to \mathcal{A} \to \mathcal{T}^* \to \Lambda^2 \mathcal{T}^* \to \cdots$$

with all sheaf homomorphisms given by the exterior derivative, is called the *de Rham complex.*

6.14. Proposition. *The de Rham complex is exact as a sequence of sheaves except at the initial stage, and the kernel of $\mathcal{A} \to T^*$ is the constant sheaf $\mathbb{R}$.*

Proof. Clearly it is enough to show that if U is an open cube in $\mathbb{R}^n$, then any form

$$\omega = \sum_{i_1 < \cdots < i_r} a_{i_1,\ldots,i_r} dx_{i_1} \wedge \cdots \wedge dx_{i_r}$$

on U satisfying $d\omega = 0$ can be written as $d\alpha$ where α is an $(r-1)$-form. We will generalise this a little in order to set up a suitable induction. The generalisation we intend is to let $a_{i_1,\ldots,i_r}$ to be not merely differentiable functions on U, but differentiable functions on $V \times U$ where V is a parameter space assumed to be an open cube in $\mathbb{R}^m$. Then we claim that the form α can be chosen to depend differentiably on the parameter space as well, in the sense that the coefficient functions are also differentiable on $V \times U$. When we take a form in n variables with coefficients which depend on m other variables as well, the notation d for exterior derivative could be somewhat ambiguous. To obviate this, we will denote by d_n the exterior derivative when α is treated as a form on an open set of $\mathbb{R}^n$, perhaps depending on some parameters. Now we want to use induction on n. Any form as above (abbreviating the parameters to t) is best written as

$$\omega(t) = \omega_1(t, x_n) \wedge dx_n + \omega_2(t, x_n)$$

where $\omega_1(t, x_n)$ and $\omega_2(t, x_n)$ are uniquely determined forms of degrees $r - 1$ and r, respectively, which depend differentiably on (t, x_n) and do not involve dx_n in the wedge part of their expressions. In other words, they are differential forms in $n - 1$ variables depending on $m + 1$ parameters. Then we get

$$
\begin{aligned}
d_n \omega(t) &= d_n \omega_1(t, x_n) \wedge dx_n + d_n \omega_2(t, x_n) \\
&= d_{n-1} \omega_1(t, x_n) \wedge dx_n + d_{n-1} \omega_2(t, x_n) + (-1)^r \frac{\partial \omega_2(t, x_n)}{\partial x_n} \wedge dx_n.
\end{aligned}
$$

Here when we write $\frac{\partial \omega}{\partial x_n}$, we mean that we differentiate all the coefficient functions in the local expression for ω. Thus we deduce that

$$d_n \omega(t) = \left(d_{n-1} \omega_1(t, x_n) + (-1)^r \frac{\partial \omega_2(t, x_n)}{\partial x_n} \right) \wedge dx_n + d_{n-1} \omega_2(t, x_n).$$

But we are given that $d_n\omega(t) = 0$, so that we obtain two equations:

$$d_{n-1}\omega_1(t,x_n) + (-1)^r \frac{\partial\omega_2(t,x_n)}{\partial x_n} = 0,$$
$$d_{n-1}\omega_2(t,x_n) = 0.$$

The second equation implies (by the induction assumption) that there exists a form $\alpha(t,x_n)$ in $n-1$ variables such that $d_{n-1}\alpha(t,x_n) = \omega_2(t,x_n)$. Then the first equation gives on substitution,

$$d_{n-1}\omega_1(t,x_n) + (-1)^r \frac{\partial}{\partial x_n} d_{n-1}\alpha(t,x_n) = 0,$$

or what is the same,

$$d_{n-1}(\omega_1(t,x_n) + (-1)^r \frac{\partial}{\partial x_n}\alpha(t,x_n)) = 0.$$

Again by the induction assumption, there exists a form $\beta(t,x_n)$ in $n-1$ variables such that

$$d_{n-1}\beta(t,x_n) = \omega_1(t,x_n) + (-1)^r \frac{\partial}{\partial x_n}\alpha(t,x_n).$$

Now let us set $\gamma(t) = \beta(t,x_n) \wedge dx_n + \alpha(t,x_n)$. Then we have

$$d_n\gamma(t) = d_{n-1}\beta(t,x_n) \wedge dx_n + d_{n-1}\alpha(t,x_n) + (-1)^{r-1}\frac{\partial\alpha(t,x_n)}{\partial x_n} \wedge dx_n$$
$$= (\omega_1(t,x_n) + (-1)^r \frac{\partial\alpha(t,x_n)}{\partial x_n}) \wedge dx_n + \omega_2(t,x_n)$$
$$\qquad + (-1)^{r-1}\frac{\partial\alpha(t,x_n)}{\partial x_n} \wedge dx_n$$
$$= \omega(t),$$

as was to be proved. In order to start the induction, we have to prove the assertion in the case $n = 1$. This means that if f is a function of x, say in the interval $(-1,+1)$, and parameters t, then there exists a function $\varphi(x,t)$ such that $\frac{\partial\varphi(x,t)}{\partial x} = f(x,t)$. Indeed the function defined by $\varphi(x,t) = \int_0^x f(x,t)dx$ is clearly differentiable in x and t and has the required property.

Finally it is obvious that on a connected open set of $\mathbb{R}^n$, if $df = 0$, i.e. $\frac{\partial f}{\partial x_i} = 0$ for all i, then f is a constant.

6.15. Remark. While the de Rham complex is exact as a sequence of sheaves, it is *not* true (in general) that the sequence

$$0 \to \mathbb{R} \to \mathcal{A}(M) \to \mathcal{T}^*(M) \to \cdots$$

is exact. In other words, it is not in general true that if a differential form ω satisfies $d\omega = 0$ (namely, a *closed* form) then it can be expressed as $d\alpha$ (namely, an *exact* form) on an arbitrary manifold M.

6.16. Example. Let us consider the manifold S^1. The differential form $\omega = X\,dY - Y\,dX$ where X, Y are restrictions of the coordinate functions on $\mathbb{R}^2$ to S^1, is clearly closed since there are no nonzero forms of degree 2 on S^1. But there is no function f on S^1 such that $df = X\,dY - Y\,dX$. For, assume that such a function exists. Locally to S^1 we can find a function θ such that $X = \cos\theta$ and $Y = \sin\theta$. Substituting in the defining equation for ω, we see that $\omega = d\theta$. Since by assumption $df = \omega$ as well, we see that $d(f - \theta) = 0$ and so f and θ differ by a constant. In particular the local function θ can be extended to the whole of S^1. It is well known and easy to prove that this is not possible.

The above example has to do with the fact that S^1 is not simply connected. This suggests that the problem of the exactness of the global de Rham complex is related to the *topology* of the manifold in question. We will take up this question for detailed discussion in Chapter 4. For now, we simply wish to point to the phenomenon of an exact sequence of sheaves

$$0 \to \mathcal{F}' \to \mathcal{F} \to \mathcal{F}'' \to 0$$

not giving rise to an *exact* sequence

$$0 \to \mathcal{F}'(M) \to \mathcal{F}(M) \to \mathcal{F}''(M) \to 0.$$

The study of this question leads to the concept of *cohomology of sheaves* which will be studied in Chapter 4.

7. Differential Operators of Higher Order

We will now turn to differential operators of higher order and set up the machinery of a symbol calculus. This will be systematically used in the subsequent chapters.

We have remarked that interesting operators on a differential manifold are often defined as maps of a tensor bundle into another, which are, in terms of their local expressions, differential operators in the usual sense. Many such operators are of first order, but there are also interesting operators of higher orders, for example the Laplacian on a Riemannian manifold. We will therefore proceed to define higher order operators, first on functions and later, from one vector bundle to another.

7.1. Definition. A *(linear) differential operator (of finite order) on functions* is an $\mathbb{R}$-linear sheaf homomorphism $\mathcal{A}^M \to \mathcal{A}^M$ which is in the algebra generated by differential operators of order ≤ 1. It is said to be of *order* $\leq k$ if it can be expressed as a linear combination of composites of k operators of order ≤ 1. The set of all such operators will be denoted by Diff^k_M or sometimes for shortness D^k_M.

7.2. Remarks.

1) Since we have seen that differential operators of order ≤ 1 have local expressions of the form $\sum f_i \frac{\partial}{\partial x_i} + \varphi$, it is clear that a differential operator of order $\leq k$ is locally of the form $\sum_{|\alpha| \leq k} f_\alpha \frac{\partial}{\partial x_\alpha}$.

2) If M is not compact, it is possible to construct differential operators of 'infinite order' by constructing operators on compact sets with orders that increase at infinity. For example, consider the open covering of $\mathbb{R}^+$ by open intervals $U_n = (n, n+2), n \in \mathbb{Z}^+$. If $\{\varphi_n\}$ is a partition of unity with respect to U_n, then $\sum \varphi_n \frac{\partial^n}{\partial x^n}$ is such an operator. But this phenomenon cannot happen if M is compact. In any case, we will only consider operators of finite order.

3) Any sheaf homomorphism of $\mathcal{A}$ into itself can be shown to be locally a differential operator. This is a theorem of Peetre.

Structure of the algebra of differential operators.

We have seen that the set of homogeneous first order operators form a Lie algebra, besides being a module over the ring of functions. We would like now to construct an analogue of the universal enveloping algebra in this case, although it is not a Lie algebra over functions. Recall first the definition of the enveloping algebra.

7.3. Definition. Let $\mathfrak{g}$ be a Lie algebra over a field k. Then *the enveloping algebra of* $\mathfrak{g}$ is the quotient of the tensor algebra of the vector space $\mathfrak{g}$ by the 2-sided ideal generated by elements of the form $X \otimes Y - Y \otimes X - [X, Y]$, with $X, Y \in \mathfrak{g}$.

In order to imitate this construction, we would like first to construct the analogue of the *tensor algebra*. As a starting point, it is better to take $\mathcal{D}^1$ than $\mathcal{T}$, since the former is well adapted to the construction of the tensor algebra. We will now explain why.

As a sheaf of $\mathbb{R}$-vector spaces, $\mathcal{D}^1$ is simply $\mathcal{A} \oplus \mathcal{T}$. For any $f \in \mathcal{A}(M)$ it is convenient to denote the corresponding element of $\mathcal{D}^1(M)$ by $m(f)$ rather than f itself (although when we judge that no confusion is likely, we may use f instead of $m(f)$). Then $\mathcal{D}^1$ is an $\mathcal{A}$-bimodule. In fact, if $f \in \mathcal{A}(U)$, and $D = m(g) + X \in \mathcal{D}^1(U)$, then we may define

$$f \circ D = m(f) \circ (m(g) + X) = m(fg) + fX,$$

$$D \circ f = (m(g) + X) \circ m(f) = m(fg + Xf) + fX.$$

The reason for the latter definition is that

$$
\begin{aligned}
(D \circ m(f))(\varphi) &= (m(g) + X)(f\varphi) \\
&= gf\varphi + (Xf)(\varphi) + fX\varphi \\
&= m(gf + Xf)\varphi + fX\varphi.
\end{aligned}
$$

It is easy to check therefore that this definition makes $\mathcal{D}^1$ an $\mathcal{A}$-bimodule.

It will turn out that the structure of the algebra of differential operators can be determined from the structure of $\mathcal{D}^1$ as an $\mathcal{A}$-bimodule on the one hand and as a Lie algebra over $\mathbb{R}$ on the other. Indeed, the construction is akin to that of the enveloping algebra of a Lie algebra.

In the case of a Lie algebra $\mathfrak{g}$ over a field k, the situation is like starting with $\mathfrak{g} \oplus k$. Let us denote inclusion of k in $\mathfrak{g} \oplus k$ by $\lambda \mapsto m(\lambda)$. We leave the following as a simple exercise to the reader.

7.4. Exercise. Consider the tensor algebra of $W = V \oplus k$. Pass to the quotient by the 2-sided ideal generated by $m(1) - 1$. Show that it is canonically isomorphic to the tensor algebra of V.

Consider $\mathcal{D}^1 \otimes \mathcal{D}^1$ where the tensor product is with respect to the right $\mathcal{A}$-module structure of the first factor and the left one of the second factor. Also using the bimodule structure of both we see that the tensor product is also a bimodule. We can iterate this construction to obtain a tensor algebra. Now we will take the quotient of this ring by the two-sided ideal spanned by $m(1) - 1$ to obtain an $\mathbb{R}$-algebra. This is the algebra which we sought to construct. We will denote it by $\mathcal{C}(M)$.

It is easy to describe this algebra by a universal property.

7.5. Proposition. *Let B be any $\mathbb{R}$-algebra containing $\mathcal{A}$. Then B has a natural bimodule structure over $\mathcal{A}$. Any bimodule homomorphism P of $\mathcal{D}^1$ into B which is the identity on $\mathcal{A}$ gives rise to a unique $\mathbb{R}$-algebra homomorphism of $\mathcal{C}$ into B which is the identity on $\mathcal{A}$ and coincides with P on $\mathcal{D}^1$.*

Proof. In fact, we will define inductively, for each r, a bimodule homomorphism $P^{(r)}$ from $\bigotimes^r(\mathcal{D}^1)$ into B. Taking $P^{(0)}$ to be the identity map on $\mathcal{A}$, and assuming that $P^{(r)}$ has been defined, we notice that the map $\mathcal{D}^1 \times \bigotimes^r \mathcal{D}^1 \to B$ given by $(Y, Z) \mapsto P(Y)P^{(r)}(Z)$ is balanced. Hence it goes down to a map $P^{(r+1)} : \bigotimes^{r+1} \mathcal{D}^1 \to B$ as claimed. It is evidently a homomorphism from the $\mathbb{R}$-algebra $T(\mathcal{D}^1)$ into B. From the assumption, we see that $m(1) - 1$ is in the kernel of this map and so induces a map of $\mathcal{C}$ into B. It is unique since $\mathcal{D}^1$ and $\mathcal{A}$ together generate $T(\mathcal{D}^1)$.

7.6. Remark. We could have restated the assumption on P by saying that it is a left $\mathcal{A}$-linear map from $\mathcal{T}$ into B which satisfies

$$P(X)f = Xf + fP(X).$$

Thus we also have a characterisation of $\mathcal{C}$-modules. It is an $\mathcal{A}$-module E together with an $\mathbb{R}$-bilinear map $\mathcal{T}(M) \times E \to E$ denoted $(X, s) \mapsto \nabla_X m$ such that

7.7. $\nabla_{fX}(s) = f\nabla_X s, \text{ and } \nabla_X(fs) = (Xf).s + f\nabla_X s.$

We have constructed, for any open set U, such an algebra $\mathcal{C}(U)$ (although for notational convenience we dropped the parenthetical U). It is easy to define compatible restriction maps and obtain a presheaf $\mathcal{C}$. We will denote the image of any vector field X in $\mathcal{C}$ by ∇_X and the image of any function f by $m(f)$.

7.8. Definition. We call $\mathcal{C}$ the *connection algebra* of M.

Notice that this is *not* an algebra over $\mathcal{A}$, since the right and left multiplications by functions are not the same. The $\mathbb{R}$-subsheaves $\mathcal{C}_r$ defined above as images of $\bigotimes^r \mathcal{D}^1$ are all submodules for the $\mathcal{A}$-biModule structure. Moreover, it is clear from the definition that these satisfy the condition

$$\mathcal{C}_r.\mathcal{C}_s \subset \mathcal{C}_{r+s}.$$

This is paraphrased by saying that the algebra $\mathcal{C}$ is actually a *filtered algebra*, of which we recall the definition below. It is easily verified that $\mathcal{C}_r$ are sheaves.

7.9. Definition. An algebra A together with subspaces $F_r(A), r \in \mathbb{Z}^+$ is said to be a *filtered algebra* if $F_r(A) \subset F_{r+1}(A)$ for all $r \in \mathbb{Z}^+$ and $F_r(A).F_s(A) \subset F_{r+s}(A)$ for all $r, s \in \mathbb{Z}^+$.

Every filtered algebra gives rise to a graded algebra as follows. Consider the direct sum $\bigoplus(F_r(A)/F_{r-1}(A))$. It is an algebra under the natural definition which follows. Suppose $v \in Gr_r(A) = F_r(A)/F_{r-1}(A)$ and $w \in Gr_s(A)$. Then taking representatives for v, w in $F_r(A), F_s(A)$, and multiplying them in A, we obtain an element of $F_{r+s}(A)$. Its image in $Gr_{r+s}(A)$ is independent of the choices made and this multiplication defines an algebra operation. This is called the *associated graded algebra* and is denoted by $Gr(A)$.

Let us now consider the graded $\mathbb{R}$-algebra associated to the filtered algebra $\mathcal{C}$. Its 0th graded component, $\mathcal{A}$, is a subalgebra. Indeed, we claim it actually commutes with all elements of $Gr(\mathcal{C})$. Let us prove this assertion by induction. Since $\mathcal{A}$ is itself commutative, we can start the induction. Assume that for every function f, the corresponding element $m(f)$ commutes with all elements of $\mathcal{C}_r$ modulo $\mathcal{C}_{r-1}$. Any element α of $\mathcal{C}_{r+1}$ is a linear combination of images of elements of the form $\beta \otimes (X + m(g))$, with $\beta \in \mathcal{C}_r$. Then we have

$$
\begin{aligned}
m(f)&(\beta \otimes (X + m(g))) - (\beta \otimes (X + m(g)))m(f) \\
&\equiv\ m(f)(\beta \otimes X) - (\beta \otimes X)m(f)(\mathrm{mod}\ \mathcal{C}_r) \\
&=\ m(f)(\beta \otimes X) - \beta \otimes (m(f)X + m(Xf)) \\
&\equiv\ ((m(f)\beta - \beta m(f)) \otimes X (\mathrm{mod}\ \mathcal{C}_r),
\end{aligned}
$$

but by the induction assumption $m(f)\beta - \beta m(f) \in \mathcal{C}_{r-1}$. This proves our claim. Notice also that the first graded component of $Gr(\mathcal{C})$ is canonically $\mathcal{D}^1/\mathcal{A} = \mathcal{T}$. Hence we have a canonical algebra homomorphism of the tensor algebra over $\mathcal{A}$ of $\mathcal{T}$ into $Gr(\mathcal{C})$. We claim that this homomorphism is actually an isomorphism. It is obvious from the definition that $\mathcal{C}$ is generated as an $\mathbb{R}$-algebra by $\mathcal{C}_1$ and $\mathcal{A}$ so that $Gr(\mathcal{C})$ is generated as an $\mathcal{A}$-algebra by its first graded component, namely $\mathcal{T}$. This implies that the above homomorphism is surjective.

Since $\mathcal{C}_r$ are all sheaves and all these maps are compatible with restrictions to open sets, it is enough to check the isomorphism statement over open sets in $\mathbb{R}^n$.

We will define a $\mathcal{C}$-module structure on a specific $\mathcal{A}$-module, namely the tensor algebra Q of $\mathcal{T}$. We have shown that we need only to define an action ∇_X on Q subject to equation 7.7. We define $\nabla_X(\sum f_\alpha X_\alpha)$ to be $X \otimes \sum f_\alpha X_\alpha + \sum (X f_\alpha) X \alpha$. Here α is a multiindex $(i_1, \ldots, i_r)$, X_i denotes $\frac{\partial}{\partial x_i}$ and X_α denotes the element $X_{i_1} \otimes \cdots \otimes X_{i_r}$, in Q. That this satisfies the two conditions 7.7 is obvious. It is also clear that if $F_i Q$ denotes the filtration on Q associated to the natural gradation on the tensor algebra, then we have $\mathcal{C}_r . F_s Q \subset F_{r+s} Q$. Hence we have an induced action of the associated graded algebra $Gr(\mathcal{C})$ on the tensor algebra Q. Since we have a natural homomorphism of the tensor algebra Q into Gr, we therefore have an action of Q on itself. This action is by definition the juxtaposition action, namely the multiplication on the left in the algebra structure of Q. Since this action is faithful, we conclude that the map $Q \to Gr(\mathcal{C}(M))$ is injective. We have therefore proved the following.

7.10. Theorem. *The filtered $\mathbb{R}$-algebra $\mathcal{C}$ has as its associated graded algebra an $\mathcal{A}$-module which is in fact the tensor algebra of $\mathcal{T}$.*

We shall go on to the definition of the algebra $\mathcal{D}$ of differential operators. Consider the two-sided ideal $\mathcal{I}$ in $\mathcal{C}$ generated by elements of the form $R(X, Y) = \nabla_X \nabla_Y - \nabla_Y \nabla_X - \nabla_{[X,Y]}$ with X, Y vector fields. The quotient of $\mathcal{C}$ by $\mathcal{I}$ gives rise to an $\mathbb{R}$-algebra. Again, since this can be done for all U, it follows easily that we have a presheaf. The resulting sheaf of algebras $\mathcal{D}$

comes with a filtration $F_k(\mathcal{D})$ making it a filtered algebra. Now the sheaf $\mathcal{A}$ may be made into a $\mathcal{C}$-module by defining $X.f$ to be Xf, for all vector fields X. It is clear that this passes down to a $\mathcal{D}$-module structure on $\mathcal{A}$. Under this action, elements of $F_k(\mathcal{D})$ act as $\mathbb{R}$-homomorphisms of $\mathcal{A}$ which are linear combinations of composites of k or less vector fields. In other words, we have given a sheaf homomorphism $F_k(\mathcal{D}) \to \mathrm{Diff}^k$. In order to check that this is an isomorphism it is enough to verify it on sections over any open domain U in $\mathbb{R}^n$. It is surjective by definition, and in order to prove injectivity, we note that modulo the relations we have imposed, any element of $\mathcal{F}_k(\mathcal{D})(U)$ can be written as

$$\sum f_{i_1,\ldots,i_n}\left(\tfrac{\partial}{\partial x_1}\right)^{i_1} \cdots \left(\tfrac{\partial}{\partial x_n}\right)^{i_n}, \qquad i_1 + \cdots + i_n \le k.$$

The image in $\mathrm{Diff}^k(U)$ is the differential operator on functions given by the above expression. Thus we have only to check that it is zero if and only if the action of the operator on all functions is 0. But this is easily seen by looking at its action on functions of the form $\Pi(x_i - a_i)^{i_r}$, where $a = (a_1, \ldots, a_n)$ is some point in U at which the coefficient functions are nonzero. Thus we have proved the following theorem.

7.11. Theorem. *The algebra $\mathcal{D}$ is naturally isomorphic to the quotient of the connection algebra $\mathcal{C}$ by the two-sided ideal generated by elements of the form $R(X,Y) = \nabla_X\nabla_Y - \nabla_Y\nabla_X - \nabla_{[X,Y]}$ with X,Y vector fields.*

7.12. Exercise. Show that left invariant differential operators on a Lie group form an algebra and that it is canonically isomorphic to the universal enveloping algebra of its Lie algebra.

7.13. The symbol sequence.

Consider the graded algebra associated to the filtered algebra $\mathcal{D}$. As in the case of $Gr(\mathcal{C})$, we see that the left and right $\mathcal{A}$-module structures on $\mathcal{D}$ pass down to identical structures on $Gr(\mathcal{D})$, making it actually an $\mathcal{A}$-algebra. Moreover, in view of the fact that $R(X,Y)$ are all 0 in this algebra, one also sees that it is a *commutative* algebra. We have a natural inclusion of $\mathcal{T}$ in $\mathcal{D}$ as the space of homogeneous first order operators and hence we also get a homomorphism of $\mathcal{T}$ into $Gr(\mathcal{D})$. By the universal property of symmetric algebras, this induces an algebra homomorphism of the symmetric algebra $\mathcal{S}(T)$ into $Gr(\mathcal{D})$, which is obviously onto.

7.14. Theorem. *The algebra $\mathcal{D}$ of differential operators on a differential manifold M is a filtered algebra whose associated graded algebra is canonically isomorphic to the symmetric algebra of $\mathcal{T}$.*

Proof. As in the proof of Theorem 7.10, we have only to check that the natural map $S^k(\mathcal{T}) \to \mathcal{D}^k/\mathcal{D}^{k-1}$ is an isomorphism on a domain U in $\mathbb{R}^n$. But then we have proved that $\mathcal{D}^k(U)$ has an $\mathcal{A}(U)$-basis consisting of $(\frac{\partial}{\partial x_1})^{i_1} \cdots (\frac{\partial}{\partial x_n})^{i_n}, i_1+\cdots+i_n \leq k$, and hence the images of $(\frac{\partial}{\partial x_1})^{i_1} \cdots (\frac{\partial}{\partial x_1})^{i_k}$, $i_1+\cdots+i_n = k$, in $\mathcal{D}^k/\mathcal{D}^{k-1}$ form a basis. But clearly these are images of a basis in $S^k(\mathcal{T})$ under the natural map we have given above. This completes the proof.

7.15. Definition. In particular, we have shown that there is an exact sequence of vector bundles:

$$0 \to \mathcal{D}^{k-1} \to \mathcal{D}^k \to S^k(\mathcal{T}) \to 0.$$

It is called the *symbol sequence*. If D is a differential operator on an open set U of order $\leq k$, then its image in $\mathcal{S}^k(\mathcal{T})(U)$ is called its *symbol* or *kth order symbol*.

7.16. Remarks.

1) If we consider a differential operator of order $\leq k$, and take its image modulo lower order operators, it is essentially the same as considering its kth order terms. The above sequence says in effect that the 'highest order term' of a differential operator makes invariant sense only as a section of $S^k(T)$.

2) A differential operator D satisfying $D(1) = 0$ may be said to be an *operator without constant term*. In particular an operator of order 1 without constant term is a *homogeneous operator*. For higher orders, of course these two notions are different, and while an operator without constant term makes sense in any manifold, there is no invariant way of defining a homogeneous operator in general.

3) An operator of order at most 1 is of the form $X + m(f)$ where X is a vector field and f is a function. Its symbol is X.

7.17. Differential operators on vector bundles.

We have seen a few examples of differential operators on manifolds, but they are rarely on functions, or even on systems of functions. They are often defined on sheaves of tensors on a differential manifold. We are therefore obliged to generalise the notion of a differential operator. In fact, we will define a differential operator (of order $\leq k$) from one locally free sheaf $\mathcal{E}$ into another, say $\mathcal{F}$. This is done by considering an ($\mathbb{R}$-linear) sheaf homomorphism, which has the following property. Whenever s is a section of $\mathcal{E}$ and A a section of $\mathcal{F}^*$, the map $f \mapsto \langle D(fs), A \rangle = A(D(fs))$ from functions to functions is a differential operator (of order $\leq k$). While this is a practical definition, the theoretically satisfactory definition would be that it is a section of $\mathcal{F} \otimes \mathcal{D} \otimes \mathcal{E}^*$, where the first tensor product is with respect to the left, and the second with respect to the right $\mathcal{A}$-module structure on $\mathcal{D}$. A slightly different (but obviously equivalent) formulation is the following.

7.18. Definition. A right $\mathcal{A}$-linear homomorphism of $\mathcal{E}$ into $\mathcal{F} \otimes \mathcal{D}^k$ is called a *differential operator from $\mathcal{E}$ to $\mathcal{F}$ of order $\leq k$.*

Using the left $\mathcal{A}$-linear map $\mathcal{D}^k \to \mathcal{A}$ which associates to any D its constant term $D(1)$ we can associate to any differential operator from $\mathcal{E}$ to $\mathcal{F}$ an $\mathbb{R}$-linear homomorphism $\mathcal{E} \to \mathcal{F}$. We will denote the sheaf of such operators by $\mathcal{D}(E, F)$. In the case when E and F are trivial, we refer to such an operator as a *system*. If D is a differential operator from E to F of order $\leq k$, its kth order symbol is the $\mathcal{A}$-linear homomorphism $\mathcal{E} \to \mathcal{F} \otimes S^k(\mathcal{T})$ obtained by using the symbol map $D^k \to S^k(T)$.

7.19. Remark. One can also define inductively an $\mathbb{R}$-linear map $D : E \to F$ to be a differential operator of order at most k, by requiring that for every function f, the operator $f \mapsto [D, f] = D \circ m(f) - m(f) \circ D$ be an operator of order at most $k - 1$. If $\{f_i\}$, $1 \leq i \leq k$, are functions, then

$$[\dots [[D, f_1,], f_2], \dots, f_k]$$

is a 0th order operator, namely an $\mathcal{A}$-linear homomorphism $E \to F$. This function depends symmetrically on the f_i's. For example, this follows from the equality $[[D, f_1], f_2] + [[f_2, D], f_1] = 0$ by the Jacobi identity. Moreover, assume that all the f_i's vanish at a point $m \in M$. Then evaluation of this bracketed function at m goes down to a map $S^k(T_m^*) \to \operatorname{Hom}(E_m, F_m)$. In other words we get a homomorphism $\tilde{\sigma}(D)$ of $S^k(T^*)$ into $\operatorname{Hom}(E, F)$. Associated to this, we get also a section of $S^k(T) \otimes \operatorname{Hom}(E, F)$. This can be easily seen to be the symbol σ as we have defined. We will call either of these the symbol, but usually reserve the notation $\tilde{\sigma}$ for one and σ for the other.

7.20. Exercise. Show that the above definition coincides with the symbol of a differential operator as defined in 7.15 on functions and hence also on bundles as in 7.18.

Computation of symbols.

We will now give some examples of computation of symbols of first order operators.

7.21. Examples.

1) Consider the exterior derivative $d : \Lambda^i(T^*) \to \Lambda^{i+1}(T^*)$. From its local expression it is clear that it is a first order operator. In any case, let α be an i-form and $X_1, \dots, X_{i+1}$ be vector fields providing a linear form, namely $\omega \mapsto \omega(X_1, \dots, X_{i+1})$ on $\Lambda^{i+1}(T^*)$. Then in order to see that d is a differential operator we have only to show that the map $f \mapsto d(f\alpha)(X_1, \dots, X_{i+1})$ is a differential operator. But since $d(f\alpha) = df \wedge \alpha + f\, d\alpha$, we have

$$
\begin{aligned}
d(f\alpha)(X_1, \dots, X_{i+1}) \;=\;& \sum (-1)^{r+1} df(X_r)\alpha(X_1, \dots, \hat{X}_r, \dots, X_{i+1}) \\
+\;& f(d\alpha)(X_1, \dots, X_{i+1}) \\
=\;& \sum (-1)^{r+1} (X_r f)\alpha(X_1, \dots, \hat{X}_r, \dots, X_{i+1}) \\
+\;& f\, d\alpha(X_1, \dots, X_{i+1}).
\end{aligned}
$$

Hence this is a differential operator of order 1 on functions whose symbol is the vector field $\sum(-1)^{r+1}\alpha(X_1, \dots, \hat{X}_r, \dots, X_{i+1})X_r$ and whose constant term is $d\alpha(X_1, \dots, X_{i+1})$. Now the symbol of d is a homomorphism $\sigma(d) : \Lambda^i(T^*) \to \Lambda^{i+1}(T^*) \otimes T$, and so can equally well be interpreted to be a bilinear homomorphism $\Lambda^i(T^*) \times T^* \to \Lambda^{i+1}(T^*)$. Then the above computation gives the following:

$$
\begin{aligned}
\sigma(d)(\alpha, \beta)(X_1, \dots, X_{i+1}) \;=\;& \beta\big(\sum(-1)^{r+1}\alpha(X_1, \dots, \hat{X}_r, \dots, X_{i+1})X_r\big) \\
=\;& \sum(-1)^{r+1}\alpha(X_1, \dots, \hat{X}_r, \dots, X_{i+1})\beta(X_r) \\
=\;& (\beta \wedge \alpha)(X_1, \dots, X_{i+1}).
\end{aligned}
$$

In other words, $\sigma(d)(\alpha, \beta) = \beta \wedge \alpha$.

2) Consider the Lie derivative with respect to a vector field X, say of an i-form α. Then for $f \in \mathcal{A}(M)$ and vector fields $X_r \in \mathcal{T}(M), r \leq i$, we see that the operator $f \mapsto L(X)(f\alpha)(X_1, \dots, X_i)$ is given by

$$
\begin{aligned}
f \;\mapsto\;& X(f\alpha(X_1, \dots, X_i)) - \sum_{r=1}^{i}(f\alpha)(X_1, \dots, [X, X_r], \dots, X_i) \\
=\;& (Xf)\alpha(X_1, \dots, X_i) + f(X\alpha(X_1, \dots, X_i)) \\
&- \sum_{r=1}^{i} f\alpha(X_1, \dots, [X, X_r], \dots, X_i).
\end{aligned}
$$

This map is obviously a differential operator of order 1 whose symbol is $\alpha(X_1, \dots, X_i)X$. This shows that the symbol of $L(X)$, which is a homomorphism $\Lambda^i(T^*) \to \Lambda^i(T^*) \otimes T$, is given by $\sigma(L(X))(\alpha) = \alpha \otimes X$.

7.22. Symbol of a composite of differential operators.

Let $D_1 : E \to F$ and $D_2 : F \to G$ be operators of order $\leq k, l$, respectively. By composition we obtain an operator of order $\leq k + l$ from E to G. We wish to compute the $(k+l)$th symbol of $D_2 \circ D_1$. Firstly, we remark that in the case when $E = F = G = \mathcal{A}$, this is easily done. For the symbol is the natural homomorphism $\mathcal{D}^k \to S^k(T)$, and this is multiplicative in the sense that $\sigma_{k+l}(D_2 \circ D_1) = \sigma_l(D_2).\sigma_k(D_1)$ where the multiplication on the right side is intended in the sense of symmetric algebras. In general, the symbols of D_1, D_2 are homomorphisms $\sigma : E \to F \otimes S^k(T)$, $\tau : F \to G \otimes S^l(T)$, respectively. Any two maps σ, τ as above give rise to a map $(\tau \otimes 1_{S^k(T)}) \circ \sigma :$ $E \to G \otimes S^l(T) \otimes S^k(T)$, and using the multiplication in the symmetric algebra, we get a map $E \to G \otimes S^{k+l}(T)$. This is called the *composite of the symbols* σ and τ. Thus our computation above can be restated as

7.23. Proposition. *If D_1, D_2 are differential operators $E \to F$ and $F \to G$ of orders k, l, respectively, then the $(k + l)$th order symbol of the composite differential operator $D_2 \circ D_1 : E \to G$ is the composite of the two symbols $\sigma_k(D_1)$ and $\sigma_l(D_2)$.*

7.24. Tensorisation with local systems.

If D is a differential operator from a vector bundle E to F, and L is any other vector bundle, there is no natural way of defining a differential operator $L \otimes E \to L \otimes F$. The reason is that the tensorisation is over $\mathcal{A}$, while D is not $\mathcal{A}$-linear. For the same reason, it is clear that if L is a *local system* of $\mathbb{R}$- or $\mathbb{C}$-vector spaces, then one *can* define such an operator. It is moreover clear that its symbol, considered as a map $L \otimes E \to L \otimes F \otimes S^k(T)$, is obtained as the tensor product of the symbol of D with 1_L.

In particular, if L is a local system, then one can talk of the *de Rham complex with coefficients in L*, namely

$$0 \to L \to L \otimes \mathcal{A} \to \cdots \to L \otimes \Lambda^i(T^*) \to \cdots ,$$

and clearly the de Rham sequence of sheaves remains exact after tensorisation with the local system L. Notice that if L is a local system of $\mathbb{R}$- (resp. $\mathbb{C}$-) vector spaces and $\mathcal{E}$ is a locally free $\mathcal{A}$-sheaf, then $L \otimes \mathcal{A}$ is a locally free $\mathcal{A}$-module.

Exercises

1) Determine the space of derivations of $A = \mathbb{R}[x_1, \ldots, x_n]$.

2) Show that the group B of triangular matrices, namely matrices (A_{ij}) such that $A_{ij} = 0$ whenever i is greater than j, is a Lie subalgebra of $\mathfrak{g}l(n, \mathbb{R})$. What is the corresponding Lie subgroup of $GL(n, \mathbb{R})$?

3) Let X be a connected topological space, x a point of X and $\mathcal{L}$ a local system of vector spaces over X. Show that the natural map $\mathcal{L}(X) \to \mathcal{L}_x$ is injective.

4) If E is a complex vector bundle of rank n, treated as a real vector bundle of rank $2n$, show that $E \otimes \mathbb{C}$ is isomorphic to $E \oplus \overline{E}$.

5) Show that the Plücker imbedding of the Grassmannian has nonzero differential at all points.

6) If a Lie group G acts on a differential manifold M, then there is a natural homomorphism of the Lie algebra $\mathfrak{g}$ of G into the Lie algebra of differentiable vector fields on M. When is this map injective? Compute this map for the action of $GL(n, \mathbb{R})$ on $\mathbb{R}^n$.

7) Show that the universal covering space of any connected Lie group is also a Lie group. What is its Lie algebra?

8) Show that if Y is a (differentiable) vector field and (φ_t) is its flow, then

$$\lim_{t \to 0} \left(\frac{\varphi_t^*(Y) - Y}{t} - L_X(Y) \right)$$

exists uniformly, where Y is a vector field with compact support.

9) Let ω be a 2-form on $\mathbb{R}^4$ with coordinates (x, y, z, t). Write ω in terms of coordinates and note that it gives rise to a 2-form α and a 1-form β on $\mathbb{R}^3$, depending on t. Compute $d\omega$ in terms of α and β, giving rise to a 1-form γ and a function φ on $\mathbb{R}^3$ (again depending on t).

10) Let G be a compact connected Lie group. Consider the differentiable map $g \to g^2$ of G into itself. Compute its differential at any $x \in G$ and determine at what points it is an isomorphism.

Chapter 3

Integration on Differential Manifolds

So far we have been discussing differential operators on a manifold. In order to develop calculus on it, we need also a theory of integration. Fortunately, the abstract theory of integration, particularly that of Borel measures on a locally compact space, is applicable to differential manifolds. But it is convenient to restrict oneself to what are called *densities*. These are in the first place Borel measures which are in some sense indefinitely differentiable. We will make this notion precise by requiring that Lie derivatives of measures exist.

1. Integration on a Manifold

We start with the remark that from the point of view of integration, a Borel measure on a differential manifold can be defined as follows.

1.1. Definition. A *Borel measure* on a differential manifold M is a linear form μ on the vector space $C_c^\infty(M)$ of indefinitely differentiable functions with compact support on M, which satisfies the following continuity axiom: If $\{f_k\}$ is any sequence of elements in C_c^∞ all of whose supports are contained in a compact set K and $\sup|f_k|$ tends to zero as k tends to infinity, then $\mu(f_k)$ tends to zero. The scalar $\mu(f)$ is also denoted $\int f\,d\mu$.

If $\varphi : M \to N$ is a proper differentiable map, then any Borel measure μ on M has an *image* measure $\varphi(\mu)$ in N. Indeed, if f has support in a compact subset K of N, then $f \circ \varphi$ has support in $\varphi^{-1}(K)$ which is compact by assumption. Hence we may define $(\varphi(\mu))(f)$ to be $\mu(f \circ \varphi)$ for every

$f \in C_c^\infty(N)$. That this linear form satisfies the continuity requirement is obvious.

1.2. Remark. For any compact subset K of M, one can provide the space C_K^∞ of indefinitely differentiable functions on M with support in K, with the structure of a normed topological vector space in such a way that a sequence $\{f_k\}$ tends to zero if and only if $\{\sup |f_k|\}$ tends to zero. A measure is then a continuous linear form on this space. A base for neighbourhoods of 0 in this topology is provided by $\{f \in C_K^\infty : \sup |f| < a\}$. A linear form μ is continuous at 0 in this topology if and only if there is a positive number λ such that $|\mu(f)| < \lambda \sup |f|$ for all f. The constant λ may depend on K and so it is better to denote it by λ_K.

1.3. Example. The Lebesgue measure on the Euclidean space $\mathbb{R}^n$ is a Borel measure. The continuity of the linear form follows from the 'dominated convergence theorem'.

1.4. Exercises.

1) Show that if f is a differentiable function, then the map $g \mapsto fg$ of C_c^∞ into itself is continuous.

2) Is the linear form $f \mapsto f(0)$ of $C_c^\infty(\mathbb{R}^n)$ a Borel measure?

3) Is the linear form $f \mapsto \frac{df}{dx}(0)$ of $C_c^\infty(\mathbb{R})$ a Borel measure?

1.5. Product measure.

If μ, ν are measures on differential manifolds M and N, then one can define a natural measure on the product $M \times N$. We note that it is enough to define in a consistent manner, continuous linear forms on C_A^∞ for any compact subset A of $M \times N$. Since any compact set A in $M \times N$ is contained in $p_1(A) \times p_2(A)$, it is clear that we may as well assume that A is itself of the form $K \times L$ where K, L are compact subsets of M, N, respectively.

To elements $f \in C_K^\infty$ and $g \in C_L^\infty$, associate the function on $M \times N$ taking (x, y) to $f(x)g(y)$. Clearly this induces a linear map $C_K^\infty \otimes C_L^\infty \to C_{K \times L}^\infty$. It is obvious that this map is in fact injective.

1.6. Exercise. Prove the above assertion by showing that if $\sum f_i(x)g_i(y) = 0$ with g_i linearly independent in C_L^∞, then $f_i = 0$ for all i.

The linear form on this subspace, given by the bilinear map $(f, g) \mapsto \mu(f).\nu(g)$, is easily seen to be continuous for the induced topology. If we now show that this subspace of $C_{K \times L}^\infty$ is dense, then it would follow that the linear form above extends uniquely to a continuous linear form as required.

Let f belong to $C^\infty_{K \times L}$. Given $\epsilon > 0$, $a \in K$ and $b \in L$, we can find open neighbourhoods U_a, V_b such that $|f(x, b') - f(a, b')| < \epsilon$ and $|f(a', y) - f(a', b)|$ $< \epsilon$ for all $x \in U_a$, $y \in V_b$, $a' \in K$, and $b' \in L$. Using the compactness of K and L, we choose finitely many points a_i, b_j such that $U_i = U_{a_i}, V_j = V_{b_j}$ cover K, L, respectively. Let (φ_i) and (ψ_j) be partitions of unity for these coverings. We define a function $g \in C^\infty_K \otimes C^\infty_L$ by setting $g(x, y) = \sum f(a_i, b_j)\varphi_i(x)\psi_j(y)$. Then $f(x, y) - g(x, y) = f(x, y)(\sum \varphi_i(x)\psi_j(y)) - \sum f(a_i, b_j)\varphi_i(x)\psi_j(y) = \sum (f(x, y) - f(a_i, b_j))\varphi_i(x)\psi_j(y))$. From the choice of a_i and b_j and noticing that the support of $\varphi_i(x)\psi_j(y)$ is contained in $U_i \times V_j$, we conclude that $|f(x, y) - g(x, y)| < \epsilon$ for all x, y. This proves our claim that $C^\infty_K \otimes C^\infty_L$ is dense in $C^\infty_{K \times L}$.

Thus we have the following

1.7. Proposition. *If μ and ν are Borel measures on M and N, respectively, there exists a unique Borel measure $\mu \otimes \nu$ on $M \times N$ which satisfies*

$$\int f(x)g(y)d(\mu \otimes \nu) = \int f(x)d\mu(x) \int g(y)d\nu(y)$$

for all f, g with compact support.

1.8. Lie derivatives of measures.

One can try to define the Lie derivative $L(X)(\mu)$ of a measure μ with respect to a (differentiable) vector field X as follows. The Lie derivative $L(X)(\mu)$ is the functional which takes any $f \in C^\infty_c$ to $\lim_{t \to 0} \frac{(\varphi_t^{-1}(\mu))(f) - \mu(f)}{t}$ where (φ_t) is the flow of X. Note that although φ_t may not be defined for all t, they can be defined for small t in a neighbourhood of the support of f. This is adequate for the definition above. We see that this limit exists and is equal to $\mu(\lim_{t \to 0} \frac{f \circ \varphi_t^{-1} - f}{t}) = -\mu(Xf)$. While it is easy to check that the map $f \mapsto -\mu(Xf)$ is linear, there is no reason why it should satisfy the continuity axiom. We will assume that it in fact does. Indeed, we will assume that if $X_1, \ldots, X_k$ are any number of vector fields, then the iterated Lie derivative of μ is also a measure, that is to say, the linear form $f \mapsto (-1)^k \mu(X_k \cdots X_1 f)$ satisfies the continuity axiom. These considerations motivate the following definition.

1.9. Definition. If μ is a Borel measure with the property that for any family of vector fields $X_1, \ldots, X_k$, the linear form $f \mapsto \mu(X_1 \cdots X_k f)$ on C^∞_c is a Borel measure, then we say that μ is *indefinitely differentiable*. If X is a vector field, the measure $f \mapsto -\mu(Xf)$ is called the *Lie derivative* of μ with respect to X and is denoted $L(X)(\mu)$.

If $U \subset V$ are open subsets of M, then a measure on V gives rise to a measure on U since there is a natural continuous inclusion of $C^\infty_c(U)$ in

$C_c^\infty(V)$, namely, extension of functions by setting them zero outside U. It is also obvious that differentiable measures restrict to differentiable measures. The assignment to every open set U of M, of the set of all differentiable measures on U, is thus a presheaf under these restriction maps. It is in fact a sheaf. Let $\{U_i\}$ be a locally finite family of open sets and μ_i be Borel measures on them, such that $\mu_i|U_i \cap U_j$ and $\mu_j|U_i \cap U_j$ coincide for all i, j. Then we can define a measure μ on $U = \bigcup U_i$ as follows. Let $\{\varphi_i\}$ be a partition of unity with respect to the above covering. Then for any $f \in C_c^\infty(U)$ define $\mu(f) = \sum \mu_i(f\varphi_i)$. Note that this summation is only over the set of i's such that $\operatorname{supp}(f) \cap U_i$ is nonempty. This set is finite since f has compact support. If f has compact support contained in U_j, then this summation becomes $\sum_i \mu_j(f\varphi_i)$ because $f\varphi_i$ has support inside $U_i \cap U_j$ and we know that $\mu_i|(U_i \cap U_j) = \mu_j|(U_i \cap U_j)$. But then $\sum \mu_j(f\varphi_i) = \mu_j(f \sum \varphi_i) = \mu_j(f)$. This implies that μ restricts to μ_j on U_j for all j. It is a straightforward verification that μ is a Borel measure on U, that is to say, it satisfies the continuity axiom, and also that it is differentiable if all the μ_i's are. We will denote the sheaf of all differentiable measures by $\mathcal{B}$.

In a natural way, $\mathcal{B}$ is an $\mathcal{A}$-module. If μ is a measure and h a differentiable function, then $h\mu : f \mapsto \mu(hf)$ is a linear form on C_c^∞. This satisfies the continuity axiom, for if $\{f_n\}$ is a sequence of functions with support in a compact set K and $\sup|f_n|$ tends to zero, then $\{hf_n\}$ is also a sequence of functions with support in K and $\sup|hf_n|$ tends to zero. Hence by definition $\mu(hf_n)$ tends to zero as well. Moreover, if μ is differentiable, then $h\mu$ is also differentiable. We have in fact the following formula regarding Lie derivatives of measures.

1.10. Proposition. *If μ is a differentiable measure and h a differentiable function, then $h\mu$ is also differentiable, and for every vector field X, we have*

$$L(X)(h\mu) = L(hX)(\mu) = Xh.\mu + hL(X)(\mu).$$

Moreover, if X, Y are vector fields, we have

$$L([X,Y])(\mu) = L(X)L(Y)(\mu) - L(Y)L(X)(\mu).$$

Proof. The first assertion is similar to [Ch. 2, Proposition 5.3] and can be proved similarly. In any case, using the formula $L(X)(\mu) = -\mu(Xf)$, we can check it directly:

$$
\begin{aligned}
(L(hX)(\mu))(f) &= -\mu(hX(f)) \\
&= -\mu(X(hf) - Xh.f) \\
&= (L(X)\mu)(hf) + (Xh.\mu)(f) \\
&= (h.L(X)(\mu))(f) + (Xh.\mu)(f).
\end{aligned}
$$

It is easy to check similarly that $L(X)(h\mu)$ also gives the same expression. The second assertion is also straightforward. In fact, we have

$$
\begin{aligned}
L([X,Y])(\mu)(f) &= -\mu([X,Y]f) \\
&= -\mu(XYf) + \mu(YXf) \\
&= (L(X)(\mu))(Yf) - (L(Y)(\mu))(Xf) \\
&= (L(X)(L(Y)(\mu)))(f) - (L(Y)(L(X)(\mu)))(f) \\
&= ((L(X)L(Y) - L(Y)L(X))(\mu))(f).
\end{aligned}
$$

We will hereafter mean by $\mathcal{B}$ the $\mathcal{A}$-module of differentiable Borel measures.

1.11. Example. Let M be a domain in a real vector space V. Then any (translation) invariant measure μ is infinitely differentiable. In fact, if X denotes the vector field $\partial_v, v \in V$, then its flow is given by translations by tv. Hence the Lie derivative $L(X)$ of the measure μ exists and is actually 0. From this we conclude that the Lie derivative of μ, with respect to any vector field of the form $f\partial_v$ with f differentiable, is actually $(\partial_v f)\mu$. In particular the Lie derivative with respect to a differentiable vector field of any measure of the form $h\mu$ with h a differentiable function, is again a measure of the same form.

1.12. Remark. If μ is any measure on a real vector space V such that $L(\partial_v)\mu = 0$ for all $v \in V$, then it is clear that μ is invariant under translations.

1.13. Theorem. *An invariant measure on a real vector space V exists and is unique up to a scalar factor.*

Proof. We will prove the uniqueness by giving an injective map H of the space of invariant measures into the one-dimensional space $\Lambda^n(V^*)$ where n is the dimension of V. We interpret the latter as the space of alternating n-linear forms on V. In order to give this map, we choose one of the two connected components of $\Lambda^n(V) \setminus \{0\}$. Then for any $v_1, \ldots, v_n \in V$ we define $H(\mu)(v_1, \ldots, v_n)$ to be $\pm\mu([v_i])$, where $[v_i]$ denotes the cube formed by 0 and the v_i's. This set can be defined as the convex closure of all the vectors of the form $\sum v_j$, with the summation running through any subset of $\{1, \ldots, n\}$. Since the cube is compact, its measure is finite. Note that if the v_i's are linearly dependent, the set $[v_i]$ is contained in a proper subspace. It cannot have nonzero measure, since a compact set contains the disjoint union of infinitely many translates of $[v_i]$. In particular, if two of the v_i's are equal, then H takes the value 0. If they are linearly independent and μ is nontrivial, then $\mu([v_i])$ is nonzero, since any open set can be covered by countably many translates of $[v_i]$. We assign the sign 1 (resp. -1) in our definition of H, if $v_1 \wedge \cdots \wedge v_n$ belongs to (resp. does not belong to) the chosen component.

It is clear that if v_1 is replaced by a positive integral multiple rv_1, then $[rv_1, \ldots, v_n]$ is the union of $[(r-1)v_1, \ldots, v_n]$ and the translate of $[v_1, \ldots, v_n]$ by $(r-1)v_1$, the intersection being of measure zero. Hence one concludes that $\mu([rv_1, \ldots, v_n]) = r\mu([v_1, \ldots, v_n])$ by induction. It is clear that the same equality persists for rational values of r as well. If r is a positive real number, we can approximate it by an increasing sequence $\{a_k\}$ of positive rational numbers and notice that $[rv_1, \ldots, v_n]$ is the increasing union of $[a_k v_1, \ldots, v_n]$ and so the same equality is valid for r as well. Finally, the signs in our mapping are the same for all these cubes and so it follows that $H(\mu)(\lambda v_1, \ldots, v_n) = \lambda H(\mu)(v_1, \ldots, v_n)$ for all real λ, since multiplying v_1 by -1 changes the sign in our definition, but does not change the measure. By a similar elementary argument, one can also check that it is additive in each variable. Thus $H(\mu)$ is an alternating n-form on V. We have given an injective linear map of the space of invariant measures into $\Lambda^n(V^*)$. As for the existence, we identify V with $\mathbb{R}^n$ and take the Lebesgue measure on $\mathbb{R}^n$ which is of course left invariant.

1.14. Exercise. Prove the above assertion that the map $(v_1, \ldots, v_n) \mapsto \pm\mu([v_i])$ is additive in each variable.

1.15. Corollary. *The set of all measures in $\mathcal{B}$ in a domain U in $\mathbb{R}^n$ satisfying $L_X(\mu) = 0$ for all vector fields of the form $X = \frac{\partial}{\partial x_i}$, is one-dimensional.*

Proof. Let μ be a nonzero measure as above. There is no loss in generality in assuming that U is a unit disc. Then by looking at the flow, we deduce that for any open subset V of U and $a \in \mathbb{R}^n$ such that the translated set $W = T_a(V)$ is contained in U, we have $T_a(\mu|V) = \mu|W$. Consider the open covering $\{T_x U\}$ of $\mathbb{R}^n$ obtained by translating U by all points of $\mathbb{R}^n$. Provide each of these open sets with the translated measure $T_x(\mu)$. We will get a global measure on $\mathbb{R}^n$, if we verify that they agree on the intersections. Let $z \in T_x U \cap T_y U = U_{x,y}$. Then $T_{-x} U_{x,y}$ and $T_{-y} U_{x,y}$ are both contained in U, and T_{x-y} translates one to the other. From this and the translation invariance of μ, we check that $T_x(\mu)$ and $T_y(\mu)$ coincide on the intersection. Thus there is a global measure on $\mathbb{R}^n$ which restricts to $T_x(U)$ as $T_x(\mu)$. Clearly this measure is translation invariant. Our assertion therefore follows from Theorem 1.13.

From the above, one can surmise that the sheaf $\mathcal{B}$ of differentiable measures as a sheaf of $\mathcal{A}$-modules, is very close to the sheaf $\mathcal{K}$ of n-differentials on M (where n is the dimension of M). For one thing, the Lie derivatives of n-forms behave very much the same way.

1.16. Proposition. *Let ω be an n-form, X a differentiable vector field and h a differentiable function. Then we have*

$$(L(hX))(\omega) = L(X)(h\omega) = Xh.\omega + hL(X)(\omega).$$

Also we have

$$L([X,Y])(\omega) = L(X)L(Y)(\omega) - L(Y)L(X)(\omega)$$

for any two vector fields X, Y.

Proof. We use the formula $d\iota_X + \iota_X d = L(X)$, proved in [Ch. 2, 6.7, ii)]. Applying this on ω, and using the fact that the exterior derivatives of n-forms are zero, we get $(L(hX))(\omega) = d\iota_{hX}(\omega) = d(h\iota_X)(\omega) = dh \wedge (\iota_X\omega) + hd\iota_X\omega$. Using the fact that ι_X is an odd derivation, we may write this as $-\iota_X(dh \wedge \omega) + \iota_X(dh)\omega + hL(X)(\omega)$. But $dh \wedge \omega$ is clearly zero and $\iota_X(dh) = Xh$, proving one of the equalities above. The remaining statements have already been proved in [Ch. 2, 5.8].

1.17. Remark. We will again consider any real vector space V of dimension n as a differential manifold. Note first that the sheaf $\Lambda^n(\mathcal{T}^*)$ is in this case simply $\mathcal{A} \otimes_{\mathbb{R}} \Lambda^n(V^*)$. We may also tensor with $\mathcal{A}$, the one-dimensional space of invariant measures. The latter sheaf can be identified with the subsheaf of $\mathcal{B}$ consisting of measures of the form $f.\mu$ where f is a differentiable function and μ is an invariant measure. This isomorphism between the space of invariant measures and the space of n-differentials proved in Theorem 1.13 yields an isomorphism of these two sheaves as well. Since the Lie derivatives with respect to vector fields of the type ∂_v are zero on both invariant forms and invariant measures, the above formulae show that this isomorphism is compatible with the operation of Lie derivatives. However, one should note that the isomorphism depends on the choice of a connected component of $\Lambda^n(V) \setminus \{0\}$. Such a choice is called an *orientation* of the vector space V.

2. Sheaf of Densities

We will take up the relationship between $\mathcal{K}$ and $\mathcal{B}$ in general now.

2.1. Definition. An $\mathcal{A}$-homomorphism of the sheaf $\mathcal{K} = \Lambda^n(\mathcal{T})^*$ into $\mathcal{B}$ is said to be *flat* if it commutes with Lie derivatives with respect to all (differentiable) vector fields.

2.2. Proposition. *If M is connected, any two nonzero flat homomorphisms of $\mathcal{K}$ into $\mathcal{B}$ differ by a nonzero scalar factor. If M is a domain in $\mathbb{R}^n$, then there does exist a nonzero flat homomorphism.*

Proof. It is enough to prove the first statement also when M is a domain in $\mathbb{R}^n$. For any invariant form ω and any $v \in \mathbb{R}^n$, we have $L(\partial_v)\omega = 0$. If A is a flat homomorphism of $\mathcal{K}$ into $\mathcal{B}$, then clearly we also have $L(\partial_v)(A\omega) = 0$. In other words, a flat homomorphism takes invariant forms to invariant measures and is therefore determined up to a scalar factor. On the other hand, we have observed that the $\mathcal{A}$-homomorphism which maps any nonzero invariant form ω to the Lebesgue measure is actually flat.

2.3. Remark. We have seen that any domain in a vector space V admits a flat homomorphism $\mathcal{K} \to \mathcal{B}$. This is even canonical once one has chosen an orientation in V. The question arises whether any differential manifold admits a flat homomorphism. It is not so for the following reason. We may cover the manifold with coordinate open sets, and in each of these open sets which are diffeomorphic to domains in $\mathbb{R}^n$, take a flat homomorphism. These may not patch together since the flat homomorphism that we have given above for domains in $\mathbb{R}^n$, depends on the particular coordinate system. Two different choices of coordinate systems may give rise to flat homomorphisms which are negatives of each other.

The assignment, to any open set U, of the set of all flat homomorphisms $\mathcal{K}_U \to \mathcal{B}_U$ is a sheaf of $\mathbb{R}$-modules. Actually we may consider the set of all flat homomorphisms which coincide with an integral multiple of any standard flat homomorphism on a coordinate subset. Since the two standard isomorphisms are negatives of each other, it is a local system of rank 1 over integers, which we will denote by OZ_M.

2.4. Definition. The sheaf OZ_M is called the local system of *twisted integers*. If this local system is isomorphic to the trivial local system $M \times \mathbb{Z}$, then the manifold M is said to be *oriented*. In that case, there are two trivialisations and each is called an *orientation* on M.

2.5. Remarks.

1) Any connected manifold is either not oriented or has two orientations.

2) Any orientation of a real vector space V (namely, a choice of a component of $\Lambda^n(V) \setminus \{0\}$), gives rise to an orientation on the *differential manifold V*.

3) If M and N are two differential manifolds, then the sheaf $\mathcal{K}_{M \times N}$ is the tensor product of the sheaves $\mathcal{K}_M$ and $\mathcal{K}_N$ pulled back to $M \times N$. Also the operation of taking the product of two measures yields an isomorphism of the tensor product of $\mathcal{B}_M$ and $\mathcal{B}_N$ pulled back to $M \times N$, with $\mathcal{B}_{M \times N}$. Hence we also get an isomorphism of local systems $OZ_{M \times N} \sim p_1^*(OZ_M) \otimes p_2^*(OZ_N)$. In particular, if M and N are oriented, then so is $M \times N$.

4) The local system OR of all flat homomorphisms is obviously $OZ \otimes_{\mathbb{Z}} \mathbb{R}$.

2.6. Definition. The image of $\mathcal{K}$ in $\mathcal{B}$ by any nonzero flat homomorphism is called the *sheaf of densities*. We will denote the sheaf of densities by $\mathcal{S}_M$ or simply $\mathcal{S}$.

By definition then we have the canonical isomorphism $\mathcal{K} \otimes_{\mathbb{Z}} OZ \simeq \mathcal{S}$.

In the case of domains in $\mathbb{R}^n$, sections of $\mathcal{S}$ are simply measures of the form $f\,dx$ where f is a differentiable function and dx is the Lebesgue measure. Hence in a manifold these are measures which in any coordinate system can be expressed as $f\,dx$ as above.

2.7. Remark. It will turn out as a consequence of a theorem of Sobolev that $\mathcal{S}$ is the same as $\mathcal{B}$ (see [Ch. 8, Remark 2.6], but we do not need it here.

Since our definition is intrinsic, it follows that if U and V are domains in $\mathbb{R}^n$, and φ a diffeomorphism of U with V, then the pull-back of a density $h\,dx$ is again of the form $g\,dx$. The formula making this explicit is called 'the change of variable' formula and will be derived presently (Corollary 2.9).

Suppose that U is a manifold with an étale map into an oriented manifold M (for example $\mathbb{R}^n$). Then the tangent sheaf on M pulls back to the tangent sheaf on U. Hence the sheaf $\mathcal{K}$ of n-differentials also pulls back to $\mathcal{K}$. It is also easy to check that $\mathcal{S}$ on M pulls back to $\mathcal{S}$ on U. In particular, the standard flat isomorphism on M gives rise to a flat isomorphism on U. In other words, any orientation on M induces an orientation on U. (Is it also true that if U is oriented, then M is oriented?)

Let $i_t : U \to M$ be a continuous family of étale maps of U into M. Formally this means that there is a differentiable map $i : (0,1) \times U \to M$ whose restriction i_t to each $\{t\} \times U$ is an étale map of U into M. Then i_t gives an orientation on U for each t. Since t varies continuously, it is natural to expect that all these orientations on U are the same. To check it rigorously, we proceed as follows. Consider the differentiable map $(0,1) \times U \to (0,1) \times M$ taking (t,x) to $(t, i_t(x))$. This map is clearly an étale map into $(0,1) \times M$. Thus we see that the pull-back of $OZ_{(0,1) \times M}$ comes with a trivialisation. Its restriction to each $\{t\} \times U$ is the local system of twisted integers on U. Hence the orientation on $(0,1) \times M$ gives rise to one on each of $\{t\} \times U$. Since the fibres of the local system are discrete, all these sections coincide, i.e. all these orientations are the same.

2.8. Proposition. *If $\varphi : U \to V$ is any diffeomorphism of a domain in $\mathbb{R}^n$ onto another, then φ is compatible with the orientations on the two open domains if and only if the sign of $\det(d\varphi)$, at one (and hence every) point of U is positive.*

Proof. We may assume, using a translation of U, that $0 \in U$ and then by composing by a translation of V, that $\varphi(0) = 0$. Let $L(\varphi)$ be the restriction to U of the linear map of $\mathbb{R}^n$ into itself given by the differential $d\varphi$ of φ at 0. Then we have a family of maps $U \to \mathbb{R}^n$ given by $\varphi_t = tL(\varphi) + (1-t)\varphi$. Clearly these maps are all étale. At $t = 0$ this gives φ and at $t = 1$, it is $L(\varphi)|U$. From this we deduce that the orientations on U induced by φ and $L(\varphi)$ are the same. Since we have identified the orientations of a vector space V and the differentiable manifold V, our assertion follows from the obvious fact that a linear transformation of a vector space preserves the orientation if and only if the induced map on top exterior power preserves each component, that is to say, it has positive determinant.

Since we know how a diffeomorphism of a domain U of $\mathbb{R}^n$ with another acts on $\Lambda^n(T^*)$, we can deduce how it acts on $\mathcal{S}$.

2.9. Corollary (Change of variable formula). *If φ is a diffeomorphism of a domain in $\mathbb{R}^n$ into another domain D, then for any function f on D with compact support, we have*

$$\int f(\varphi(x))|\det(d\varphi)|dx = \int f(y)dy.$$

Proof. Using the isomorphism $\mathcal{K} \otimes OZ \to \mathcal{S}$ and the compatibility of the maps on the three sheaves induced by φ with the above isomorphism, we can compute the induced map on $\mathcal{S}$ by computing the induced maps on $\mathcal{K}$ and OZ. Now the isomorphism given rise to on the tangent bundle by φ is by definition $d\varphi$. Hence the induced map on $\mathcal{K} = \Lambda^n(\mathcal{T}^*)$ is multiplication by $\det(d\varphi)^{-1}$. On the other hand we have computed the induced map on OZ above. Putting these together, we see that the induced map (which is the pull-back of measures) is given by multiplication by $|\det(d\varphi)|^{-1}$. The formula claimed gives the image measure.

2.10. Remark. A trivialisation of OZ gives rise to a trivialisation of the local system OR. It is easy to see that if OR is trivial then OZ is so, as well. We could therefore have defined orientation as a trivialisation of OR, considering two trivialisations equivalent if they differ by a positive real multiple. Moreover, any trivialisation of OR leads to a trivialisation of the $\mathcal{A}$-module $\mathcal{A} \otimes OR$. Again, the converse is easy to see. Thus an orientation gives a canonical isomorphism $\mathcal{K} = \Lambda^n(T^*) \to \mathcal{S}$. On the other hand, $\mathcal{S}$

is always trivial (although there is no canonical trivialisation). In fact, we may cover the manifold with a locally finite family of coordinate open sets (U_i), take the Lebesgue measure μ_i in each of them, and consider $\sum \varphi_i \mu_i$, where (φ_i) is a partition of unity for the covering. This gives a section of $\mathcal{S}$ which is nonzero everywhere. Thus M is orientable if and only if the bundle $\Lambda^n(T^*)$ is trivial.

2.11. Examples.

1) We have seen that any domain in $\mathbb{R}^n$ comes with a canonical orientation.

2) If the tangent bundle of a differential manifold is trivial, then the canonical bundle is also trivial. Hence the manifold is orientable. In particular, any Lie group is orientable.

3) Consider a connected closed submanifold H of $\mathbb{R}^n$ of dimension $n-1$. In view of the exact sequence

$$0 \to T_H \to T_{\mathbb{R}^n}|H \to \mathrm{Nor}(H, \mathbb{R}^n) \to 0$$

and the fact that the normal bundle is one-dimensional and the middle term is trivial, we conclude that $\Lambda^{n-1}(T_H^*)$ is trivial if and only if the normal bundle is trivial. We recall that this is indeed the case. There is a differentiable function f such that all partial derivatives of f do not vanish simultaneously at any point of the hypersurface H given by $f = 0$ [Ch. 2, Example 2.6]. The vector field $v_f = \sum \frac{\partial f}{\partial x_i} \frac{\partial}{\partial x_i}$ on $\mathbb{R}^n$ can be restricted to H_f to give a section of $T_{\mathbb{R}^n}|H_f$. Now the tangent space at a point $(a_1, \ldots, a_n) \in H_f$ is given by the subspace $\{\sum y_i \frac{\partial}{\partial x_i} : \sum y_i \frac{\partial f}{\partial x_i}(a) = 0\}$. The vector v_f is therefore nonzero at all points of H_f. The image of this section in the normal bundle is nonzero everywhere and gives a trivialisation. Thus we conclude that H_f gets a canonical orientation. Note that if we replaced f by $-f$, we would get the opposite orientation on H_f. In particular, if f is the quadratic polynomial $\sum x_i^2 - 1$, the hypersurface H_f is the sphere S^{n-1}. The vector field we have given is the outward radial vector at every point of S^{n-1} and as such it is never tangential to S^{n-1}.

Let M be a differential manifold and (U, x) a local coordinate system on it. Then U has a canonical orientation. If (V, y) is another local coordinate system, then we get two orientations on $U \cap V$. They are related by ± 1 (if $U \cap V$ is connected, which we shall assume). It is clear from our analysis above that these orientations are the same if and only if the determinant of the transformation matrix is positive. From this we conclude the following.

2.12. Proposition. *The local system OZ on a connected manifold M is trivial, i.e., the manifold is orientable, if and only if there exists a system of* *coordinate charts covering M such that on the overlaps the transformations of the coordinate systems have positive Jacobians. Moreover, any orientation gives rise to a special system of oriented charts as above.*

2.13. Example. The real projective space, $\mathbb{RP}^n$, is orientable if and only if n is odd. In fact, we have a natural map $S^n \to \mathbb{RP}^n$. Noting that this is an étale map, we see that the pull-back of $OR_{\mathbb{RP}^n}$ is OR_{S^n} which is trivial as we have seen above. The antipodal involution also acts on Or_{S^n}, If this local system is the pull-back of a trivial system on $\mathbb{RP}^n$, then the involution preserves the trivialisation on S^n. Hence $\mathbb{RP}^n$ is orientable if and only if the antipodal involution preserves the orientation on S^n. Using the radial vector we see that it preserves the orientation of S^n if and only if it preserves that of $\mathbb{R}^{n+1}$. Note that this map is given by $(x_1, \ldots, x_{n+1}) \mapsto (-x_1, \ldots, -x_{n+1})$ which preserves the orientation in $\mathbb{R}^{n+1}$ if and only if n is odd.

2.14. Exercises.

   1) Show that $\mathbb{RP}^2$ cannot be imbedded as a submanifold of $\mathbb{R}^3$.

   2) Let V be a real vector space of dimension n and $\mathrm{Grass}_k(V)$ the Grassmannian of r-dimensional spaces in V. When is it orientable?

2.15. Remark. The vector bundle S associated to $\mathcal{S}$ has an interesting interpretation. For $a \in M$, the fibre S_a is isomorphic to $\mathcal{S}_a/\mathcal{M}_a\mathcal{S}_a$ where $\mathcal{M}_a$ is the ideal of functions f vanishing at a. Any element μ of $\mathcal{S}_a$ can be represented by a measure $h(x)dx$ in a coordinate chart $(U, x_1, \ldots, x_n)$ around a. Consider the map which associates to μ the invariant measure $h(a)dx$ on the tangent space $T_a(M)$. Note that since a choice of coordinate system $(x_1, \ldots, x_n)$ has been made, there is a canonical basis $(\frac{\partial}{\partial x_1}, \ldots, \frac{\partial}{\partial x_n})$ of $T_a(M)$ and hence one can talk of the Lebesgue measure dx on $T_a(M)$ as well. If a different coordinate system $(y_1, \ldots, y_n)$ (again with a as origin) is used, then $h(x)dx$ pulls back to $h(\varphi(x))|\det(\frac{\partial \varphi_i}{\partial x_j})|dy$. Under our map it induces the measure $h(\varphi(0))|\det(\frac{\partial \varphi_i}{\partial x_j})(0)|dy$ on $T_a(M)$. But this is simply the pull-back by the linear transformation $d\varphi$ (the differential $T_a(M) \to T_a(M)$ of the map φ) of the invariant measure $h(0)dx$. Thus we have

2.16. Proposition. *The vector space $S_a, a \in M$, may be canonically identified with the space of all invariant measures on $T_a(M)$.*

Proof. We have already shown that there is a well-defined map of the stalk $\mathcal{S}_a$ of $\mathcal{S}$ at a, into the set of invariant measures on $T_a(M)$, namely that which in any coordinate chart around a point $a \in M$ is given by $h(x)dx \mapsto h(a)dx$. Clearly $\mathcal{M}_a\mathcal{S}_a$ maps to 0 under this map and induces an isomorphism of S_a with the space of invariant measures on $T_a(M)$.

2.17. Remark. An important class of manifolds which are orientable is that of complex manifolds. In fact, we may take the complex coordinate charts as a special system of coordinates. Then on the overlaps, the coordinate transformations are given by *holomorphic* transformations $\varphi_i(z)$ of an open set in $\mathbb{C}^n$. Considering them as real transformations, we get the Jacobian matrix

$$\begin{pmatrix} \dfrac{\partial \xi_i}{\partial x_j} & \dfrac{\partial \xi_i}{\partial y_j} \\[2mm] \dfrac{\partial \eta_i}{\partial x_j} & \dfrac{\partial \eta_i}{\partial y_j} \end{pmatrix}$$

where ξ_i, η_i are respectively the real and imaginary parts of $\varphi_i(z)$ and x_i, y_i are real and imaginary parts of z_i. But since φ_i are holomorphic, we have the Cauchy-Riemann equations $\dfrac{\partial \xi_i}{\partial x_j} = \dfrac{\partial \eta_i}{\partial y_j}$ and $\dfrac{\partial \xi_i}{\partial y_j} = -\dfrac{\partial \eta_i}{\partial x_j}$. In order to compute its determinant, we treat the matrix as a complex matrix and conjugate it by

$$\begin{pmatrix} 1 & -\sqrt{-1} \\ -\sqrt{-1} & 1 \end{pmatrix}$$

to obtain

$$\begin{pmatrix} \dfrac{\partial \xi_i}{\partial x_j} + \sqrt{-1}\dfrac{\partial \xi_i}{\partial y_j} & 0 \\[2mm] 0 & \dfrac{\partial \xi_i}{\partial x_j} - \sqrt{-1}\dfrac{\partial \xi_i}{\partial y_j} \end{pmatrix}$$

whose determinant, namely $|\det(\frac{\partial \varphi_i}{\partial z_j})|^2$, is clearly positive.

3. Adjoints of Differential Operators

In analysis, it is sometimes convenient to work with vector-valued measures, namely measures with values in a vector space. From a global perspective, it is expedient to have a notion of a vector-bundle-valued density on a differential manifold.

3.1. Definition. Let E be a differentiable vector bundle on a differential manifold M. Then a section of $\mathcal{E} \otimes \mathcal{K} \otimes_{\mathbb{R}} OR = \mathcal{E} \otimes_A \mathcal{S}$ is called an *E-valued density*.

3.2. Twisted forms.

We have seen that if D is a differential operator from E to F, where E and F are vector bundles, then we may also define an operator $E \otimes \mathcal{L} \to F \otimes \mathcal{L}$ where $\mathcal{L}$ is any local system. All the operators we defined on forms, such as exterior derivation, Lie derivative, inner product and the corresponding formulae are valid therefore for sections of the bundles obtained by tensoring a tensor bundle with $\mathcal{L}$. In particular, we may tensor the de Rham complex by OR and get a 'twisted de Rham complex'. Sections of the bundles $\Lambda^i(T^*) \otimes OR$ will be called *twisted forms*. Thus by definition, a density is a twisted n-form. We may also take an exterior product of a differential form and a twisted form to get a twisted form, or of two twisted forms resulting in an ordinary form.

3.3. Theorem (Stokes). *If μ is a density with compact support of the type $d\alpha$ where α is a twisted $(n-1)$-form, then $\int_M \mu = 0$.*

Proof. Let (U_i) be a locally finite coordinate covering of M and (g_i) a differentiable partition of unity with respect to (U_i). If we set $\mu_i = d(g_i \alpha)$, then clearly $\sum \mu_i = d(\sum g_i \alpha) = d\alpha = \mu$. So it is enough to show that $\int \mu_i = 0$. In other words, we may assume that $\mu = d\alpha$ with support of α contained in a coordinate neighbourhood and show that $\int_M \mu = 0$. If $(x_1, \ldots, x_n)$ is a coordinate system, we have seen that the measure μ may be represented by $f\,dx$ where dx is the Lebesgue measure and f a differentiable function with compact support. Having trivialised OR over the coordinate neighbourhood, we may assume that α is given by $\sum f_i dx_1 \wedge \cdots \wedge \widehat{dx_i} \wedge \cdots \wedge dx_n$. Hence $d\alpha$ corresponds to the measure $(\sum (-1)^{i-1} \frac{\partial f_i}{\partial x_i}) dx$, where f_i are differentiable functions with compact support. Let us compute $\int_U \frac{\partial f_i}{\partial x_i} dx$. We may assume that the support of f_i is contained in an open cube with its closure C a subset of U. Then $\int_U \frac{\partial f_i}{\partial x_i} dx = \int_C \frac{\partial f_i}{\partial x_i} = \int (f_i(a) - f_i(b)) dx'$ where dx' is the Lebesgue measure in $\mathbb{R}^{n-1}$ and a, b the maximum and minimum of the ith coordinates of points in C. But $f_i(a) = f_i(b) = 0$, proving our assertion.

If X is any vector field, the operator $\mu \mapsto -L(X)\mu$ of $\mathcal{S}$ into itself, is called the *adjoint* $\mathrm{adj}(X)$ of X. We will compute $\mathrm{adj}(X)$ in local coordinates. Let $(x_1, \ldots, x_n)$ be a local coordinate system, and let $X = \sum a_i(x) \frac{\partial}{\partial x_i}$. It is enough to compute the effect of $\mathrm{adj}(X)$ on the Lebesgue measure dx. In fact, using Proposition 1.10, we get

$$L(X)(dx) = \sum L\left(a_i \frac{\partial}{\partial x_i}\right)(dx) = \left(\sum \frac{\partial a_i}{\partial x_i}\right) dx$$

and again,

$$L(X)(f dx) = \left(\sum a_i \frac{\partial f}{\partial x_i} + f \sum \frac{\partial a_i}{\partial x_i}\right) dx = \left(\sum \frac{\partial}{\partial x_i}(a_i f)\right) dx.$$

Hence if we trivialise $\mathcal{S}$ over U, i.e. identify differentiable functions f with densities $f dx$, then $\mathrm{adj}(X)$ is the operator $-\sum \frac{\partial}{\partial x_i} \circ m(a_i)$. Incidentally, this shows that $\mathrm{adj}(X)$ is also a differential operator of order ≤ 1.

We will next define for any $D \in \mathcal{D}_M^1$, its adjoint as a differential operator $\mathrm{adj}(D) : \mathcal{S} \to \mathcal{S}$ of order ≤ 1. If $D = m(f)$, then we define $\mathrm{adj}(D)$ to be multiplication by f. If $D = X \in \mathcal{T}(M)$, then we have defined $\mathrm{adj}(D)$ above, namely the operator $\mu \mapsto -L(X)(\mu)$. We wish to extend this definition to differential operators of arbitrary orders. In fact, the map $D \mapsto \mathrm{adj}(D)$ is to be an antihomomorphism of the algebra $\mathcal{D}(M)$ into the algebra $\mathcal{D}(\mathcal{S}, \mathcal{S})$ in the sense that $\mathrm{adj}(D_1 \circ D_2) = \mathrm{adj}(D_2) \circ \mathrm{adj}(D_1)$. In view of the universal property of the algebra $\mathcal{D}$ we have only to show that

$$\begin{aligned}
\mathrm{adj}(X \circ m(f)) &= m(f) \circ \mathrm{adj}(X), \\
\mathrm{adj}(m(f) \circ X) &= \mathrm{adj}(X) \circ m(f), \\
\mathrm{adj}(X_1 \circ X_2 - X_2 \circ X_1) &= \mathrm{adj}(X_2) \circ \mathrm{adj}(X_1) - \mathrm{adj}(X_1) \circ \mathrm{adj}(X_2).
\end{aligned}$$

As for the first equality, we have $\mathrm{adj}(X \circ m(f))(\mu) = \mathrm{adj}(m(f)X + m(Xf))(\mu)$ $= -L(fX)(\mu) + (Xf)\mu = -fL(X)(\mu)$. The second equality is proved similarly. The last formula follows from Proposition 1.10. Since all these formulas are only to be verified locally, one may also verify them from our explicit local computation of $\mathrm{adj}(X)$, given above.

3.4. Proposition. *Let E, F, G be differentiable vector bundles on a differential manifold M. Then there is a natural isomorphism $D \to \mathrm{adj}(D)$ of $\mathcal{D}^k(E, F)$ with $\mathcal{D}^k(\mathrm{adj}(F), \mathrm{adj}(E))$, where $\mathrm{adj}(E) = E^* \otimes S$. If $D \in \mathcal{D}^k(E, F)$, then $\mathrm{adj}(\mathrm{adj}(D))$ belongs to $\mathrm{Diff}_k(E, F)$. With this identification, we have $\mathrm{adj}(\mathrm{adj}(D)) = D$. Moreover, if $D_1 \in \mathcal{D}^k(E, F)$ and $D_2 \in \mathcal{D}^l(F, G)$, then $\mathrm{adj}(D_2 \circ D_1) = \mathrm{adj}(D_1) \circ \mathrm{adj}(D_2)$.*

Proof. We note first that the map $D \to \mathrm{adj}(\mathrm{adj}(D))$ of $\mathcal{D}^k$ into itself is actually an algebra homomorphism and so, in order to show that it is the identity map, we have only to check that $\mathrm{adj}(\mathrm{adj}(X)) = X$ for any vector field X and that $\mathrm{adj}(\mathrm{adj}(m(f))) = m(f)$ for all functions f. The latter is obvious, since $\mathrm{adj}(m(f))$ is simply multiplication by f and so is $\mathrm{adj}(\mathrm{adj}(m(f)))$. It is enough to check the first assertion locally, and so we may assume that $X = \frac{\partial}{\partial x_i}$, where $(x_1, \ldots, x_n)$ is a local coordinate system. Then with the obvious trivialisation of $\mathcal{S}$ we have seen that $\mathrm{adj}(\frac{\partial}{\partial x_i}) = -\frac{\partial}{\partial x_i}$ and hence $\mathrm{adj}(\mathrm{adj}((\frac{\partial}{\partial x_i}))) = \frac{\partial}{\partial x_i}$, proving our assertions in the case $E = F = \mathcal{A}$.

The general case follows on remarking that the map $\mathcal{D}^k(E, F) \to \mathcal{D}^k(E, F)$ given by $D \mapsto \mathrm{adj}(\mathrm{adj}(D))$ is obtained by tensoring the map $D \to \mathrm{adj}(\mathrm{adj}(D))$ on $\mathcal{D}^k$ on the right by F and on the left by E. The proof of the last assertion is similar.

3.5. Remark. Since the sheaf $\mathcal{S}$ is isomorphic to $\Lambda^n(T^*) \otimes OR$, it follows that there is an identity between differential operators $E \otimes S \to F \otimes S$ and those from $E \otimes \Lambda^n(T^*)$ to $F \otimes \Lambda^n(T^*)$. Hence the adjoint might just as well have been defined as above with $\mathcal{S}$ replaced by $\Lambda^n(T^*)$. However, our insistence on using $\mathcal{S}$ is explained by the following assertion.

3.6. Proposition (Integration by parts). *Let $D : E \to F$ be a differential operator of order $\leq k$ and $\mathrm{adj}(D)$ its adjoint. If $s \in \mathcal{E}(M)$ and $t \in (\mathrm{adj}(F))(M)$, at least one of them having compact support, then*

$$\int_M \langle Ds, t \rangle = \int_M \langle s, \mathrm{adj}(D)t \rangle.$$

Here we use the duality to define $\langle Ds, t \rangle$ and $\langle s, \mathrm{adj}(D)t \rangle$ as elements of $\mathcal{S}(M)$.

Proof. We will first show that this is the case when $E = F = \mathcal{A}$. The assertion is clear if D is of order 0, namely $D = m(g)$ for some $g \in \mathcal{A}(M)$. We will assume to start with that D is a vector field X. In this case, we have to show that if f is a differentiable function and μ is a density, then $\int (Xf)\mu = \int f(-L(X)\mu)$. But $fL(X)(\mu) = L(X)(f\mu) - (Xf)\mu$ by Proposition 1.10, so that we are reduced to showing that $\int L(X)(f\mu) = 0$. In other words, we need to show that if ν is any density with compact support, then $\int L(X)(\nu) = 0$. But $L(X) = d\iota_X + \iota_X d$ and since $d\nu = 0$, we have $\int_M L(X)\nu = \int_M d\iota_X\nu = 0$ by Stokes' Theorem 3.3. Now we prove our assertion for higher order operators by induction on the order. We have shown above that if $D \in \mathcal{D}^1(M)$, then the formula holds. It is enough to prove it for operators of the form $X_1 \cdots X_k$, where $X_i \in \mathcal{T}(M)$. We have then

$$\int ((X_1 \cdots X_k)f)\mu \;=\; \int (X_2 \cdots X_k f)(\mathrm{adj}(X_1)\mu)$$

$$=\; \int f \, \mathrm{adj}((X_1 \cdots X_k))(\mu)$$

as was to be proved. This concludes the proof when $E = F = \mathcal{A}$.

The general case is easily reduced to the above. Firstly, note that the assertion follows easily from the above if both E and F are trivial. Secondly, if s has compact support K, then Ds also has the same support, and so both sides depend only on $t|K$. Replacing t by φt where φ is a function vanishing outside a compact neighbourhood of K, we may assume that all the sections s, t, Ds, D^*t have compact support. Hence we may cover this support by a finite open covering (U_i) such that E and F are both trivial on each U_i. We may then use a partition of unity to write out both s and t as a sum of sections whose support lie in one of the U_i's. This reduces the problem to the case when both E and F are trivial.

3.7. Remark. The formula above could have been *used* to define the adjoint of a differential operator. The advantage of our definition is that it is local, and it is *a priori* clear that $\mathrm{adj}(D)$ is also a differential operator. In any case, $\mathrm{adj}(D)$ is uniquely defined by the above formula. For if D_1, D_2 are two operators satisfying

$$\int \langle Ds, t \rangle = \int \langle s, D_i t \rangle, \qquad i = 1, 2,$$

then for any $t \in (\mathcal{F}^* \otimes \mathcal{S})(M)$, we have $\int_M \langle s, (D_1 - D_2)t \rangle = 0$ for all $s \in \mathcal{E}(M)$. By replacing s by fs, we see that the measure $\langle s, (D_1 - D_2)t \rangle$ is itself 0. But this implies that $(D_1 - D_2)t = 0$ by the duality between E and E^*.

3.8. Examples.

1) Consider the exterior derivative $d : \Lambda^i(T^*) \to \Lambda^{i+1}(T^*)$. We may identify $\mathrm{adj}(\Lambda^i(T^*)) = \Lambda^i(T) \otimes \Lambda^n(T^*) \otimes OR$ with $\Lambda^{n-i}(T^*) \otimes OR$. Thus the adjoint of d is an operator $\mathrm{adj}(d) : \Lambda^{n-i-1}(T^*) \otimes OR \to \Lambda^{n-i}(T^*) \otimes OR$. We claim that this is simply $(-1)^{i+1}d$ where d is again the exterior derivative, now on twisted forms. In fact it would be enough to check that if ω is a twisted $(n - i - 1)$-form and α is an i-form, then

$$\int_M (d\alpha \wedge \omega) = (-1)^{i+1} \int_M \alpha \wedge d\omega.$$

But then $d\alpha \wedge \omega + (-1)^i \alpha \wedge d\omega = d(\alpha \wedge \omega)$ so that we have only to check that $\int d(\alpha \wedge \omega) = 0$, which is a consequence of Stokes' formula.

2) If X is a vector field on M, then the Lie derivative $L(X) : \Lambda^i(T^*) \to \Lambda^i(T^*)$ is a differential operator. As above, $\mathrm{adj}(L(X))$ may be considered an operator $\Lambda^{n-i}T^* \otimes OR \to \Lambda^{n-i}T^* \otimes OR$. This is indeed $-L(X)$, for we have only to check that

$$\int_M L(X)\alpha \wedge \omega = \int_M \alpha \wedge (-L(X)\omega),$$

for all $\alpha \in \Lambda^i T^*$ and $\omega \in \Lambda^{n-i}T^* \otimes OR$. In other words, we need to check that $\int L(X)(\alpha \wedge \omega) = 0$. But then $L(X) = d\iota_X + \iota_X d$ and $d(\alpha \wedge \omega) = 0$ so that our assertion follows again from Stokes' formula. In this case, it is also easy to prove our assertion directly from the definition.

3.9. The adjoint of an operator with complex coefficients.

The definition of the adjoint of an operator has to be slightly modified when E and F are complex vector bundles. If D is an operator $E \to F$, then its *conjugate* $\overline{D}$ can be defined as an operator $\overline{E} \to \overline{F}$. In fact, if $s \in \overline{\mathcal{E}}(M)$ then $\overline{D}s$ is defined to be $\overline{D\overline{s}}$. The conjugate of the operator $F^* \otimes \Lambda^n T^* \otimes$

$OR \to E^* \otimes \Lambda^n T^* \otimes OR$, which can be defined as in the real case, is called the *adjoint* $\mathrm{adj}(D)$ of D. It is an operator

$$\overline{F}^* \otimes \Lambda^n T^* \otimes OR = \mathrm{adj}(F) \to \mathrm{adj}(E) = \overline{E}^* \otimes \Lambda^n T^* \otimes OR.$$

3.10. Symbol of the adjoint.

If D is a differential operator $E \to F$ of order $\leq k$, then its (kth order) symbol is a vector bundle homomorphism $E \to F \otimes S^k(T)$. It may be thought of as associating to any $v \in T^*_m$ a homomorphism $\sigma(D)_v : E \to F$. Now $\mathrm{adj}(D)$ is a differential operator $F^* \otimes \mathcal{S} \to E^* \otimes \mathcal{S}$ of order $\leq k$. The symbol of the latter assigns therefore to any $v \in T^*_m$, a homomorphism $F^* \otimes \mathcal{S} \to E^* \otimes \mathcal{S}$. We claim that it is essentially $(-1)^k$ times the conjugate transpose of $\sigma(D)_v$. In fact, $\mathcal{S}$ is a line bundle and so the symbol $\sigma(\mathrm{adj}(D))$ may equally well be considered as assigning to v, a homomorphism $F^*_m \to E^*_m$. It is in this sense that we claim that $\sigma(\mathrm{adj}(D))_v = (-1)^k \overline{\sigma(D)}^{\,t}_v$. To check this we first note that by the definition of the adjoint operator it is enough to check the assertion in the case $E = F = \mathcal{A}$. Moreover, by the formula for symbols of composites [Ch. 2, 7.2], it is enough to prove it for $D = m(f), f \in \mathcal{A}(M)$, and $D = X \in \mathcal{T}(M)$. In the former case, the assertion is obvious, while in the latter case, the claim follows from our computation of $\sigma(L(X))$ in [Ch. 2, 7.21].

3.11. Remark. Although $\mathcal{S}$ is globally trivial, we have not defined the adjoint of an element in $\mathcal{D}$ as an element in $\mathcal{D}$. In fact, in order to do so, we would have to choose a global trivialisation of $\mathcal{S}$. After such a trivialisation, we may define the adjoint of a differential operator $E \to F$ to be a differential operator $\overline{F}^* \to \overline{E}^*$. However, this definition of the adjoint would depend on the chosen trivialisation of $\mathcal{S}$. Moreover, we can choose a Hermitian metric along the fibres of E and F to define isomorphisms of F, E with $\overline{F}^*, \overline{E}^*$, respectively, and actually define the adjoint as an operator $F \to E$. If we do this, then we may compute the symbol of the adjoint as follows. If a linear map $\sigma : E \to F \otimes S^k(T)$ is given, for every $v \in T^*_v$, we get an associated map $\sigma_v : E_v \to F_v$. Its conjugate transpose (using the Hermitian metrics along the fibres) is a linear map $F_v \to E_v$. We define the *adjoint symbol* $\mathrm{adj}(\sigma)$ of σ to be the linear map $F \to E \otimes S^k(T^*)$ given by setting $\mathrm{adj}(\sigma)_v$ to be $(-1)^k$ times the conjugate transpose of σ_v. Then our discussion above gives the computation of the symbol of the adjoint as saying $\sigma(\mathrm{adj}(D)) = \mathrm{adj}(\sigma(D))$.

Exercises

1) Let G be a connected Lie group and X a right invariant vector field. Compute $L(X)(\mu)$ where μ is a left invariant measure.

2) If μ is a linear form on C_c^∞ which takes nonnegative functions with compact support to nonnegative real numbers. Show that μ is then a measure.

3) Determine the image measure on a Lie group of a left invariant measure μ by the map $x \mapsto x^2$.

4) Show that the set of unitary matrices which have at least one eigenvalue with multiplicity > 1 is a set of measure zero with respect to the left invariant measure on $U(n)$.

5) Let T be the group of diagonal matrices in $U = U(n)$. Consider the map $T \times U/T \to U$ given by $(t, \bar{x}) \mapsto xtx^{-1}$. Write down the differential form on $T \times U/T$ whose image in U is the left invariant measure.

6) Consider the diffeomorphism $\mathbb{R}^+ \times S^{n-1} \to \mathbb{R}^n \setminus \{0\}$. Write down the pull-back of the Lebesgue measure and write it down as an n-differential form.

7) Prove that the tangent bundle of any manifold, considered as a differential manifold, is orientable.

8) Let G be a connected Lie group and H a closed subgroup. Find a necessary and sufficient condition for the existence of a G-invariant measure on G/H.

9) Show that there is no nonzero measure on $\mathbb{CP}^n$ which is invariant under the action of $GL(n+1)$.

10) Using the Lebesgue measure to trivialise $\mathcal{S}$ over $\mathbb{R}^n$, find the condition for a vector field $\sum f_i \frac{\partial}{\partial x_i}$ to be its own adjoint.

Chapter 4

Cohomology of Sheaves and Applications

We have mentioned that sheaves are the gadgets which allow one to pass from local data to global ones. The obstructions in this passage are kept track of, as in a log book, by the *cohomology* of the sheaves in question. We will study the definition and properties of cohomology in this chapter.

1. Injective Sheaves

Suppose $\mathcal{F}, \mathcal{F}'$ and $\mathcal{F}''$ are sheaves of abelian groups on a topological space X and we have homomorphisms $\mathcal{F}' \to \mathcal{F}$ and $\mathcal{F} \to \mathcal{F}''$. Then we say that the sequence

$$\mathcal{F}' \to \mathcal{F} \to \mathcal{F}''$$

is *exact* if the associated sequence at the stack level, namely

$$\mathcal{F}'_x \to \mathcal{F}_x \to \mathcal{F}''_x,$$

is exact, for all $x \in X$.

Note that to say that $\mathcal{F} \to \mathcal{G} \to 0$ is exact means that at the stalk level $\mathcal{F}_x \to \mathcal{G}_x$ is surjective. In long hand, this is equivalent to saying that if $s \in \mathcal{G}(U)$, where U is a neighbourhood of x, then there exist an open neighbourhood $V \subset U$ of x and an element t of $\mathcal{F}(V)$ such that the image of t in $\mathcal{G}(V)$ is $res_{UV}(s)$.

If $\mathcal{F}$ is a subsheaf of $\mathcal{G}$ then one may define the *quotient sheaf* by taking the quotient of the étale space of $\mathcal{G}$ under the equivalence relation: $v \sim w$ if they are in the same stalk $\mathcal{G}_a$ and $v - w$ belongs to $\mathcal{F}_a$. Note however that although we can define a presheaf by associating to any open set U the group $\mathcal{G}(U)/\mathcal{F}(U)$, it is not a sheaf in general.

1.1. Exercise. Consider the constant subsheaf $\mathbb{Z}$ over the real line. Let I be the subsheaf which associates sections which vanish at 0 and 1. Show that the quotient presheaf defined above is not a sheaf.

Of course the sheaf associated to the quotient presheaf is the quotient sheaf as we have defined. This also gives an example of the fact that if

$$0 \to \mathcal{F}' \to \mathcal{F} \to \mathcal{F}'' \to 0$$

is an exact sequence of sheaves on X, it is *not* true in general that the induced sequence

1.2. $\qquad\qquad 0 \to \mathcal{F}'(X) \to \mathcal{F}(X) \to \mathcal{F}''(X) \to 0$

is exact (see [Ch. 2, 6.16] for another example). The problem is only at the right end of the sequence. More precisely we have the following.

1.3. Proposition. *If $0 \to \mathcal{F}' \to \mathcal{F} \to \mathcal{F}''$ is an exact sequence of sheaves of abelian groups over a space X, then the induced sequence*

$$0 \to \mathcal{F}'(X) \to \mathcal{F}(X) \to \mathcal{F}''(X)$$

is also exact.

Proof. For a section s of $\mathcal{F}'$ to vanish as a section of $\mathcal{F}$, we must have that $s_x \in \mathcal{F}_x$ is 0 for all $x \in X$. But $\mathcal{F}'_x \to \mathcal{F}_x$ is injective, which implies that s_x is also zero as an element of $\mathcal{F}'_x$ and hence that $s = 0$. This proves the injectivity of $\mathcal{F}'(X) \to \mathcal{F}(X)$. If a section $s \in \mathcal{F}(X)$ induces 0 on $\mathcal{F}''(X)$, then $s_x \in \mathcal{F}_x$ maps to 0 in $\mathcal{F}''_x$ for all $x \in X$. So s_x belongs to $\mathcal{F}'_x$ for all $x \in X$, in view of the exact sequence

$$0 \to \mathcal{F}_x \to \mathcal{F}'_x \to \mathcal{F}''_x.$$

This means however that s is actually a section of $\mathcal{F}$ as claimed.

The *cohomology* of a sheaf measures the nonexactness of sequence 1.2. We first define some classes of sheaves for which the exactness is nevertheless true and will define cohomology in terms of the deviation of the given sheaf from such sheaves.

1.4. Definition. A sheaf $\mathcal{F}$ of modules over a sheaf $\mathcal{O}$ of rings on X is said to be *injective* if any homomorphism of a subsheaf $\mathcal{G}'$ of a sheaf $\mathcal{G}$ into $\mathcal{F}$ can be extended to a homomorphism of the whole of $\mathcal{G}$ into $\mathcal{F}$.

1.5. Definition. A sheaf $\mathcal{F}$ of abelian groups is said to be *flabby* (resp. *soft*) if any section of $\mathcal{F}$ over an open (resp. closed) subset of X can be extended to a section over the whole of X.

1.6. Proposition. *Any injective sheaf is flabby.*

Proof. Let U an open subset of X. We will construct in a natural way a subsheaf $\mathcal{J}_U$ of $\mathcal{O}$ such that $\mathrm{Hom}(\mathcal{J}_U, \mathcal{F}) = \mathcal{F}(U)$, for any sheaf $\mathcal{F}$. Then we may interpret any section s of a sheaf $\mathcal{I}$ over U to be a homomorphism $\mathcal{J}_U \to \mathcal{I}$. If $\mathcal{I}$ is injective, this homomorphism can be extended to a map $\mathcal{O} \to \mathcal{I}$, that is to say, the section s can be extended to the whole of X. Thus the proof will be complete if we show the following.

1.7. Lemma. *i) Let $\mathcal{O}$ be a sheaf of rings on X and $\mathcal{F}$ an $\mathcal{O}$-module over an open subset $U \subset X$. Then there exists a sheaf $\mathcal{F}_U$ on X which restricts to $\mathcal{F}$ on U and has zero stalks outside U.*

ii) If $\mathcal{G}$ is any sheaf on X, then the obvious restriction map $\mathrm{Hom}(X, \mathcal{F}_U, \mathcal{G}) \to \mathrm{Hom}(U, \mathcal{F}, \mathcal{G}|U)$ is an isomorphism.

iii) If $\mathcal{F}$ is already the restriction to U of a sheaf $\mathcal{S}$ on X, then $\mathcal{F}_U$ is a subsheaf of $\mathcal{S}$.

iv) In particular, taking $\mathcal{S}$ to be $\mathcal{O}$, we get a subsheaf $\mathcal{J}_U$ of $\mathcal{O}$ such that $\mathrm{Hom}(\mathcal{J}_U, \mathcal{G}|U) = \mathcal{G}(U)$ for all $\mathcal{O}$-modules $\mathcal{G}$.

Proof. i) Define $\mathcal{F}_U(V)$ for any open subset V of X by

$$\mathcal{F}_U(V) = \{s \in \mathcal{F}(U \cap V) : \text{support of } s \text{ is a closed subset of } V\}.$$

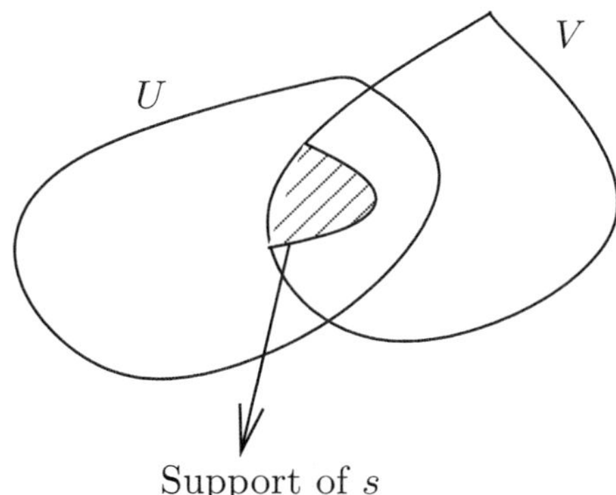

If $W \subset V$, then we can restrict s to $U \cap W$. The support of this restriction is simply $\mathrm{supp}(s) \cap W$, which is a closed subset of W. This makes our assignment a presheaf. Since $\mathcal{F}$ is a sheaf on U, in order to check that $\mathcal{F}_U$ is a sheaf, we have only to show that if (V_i) is an open covering of V and $s \in \mathcal{F}(U \cap V)$, then the support of s is closed in V if and only if the support of the restriction of s to $U \cap V_i$ is closed in V_i for every i. But this is obvious. If $V \subset U$, then $\mathcal{F}_U(V) = \mathcal{F}(V)$ by definition. Hence $\mathcal{F}_U|U = \mathcal{F}$. On the other hand if $x \notin U$ and $s_x \in (\mathcal{F}_U)_x$, then s_x is represented by $s \in \mathcal{F}(U \cap V)$ with V a neighbourhood of x. The support of s in $U \cap V$ is closed in V and its complement N contains x. The restriction of s to $U \cap N$ is zero. Hence $s_x = 0$, proving that $(\mathcal{F}_U)_x = 0$ for all $x \notin U$.

ii) It is obvious from i) that the restriction map in ii) is injective, for a homomorphism $f : \mathcal{F}_U \to \mathcal{G}$ satisfying $f_x = 0$ for all $x \in U$ is clearly 0 everywhere. To prove surjectivity, suppose $f : \mathcal{F} \to \mathcal{G}|U$ is a homomorphism. Then f yields homomorphisms $\mathcal{F}(U \cap V) \to \mathcal{G}(U \cap V)$ for all open sets V. If a section s of $\mathcal{F}$ over $U \cap V$ has support closed in V, then so has $f(s)$. Hence $f(s)$ can be viewed as an element $\widetilde{f(s)}$ of $\mathcal{G}(V)$ obtained by setting $\widetilde{f(s)}_x = 0$ if $x \in V \setminus U$ and $\widetilde{f(s)}_x = f(s)_x$ if $x \in U \cap V$. It is easy to see that $s \mapsto \widetilde{f(s)}$ gives a homomorphism $\mathcal{F}_U \to \mathcal{G}$ which restricts to f.

iii) If $\mathcal{F} = \mathcal{S}|U$, then the identity map of $\mathcal{S}|U$ gives rise to a map $\mathcal{F}_U \to \mathcal{S}$ by the bijection in ii), in which we take $\mathcal{F} = \mathcal{S}$. This is the inclusion which makes $\mathcal{F}_U$ a subsheaf of $\mathcal{S}$ in this case.

iv) is a particular case of the above.

On the other hand, it is obvious that over a paracompact space X, flabby sheaves are soft. For, by [Ch. 1, 2.1], any section over a closed subset can be extended to a neighbourhood and hence to the whole space if the sheaf is flabby.

1.8. Proposition. *If $\mathcal{F}$ is a flabby sheaf, then an exact sequence*

$$0 \to \mathcal{F} \to \mathcal{A} \to \mathcal{B} \to 0$$

of sheaves leads to an exact sequence

$$0 \to \mathcal{F}(X) \to \mathcal{A}(X) \to \mathcal{B}(X) \to 0.$$

In particular, this is true for $\mathcal{F}$ injective.

Proof. In view of Proposition 1.3, the crucial thing to check is that any section s of $\mathcal{B}$ can be lifted to a section t of $\mathcal{A}$. Consider the set of all pairs (U, τ) where U is an open set and $\tau \in \mathcal{A}(U)$ lifts s over U. This set can be partially ordered by defining $(U, \tau) < (U', \tau')$ if $U \subset U'$ and the restriction to U of τ' is τ. It is obvious that it is nonempty (since $\mathcal{A}_x \to \mathcal{B}_x$ is surjective) and forms an inductive set. Hence by Zorn's lemma, there exists a maximal element (U_0, τ_0). We have now to show that $U_0 = X$. If not, let $x \notin U_0$. There exist a neighbourhood V of x and a section τ of $\mathcal{A}$ over V lifting s. We wish to show that there exists a section over $U_0 \cup V$ lifting s and restricting to τ_0. This would prove our assertion in view of our maximality assumption. But τ_0 and τ may not coincide on $U_0 \cap V$. However, the difference $\tau_0 - \tau$ is actually a section of $\mathcal{F}$ over $U_0 \cap V$, since both of them map onto s (Proposition 1.3). Since $\mathcal{F}$ has been assumed to be flabby, this section can be extended as a section t of $\mathcal{F}$ over the whole of X. We may alter τ by adding to it this extended section t of $\mathcal{F}$, and obtain a new lift τ_1 of s over V. Now τ_0, τ_1 coincide on $U_0 \cap V$, thereby yielding a lift of s over $U_0 \cup V$ extending τ_0, which contradicts the maximality of (U_0, τ_0). Thus we must have $U_0 = X$, proving our assertion.

More or less similarly, we have the following analogue for soft sheaves.

1.9. Proposition. *If $\mathcal{F}$ is a soft sheaf on a paracompact space, then an exact sequence*

$$0 \to \mathcal{F} \to \mathcal{A} \to \mathcal{B} \to 0$$

leads to an exact sequence

$$0 \to \mathcal{F}(X) \to \mathcal{A}(X) \to \mathcal{B}(X) \to 0.$$

Proof. As in (1.8), we have only to show that $s \in \mathcal{B}(X)$ can be lifted to $\mathcal{A}(X)$. Using the regularity of the space X, we see that there exist a family (C_i) of closed sets such that their interiors cover X, and also a family of sections of $\mathcal{A}(X)$ over C_i, lying over s. We may assume that (C_i) is a locally finite covering. Consider unions of the sets C_i and sections over such unions that lift s. One can check that it is an inductive family and the proof is completed as in Proposition 1.8.

1.10. Examples.

1) According to [Ch. 1, Proposition 3.13], the sheaf $\mathcal{A}$ of differentiable functions is a soft sheaf. In fact, any sheaf of $\mathcal{A}$-modules is itself soft. More generally, any sheaf $\mathcal{F}$ of $\mathcal{O}$-modules (where $\mathcal{O}$ is a soft sheaf of rings) is a soft sheaf. For, if C is a closed set, U an open neighbourhood of C and $s \in \mathcal{F}(U)$, then we have to show that there exists $\widetilde{s} \in \mathcal{F}$ which coincides with s in a neighbourhood of C. Consider a section f of $\mathcal{O}$ which is 1 on C and 0 outside a neighbourhood N with $\overline{N} \subset U$. Then fs is a section of $\mathcal{A}$ over U vanishing outside $\overline{N}$ and hence extendable to the whole of X.

2) Let X be a topological space in which every open set is paracompact. If we consider the *presheaf* $\mathcal{S}$ of singular cochains [Ch. 1, Example 1.3, 4)], it is clear that the restriction map $\mathcal{S}(X) \to \mathcal{S}(U)$ is surjective. This does not prove that the *sheaf* $\widetilde{\mathcal{S}}$ is flabby yet. However, [Ch. 1, Proposition 1.14] shows that $\mathcal{S}(U) \to \widetilde{\mathcal{S}}(U)$ is surjective. Hence any section s over U of the associated sheaf $\widetilde{\mathcal{S}}$ can be lifted to $\mathcal{S}(U)$ and then can be extended to $\mathcal{S}(X)$. Its image in $\widetilde{\mathcal{S}}(X)$ is an extension of s, proving that $\widetilde{\mathcal{S}}$ is a flabby sheaf.

1.11. Exercises.

1) The sheaf of holomorphic functions on the complex plane is not soft.

2) The constant sheaf $\mathbb{Z}$ over the real line is not soft.

3) The sheaf of differentiable functions over the real line is not flabby.

4) Over a discrete space any sheaf is flabby. Which ones are injective?

1.12. Corollary to Propositions 1.8 and 1.9.

i) *If* $0 \to \mathcal{F}' \to \mathcal{F} \to \mathcal{F}'' \to 0$ *is exact and* $\mathcal{F}$, $\mathcal{F}'$ *are flabby, then so is* $\mathcal{F}''$.

ii) *If* X *is paracompact and* $\mathcal{F}$, $\mathcal{F}'$ *are soft, then so is* $\mathcal{F}''$.

Proof. i) It is obvious from the definition that for any open subset U of $X, \mathcal{F}'|U$ is flabby as well. According to Proposition 1.8, any section s over U of $\mathcal{F}''$ can be lifted to a section of $\mathcal{F}$ over U. Since $\mathcal{F}$ is flabby this lifted section can be extended to the whole of X. Its image in $\mathcal{F}''(X)$ of course extends s, proving our assertion.

ii) is proved similarly using Proposition 1.9 instead.

2. Sheaf Cohomology

2.1. Proposition. *If $\mathcal{F}$ is a sheaf of modules over a sheaf $\mathcal{R}$ of rings on a topological space X, then there exists an injective sheaf $\mathcal{I}^0$ of $\mathcal{R}$-modules of which $\mathcal{F}$ is a subsheaf.*

Proof. We will use the fact that the corresponding statement is valid for modules, namely any module over a ring is a submodule of some injective module. For each $x \in X$, let I_x^0 be an injective $\mathcal{R}_x$-module containing $\mathcal{F}_x$ as a submodule. Define $\mathcal{I}^0(U)$ to be the direct product $\prod_{x \in U} I_x^0$, for any open subset U of X. If $V \subset U$, then there is a natural projection $\prod_{x \in U} I_x^0 \to \prod_{x \in V} I_x^0$, which is defined to be the restriction map. It is easily seen that $\mathcal{I}^0$ is a sheaf. The inclusions $\mathcal{F}_x \to I_x^0$ give rise to a map $\prod_{x \in U} \mathcal{F}_x \to \prod_{x \in U} I_x^0$. Thus we get a homomorphism $\mathcal{F} \to \mathcal{I}^0$ by composing the natural homomorphism $\mathcal{F}(U) \to \prod_{x \in U} \mathcal{F}_x$ with the above inclusion. It is easily seen that $\mathcal{I}^0$ is a sheaf of $\mathcal{R}$-modules and that this inclusion is a homomorphism. It remains to prove that $\mathcal{I}^0$ is $\mathcal{R}$-injective. Let $\mathcal{S} \subset \mathcal{T}$ be sheaves of $\mathcal{R}$-modules and $f : \mathcal{S} \to \mathcal{I}^0$ an $\mathcal{R}$-homomorphism. Then for every $x \in X$ we have the map $\mathcal{S}_x \to I_x^0$ obtained by composing f_x with the projection of $\mathcal{I}_x^0 \to I_x^0$. These can be extended to $\mathcal{R}_x$-linear maps $\mathcal{T}_x \to I_x^0$ since I_x^0 is $\mathcal{R}_x$-injective. Hence we get a map $\mathcal{T}(U) \to \prod_{x \in U} \mathcal{T}_x \to \prod_{x \in U} I_x^0 = \mathcal{I}^0(U)$. It is easy to see that this gives a homomorphism of $\mathcal{T}$ into $\mathcal{I}^0$ extending f.

If $\mathcal{F}$ is any sheaf of $\mathcal{R}$-modules, then we include $\mathcal{F}$ in an injective sheaf, say

$$0 \to \mathcal{F} \to \mathcal{I}^0 \to \mathcal{F}_1 \to 0.$$

Then we may include $\mathcal{F}_1$ in an injective sheaf $\mathcal{I}^1$ to obtain an exact sequence

$$0 \to \mathcal{F} \to \mathcal{I}^0 \to \mathcal{I}^1 \to \mathcal{F}_2 \to 0.$$

Proceeding thus, we obtain an exact sequence

$$0 \to \mathcal{F} \to \mathcal{I}^0 \to \mathcal{I}^1 \to \mathcal{I}^2 \to \cdots \to \mathcal{I}^n \to \cdots$$

where all the sheaves $\mathcal{I}^i$ are $\mathcal{R}$-injective.

We recalled in [Ch. 2, 6.9–6.12], the concept of complexes and cohomology for abelian groups and remarked that the same go through for sheaves as well. We will formally give some of these definitions here.

2.2. Definition. A sequence $\mathcal{F}^\circ = (\mathcal{F}^i)$, $i \in \mathbb{Z}$, of sheaves (of abelian groups)

$$\cdots \to \mathcal{F}^{i-1} \to \mathcal{F}^i \to \mathcal{F}^{i+1} \to \cdots$$

is called a *complex of sheaves* or simply a *complex* if the composite of any two successive homomorphisms is 0. The homomorphisms themselves are called *differentials* of the complex. All the differentials are usually denoted by d.

2.3. Definition. An exact sequence of sheaves

$$0 \to \mathcal{F} \to \mathcal{J}^0 \to \mathcal{J}^1 \to \mathcal{J}^2 \to \cdots \to \mathcal{J}^n \to \cdots$$

is said to be a *resolution* of $\mathcal{F}$, and the complex

$$\mathcal{J}^\circ : \mathcal{J}^0 \to \mathcal{J}^1 \to \mathcal{J}^2 \to \cdots$$

is called the *resolving complex*. We denote this by

$$0 \to \mathcal{F} \to \mathcal{J}^\circ.$$

If in addition, all the sheaves $\mathcal{J}^i$ are injective, we say that it is an *injective resolution*.

2.4. Remark. The idea is that for many purposes, the sheaf $\mathcal{F}$ may, with advantage, be *replaced* by the resolving injective complex. A complex of sheaves is of course a more complicated object than the sheaf itself, but if its components are all injective, then the resolving complex is in a sense better than the sheaf $\mathcal{F}$.

We just showed that any sheaf of $\mathcal{R}$-modules admits an injective resolution. The way we constructed the injective resolution is quite arbitrary, and so for the idea to have any chance of success, we need to prove some kind of uniqueness for the resolution.

2.5. Definition. A *morphism* of a complex $\mathcal{F}^\circ$ into another complex $\mathcal{G}^\circ$ is a sequence of homomorphisms $\mathcal{F}^i$ into $\mathcal{G}^i$, commuting with the differentials in the sense that the diagrams

$$
\begin{array}{ccc}
\mathcal{F}^{i-1} & \to & \mathcal{F}^i \\
\downarrow & & \downarrow \\
\mathcal{G}^{i-1} & \to & \mathcal{G}^i
\end{array}
$$

commute.

2.6. Proposition. *Let $0 \to \mathcal{F} \to \mathcal{J}^\circ$ (resp. $0 \to \mathcal{G} \to \mathcal{I}^\circ$) be any resolution of $\mathcal{F}$ (resp. an injective resolution of $\mathcal{G}$). Then any $\mathcal{R}$-homomorphism $f : \mathcal{F} \to \mathcal{G}$ can be lifted to a morphism $\varphi : \mathcal{J}^\circ \to \mathcal{I}^\circ$ of complexes such that the diagram*

$$
\begin{array}{ccc}
\mathcal{F} & \to & \mathcal{J}^\circ \\
\downarrow{\scriptstyle f} & & \downarrow{\scriptstyle \varphi} \\
\mathcal{G} & \to & \mathcal{I}^\circ
\end{array}
$$

is commutative.

Proof. Firstly, the construction of a map $\varphi^0 : \mathcal{J}^0 \to \mathcal{I}^0$ making commutative the diagram

$$
\begin{array}{ccc}
\mathcal{F} & \to & \mathcal{J}^0 \\
\downarrow{\scriptstyle f} & & \downarrow{\scriptstyle \varphi^0} \\
\mathcal{G} & \to & \mathcal{I}^0
\end{array}
$$

is straightforward. For, since $\mathcal{I}^0$ is injective, the map $\mathcal{F} \to \mathcal{I}^0$ obtained by composing f with the inclusion $\mathcal{G} \to \mathcal{I}^0$, can be extended to a homomorphism $\varphi^0 : \mathcal{J}^0 \to \mathcal{I}^0$. We can successively construct φ^i this way. For, let us suppose that $\varphi^j : \mathcal{J}^j \to \mathcal{I}^j$ has been constructed for all $j \le i - 1$ in such a way that the diagram

$$
\begin{array}{ccccccccc}
0 & \to & \mathcal{F} & \to & \mathcal{J}^0 & \to & \cdots & \to & \mathcal{J}^{i-1} \\
& & \downarrow{\scriptstyle f} & & \downarrow{\scriptstyle \varphi^0} & & & & \downarrow{\scriptstyle \varphi^{i-1}} \\
0 & \to & \mathcal{G} & \to & \mathcal{I}^0 & \to & \cdots & \to & \mathcal{I}^{i-1}
\end{array}
$$

is commutative. Note that the image of $\mathcal{J}^{i-2}$ in $\mathcal{J}^{i-1}$ is mapped into the image of $\mathcal{I}^{i-2}$ in $\mathcal{I}^{i-1}$, thanks to the commutativity of the diagram above. By the exactness of the horizontal sequences, this is the same as saying that the kernel of $\mathcal{J}^{i-1} \to \mathcal{J}^i$ is mapped into the kernel of $\mathcal{I}^{i-1} \to \mathcal{I}^i$. Again this implies that the map φ^{i-1} induces a map of $\mathrm{im}(\mathcal{J}^{i-1} \to \mathcal{J}^i)$ into $\mathrm{im}(\mathcal{I}^{i-1} \to \mathcal{I}^i)$. Now this map of $\mathrm{Im}\,\mathcal{J}^{i-1}$ into $\mathcal{I}^i$ can be extended to a map $\varphi^i : \mathcal{J}^i \to \mathcal{I}^i$, in view of the assumption that $\mathcal{I}^i$ is an injective sheaf. This proves our assertion.

There is of course no uniqueness about the extension of f to φ. However, by an argument similar to the above, we can also prove the following statement of 'uniqueness up to homotopy' about the extension φ. Homotopy is a topological notion and we will give its definition soon (4.5). At this point, we simply give the algebraic analogue and work with it.

2.7. Definition. Two morphisms $f, g : \mathcal{I}^\circ \to \mathcal{J}^\circ$ of complexes are said to be *homotopic* if there exist (for all i) homomorphisms $k^i : \mathcal{I}^i \to \mathcal{J}^{i-1}$ such that

$$d \circ k^i + k^{i+1} \circ d = f^i - g^i.$$

2.8. Exercises.

1) Let K° and L° be two complexes. Then define $K^\circ \otimes L^\circ$ by setting its ith component to be $\sum_{j+k=i} K^j \otimes L^k$ and the differential to be the map induced by $(a, b) \mapsto da \otimes b + (-1)^j a \otimes db$ for $(a, b) \in K^j \otimes L^k$. Check that this defines a complex and that there are natural homomorphisms $\sum H^j(K^\circ) \otimes H^k(L^\circ) \to H^{j+k}(K^\circ \otimes L^\circ)$.

2) Consider the complex $\mathcal{D}^\circ$ consisting of two members $\mathcal{D}^0 = \mathcal{R} \oplus \mathcal{R}$ and $\mathcal{D}^1 = \mathcal{R}$. The differential is given by $(a, b) \mapsto b - a$. Let $\mathcal{E}^\circ$ be the complex consisting of only one member $\mathcal{E}^0 = \mathcal{R}$. Note that there are two morphisms $\mathcal{D}^\circ \to \mathcal{E}^\circ$, which map (a, b) to a and b, respectively. Consider the tensor product complex $\mathcal{J} \otimes \mathcal{D}$ and show that morphisms f, g as in 2.7 are homotopic if and only if there exists a morphism $h : \mathcal{I}^\circ \to \mathcal{J} \otimes \mathcal{D}$ such that the composites of h with the two projections $\mathcal{J}^\circ \otimes \mathcal{D}^\circ \to \mathcal{J}^\circ \otimes \mathcal{E}^\circ = \mathcal{J}$ are f and g.

2.9. Proposition. *Let $0 \to \mathcal{F} \to \mathcal{J}^\circ$ and $0 \to \mathcal{G} \to \mathcal{I}^\circ$ be resolutions, with $\mathcal{I}^\circ$ injective. If $\varphi, \psi : \mathcal{J}^\circ \to \mathcal{I}^\circ$ are morphisms of complexes such that*

$$
\begin{array}{ccc}
\mathcal{F} & \to & \mathcal{J}^\circ \\
\downarrow{\scriptstyle f} & & \downarrow{\scriptstyle \varphi, \psi} \\
\mathcal{G} & \to & \mathcal{I}^\circ
\end{array}
$$

is commutative, then φ, ψ are homotopic.

Proof. Similar to that of Proposition 2.6.

Since we have defined the notion of morphisms, we also have automatically a notion of *isomorphism* of complexes. Although two injective resolutions of a sheaf need not necessarily be isomorphic, we have the following uniqueness statement.

2.10. Corollary. *If $0 \to \mathcal{F} \to \mathcal{I}^\circ$, $0 \to \mathcal{F} \to \mathcal{J}^\circ$ are two injective resolutions of a sheaf $\mathcal{F}$, then there exist morphisms $\varphi : \mathcal{I}^\circ \to \mathcal{J}^\circ$ and $\psi : \mathcal{J}^\circ \to \mathcal{I}^\circ$ of complexes inducing the identity on $\mathcal{F}$ such that $\varphi \circ \psi$ and $\psi \circ \varphi$ are both homotopic to the identities.*

Proof. The identity map $\mathcal{F} \to \mathcal{F}$ extends to morphisms $\varphi : \mathcal{I}^\circ \to \mathcal{J}^\circ$ and $\psi : \mathcal{J}^\circ \to \mathcal{I}^\circ$ by Proposition 2.6. On the other hand both $\varphi \circ \psi$ and $\psi \circ \varphi$ extend the identity maps and so are homotopic to the identities by Proposition 2.9.

We remarked at the beginning of this chapter that exactness of a complex does not imply that the corresponding sequence of sections is also exact. The notion of cohomology of a sheaf arises when we replace a sheaf by its injective resolution and then take the sequence of sections. The latter is no longer exact, but is nevertheless a *complex* of abelian groups.

2.11. Definition. Let $\mathcal{F}$ be any sheaf and $0 \to \mathcal{F} \to \mathcal{I}^\circ$ an injective resolution of $\mathcal{F}$. The cohomology groups of the complex

$$\mathcal{I}^0(X) \to \mathcal{I}^1(X) \to \cdots \to \mathcal{I}^{i-1}(X) \to \mathcal{I}^i(X) \to \mathcal{I}^{i+1}(X) \to \cdots$$

are called the *cohomology groups $H^i(X, \mathcal{F})$ of the sheaf $\mathcal{F}$.*

2.12. Remarks.

1) Definition 2.11 is meaningful since any two injective resolutions $\mathcal{I}^\circ$, $\mathcal{J}^\circ$ of $\mathcal{F}$ are homotopically equivalent and therefore yield homotopically equivalent complexes $\mathcal{I}(X)^\circ, \mathcal{J}(X)^\circ$ by (2.10).

2) Proposition 1.3, applied to the exact sequence $0 \to \mathcal{F} \to \mathcal{I}^0 \to \mathcal{I}^1$, gives the exact sequence $0 \to \mathcal{F}(X) \to \mathcal{I}^0(X) \to \mathcal{I}^1(X)$. By definition, $H^0(X, \mathcal{F})$ is the kernel of the map $\mathcal{I}^0(X) \to \mathcal{I}^1(X)$. Consequently there is a canonical isomorphism of $H^0(X, \mathcal{F})$ with $\mathcal{F}(X)$.

3) If $\mathcal{F}$ is itself an injective sheaf, then a resolving complex $\mathcal{I}^\circ$ is given by $\mathcal{I}^0 = \mathcal{F}$ and $\mathcal{I}^i = 0$ for all $i \geq 1$. Hence we conclude that $H^i(X, \mathcal{F}) = 0$ for all $i \geq 1$.

2.13. Definition. If $f : \mathcal{F} \to \mathcal{G}$ is any homomorphism, the *induced homomorphism $H^i(f) : H^i(X, \mathcal{F}) \to H^i(X, \mathcal{G})$* is defined as follows. Let $0 \to \mathcal{F} \to \mathcal{J}^\circ$, $0 \to \mathcal{G} \to \mathcal{I}^\circ$ be injective resolutions. If $\varphi : \mathcal{J}^\circ \to \mathcal{I}^\circ$ is an extension of f as in 2.6, then $H^i(\varphi(X)) : H^i(\mathcal{J}(X)) \to H^i(\mathcal{I}(X))$ is independent of the resolutions $\mathcal{J}, \mathcal{I}$ according to Proposition 2.9 and this is defined to be $H^i(f)$.

Let $0 \to \mathcal{F}_1 \to \mathcal{F}_2 \to \mathcal{F}_3 \to 0$ be an exact sequence of sheaves. Then we can actually construct injective resolutions $0 \to \mathcal{F}_i \to \mathcal{I}_i^\circ$, $i = 1, 2, 3$, in such a way that we have *an exact sequence*

$$0 \to \mathcal{I}_1^\circ \to \mathcal{I}_2^\circ \to \mathcal{I}_3^\circ \to 0$$

of complexes fitting in a commutative diagram

$$\begin{array}{ccccccccc}
& & 0 & & 0 & & 0 & & \\
& & \downarrow & & \downarrow & & \downarrow & & \\
\mathbf{2.14.} \qquad 0 & \to & \mathcal{F}_1 & \to & \mathcal{F}_2 & \to & \mathcal{F}_3 & \to & 0 \\
& & \downarrow & & \downarrow & & \downarrow & & \\
0 & \to & \mathcal{I}_1^\circ & \to & \mathcal{I}_2^\circ & \to & \mathcal{I}_3^\circ & \to & 0
\end{array}$$

The procedure that we will adopt for such a construction is the following. Choose injective resolutions $\mathcal{I}_1^\circ, \mathcal{I}_3^\circ$ arbitrarily. Then we will construct a complex $\mathcal{I}_2^\circ$ to fit in the short exact sequence. For every i, take the sheaf $\mathcal{I}_2^i = \mathcal{I}_1^i \oplus \mathcal{I}_3^i$. Since $\mathcal{I}_1^i$ are all injective, any exact sequence

$$0 \to \mathcal{I}_1^i \to \mathcal{I}_2^i \to \mathcal{I}_3^i \to 0$$

necessarily splits, and we are led to the above choice of $\mathcal{I}_2^i$ any way. The inclusion $\mathcal{I}_1^i \to \mathcal{I}_2^i$ and the surjection $\mathcal{I}_2^i \to \mathcal{I}_3^i$ are the obvious ones. We have now to define the differentials of the complex $\mathcal{I}_2^\circ$. Note that if we make $(\mathcal{I}_2^i)$ a complex that fits in diagram 2.14, then the exactness of $0 \to \mathcal{F}_2 \to \mathcal{I}_2^\circ$ follows from the exactness of the complexes $0 \to \mathcal{F}_1 \to \mathcal{I}_1^\circ$ and $0 \to \mathcal{F}_3 \to \mathcal{I}_3^\circ$. For, we have only to check this at the stalk-level and there it follows by the '5-Lemma'.

Thus the only thing to be checked is that we can define differential maps $\mathcal{I}_2^i \to \mathcal{I}_2^{i+1}$ to fit in diagram 2.14. In the following we will work at the stalk level and will leave it to the reader to verify that the induced maps on the étale spaces are continuous. We will fix a point and denote all the stalks of sheaves at that point by the corresponding Roman letters. The differential has to be defined by $d(x, y) = (d_1 x + f_i(y), d_3 y)$ for $x \in I_1^i, y \in I_3^i$, where d_1 and d_3 are the differentials of the complexes I_1° and I_3° and f_i is a homomorphism $I_3^i \to I_1^{i+1}$. The only constraint on f_i results from the requirement that $d^2 = 0$. This yields

$$0 = d^2(x, y) = d(d_1 x + f_i(y), d_3 y) = (d_1^2 x + d_1 f_i(y) + f_{i+1} d_3 y, d_3^2 y),$$

or $d_1 f_i + f_{i+1} d_3 = 0$, since we know that $d_1^2 = 0 = d_3^2$. To define the (f_i), we proceed inductively, noting that in order to define f_{i+1}, we have only to ensure that $f_{i+1}|d_3(I_3^i)$ is given by

$$f_{i+1}(y) = -d_1 f_i(x) \text{ where } d_3 x = y.$$

We will check that $y \mapsto -d_1 f_i x$, where x is any element of I_3^{i+1} with $d_3 x = y$, is a well-defined map $\mathrm{im}(d_3) \to I_1^{i+2}$. Then it would follow that f_{i+1} can be defined to be any extension of this map as a homomorphism $I_3^{i+1} \to I_1^{i+2}$, which exists since I_1^{i+2} is injective. If $d_3 x = d_3 x'$ then by the exactness of I_3°

we see that $x - x' = d_3 a$ for some $a \in I_3^i$. Then $d_1 f_i(x - x') = d_1 f_i(d_3 a) = -d_1(d_1 f_{i-1} a)$ by our inductive assumption. Thus $d_1 f_i(x) = d_1 f_i(x')$ and our proof is complete. In particular, we obtain

2.15. Proposition. *If $0 \to \mathcal{F}' \to \mathcal{F} \to \mathcal{F}'' \to 0$ is an exact sequence of sheaves, then we have a long exact sequence of cohomology groups:*

$$0 \to H^0(\mathcal{F}') \to H^0(\mathcal{F}) \to H^0(\mathcal{F}'') \to H^1(\mathcal{F}') \to H^1(\mathcal{F}) \to H^1(\mathcal{F}'')$$

$$\to \cdots \to H^i(\mathcal{F}') \to H^i(\mathcal{F}) \to H^i(\mathcal{F}'') \to H^{i+1}(\mathcal{F}') \to \cdots .$$

Proof. Let $\mathcal{I}'^\circ, \mathcal{I}^\circ, \mathcal{I}''^\circ$ be injective resolutions fitting in an exact sequence $0 \to \mathcal{I}'^\circ \to \mathcal{I}^\circ \to \mathcal{I}''^\circ \to 0$ as above. Then from Proposition 1.8, we conclude that

$$0 \to \mathcal{I}'(X)^\circ \to \mathcal{I}(X)^\circ \to \mathcal{I}''(X)^\circ \to 0$$

is an exact sequence of complexes of abelian groups. Now our assertion follows from the definition of $H^i(X, \mathcal{F})$ and the long exact sequence associated to a short exact sequence of complexes which we recall below.

2.16. Remark. If

$$0 \to A^\circ \to B^\circ \to C^\circ \to 0$$

is an exact sequence of complexes, then the two morphisms induce homomorphisms $H^i(A^\circ) \to H^i(B^\circ)$ and $H^i(B^\circ) \to H^i(C^\circ)$. Since the composite of the two morphisms above is zero, so is the composite at the cohomology level. More is true. The sequence

$$H^i(A^\circ) \to H^i(B^\circ) \to H^i(C^\circ)$$

is *exact*. In fact if $b \in B^i$ is such that $db = 0$, and its image in C^i is dx with $x \in C^{i-1}$, then lifting x to an element y in B^i, we see that $b - dy$ maps to zero in C^i. Hence it belongs to A^i. Moreover, we also have $d(b - dy) = db = 0$. In other words, the cohomology class of b comes from a class in $H^i(A^\circ)$ proving the exactness as claimed.

The more interesting fact is that there is also a map $H^i(C^\circ) \to H^{i+1}(A^\circ)$ making the sequence

$$H^i(B^\circ) \to H^i(C^\circ) \to H^{i+1}(A^\circ) \to H^{i+1}(B^\circ)$$

exact. We will just give this *connecting homomorphism* and leave out the actual checking of exactness, which is straightforward. Take an element c in C^i with $dc = 0$, representing an element of H^i. We know that there exists an element b of B^i which maps to c. Moreover, db maps to $dc = 0$. From the exactness of $0 \to A^i \to B^i \to C^i$, it follows that $x = db$ belongs to A^{i+1}. Moreover, since dx is zero in B^{i+2}, it is also zero in A^{i+2}. Thus x represents a cohomology class of A°. If we choose a different element b' mapping to c, the difference $b' - b$ belongs to A^i and hence $db' = db + d(b' - b)$ represents the same cohomology class. This class depends only on the class of c and not on c itself. To see this, we have only to check that if $c = dx$ for some $x \in C^{i-1}$, then the class we have defined is zero. In fact, if we lift x to some y in B^{i-1}, then dy gives a lift of c. Since $d(dy) = 0$, the cohomology class we associated to c is zero. Thus we have given a homomorphism $H^i(C^\circ) \to H^{i+1}(A^\circ)$.

But conventionally we take the *negative* of the above map as the connecting homomorphism. Some rationale for this convention may be found in Exercise 7) at the end of the chapter.

3. Cohomology through Other Resolutions

Although the cohomology of a sheaf was defined in terms of injective resolutions, it is seldom practical to make computations of cohomology using injectives. The reason why we can make do with other resolutions is the following fact.

3.1. Lemma. *Let $0 \to \mathcal{F} \to \mathcal{J}^\circ$ be an arbitrary resolution of the sheaf $\mathcal{F}$. Suppose $H^i(X, \mathcal{J}^j) = 0$ for $i > 0$ and all $j \geq 0$. Then the complex $\mathcal{J}(X)^\circ$ has cohomologies canonically isomorphic to those of $\mathcal{F}$.*

Proof. In fact, our argument would actually yield the following more general statement.

3.2. Lemma. *Let $0 \to \mathcal{F} \to \mathcal{J}^\circ$ be an arbitrary resolution of the sheaf $\mathcal{F}$. Suppose $H^i(X, \mathcal{J}^j) = 0$ for $i + j \leq n, i \neq 0$. Then $H^k(X, \mathcal{F})$ is canonically isomorphic to $H^k(\mathcal{J}^\circ(X))$ for $k \leq n$.*

Proof. We will prove the lemma by induction on n. We note first that for $n = 0$, the assumptions regarding vanishing of cohomology are vacuous. Thus what we have to show is that the kernel of $\mathcal{J}^0(X) \to \mathcal{J}^1(X)$ is isomorphic to $H^0(X, \mathcal{F}) = \mathcal{F}(X)$ (Remark 2.12, 2)). But this follows on applying Proposition 1.3 to the exact sequence

$$0 \to \mathcal{F} \to \mathcal{J}^0 \to \mathcal{J}^1.$$

Let us then assume the assertion valid for all $n < m$ and prove that it holds for $n = m$ as well. Let $\mathcal{K}$ be the cokernel of the inclusion $\mathcal{F} \to \mathcal{J}^0$. Then by Proposition 2.15, we have an exact sequence

$$\cdots \to H^{i-1}(\mathcal{F}) \to H^{i-1}(\mathcal{J}^0) \to H^{i-1}(\mathcal{K})$$
$$\to H^i(\mathcal{F}) \to H^i(\mathcal{J}^0) \to H^i(\mathcal{K}) \to \cdots .$$

Take $i = 1$ to get the following exact sequence (since $H^1(\mathcal{J}^0) = 0$):

$$0 \to H^0(\mathcal{F}) \to H^0(\mathcal{J}^0) \to H^0(\mathcal{K}) \to H^1(\mathcal{F}) \to 0.$$

But then from the exact sequence

$$0 \to \mathcal{K} \to \mathcal{J}^1 \to \mathcal{J}^2$$

we conclude that there is a natural isomorphism of $H^0(\mathcal{K})$ with the kernel of $\mathcal{J}^1(X) \to \mathcal{J}^2(X)$. Thus $H^1(\mathcal{F})$ is isomorphic to the quotient of the kernel of the map $\mathcal{J}^1(X) \to \mathcal{J}^2(X)$, namely $H^0(\mathcal{K})$, by the image of the map $H^0(\mathcal{J}^0) \to H^0(\mathcal{K})$, or what is the same, image of $\mathcal{J}^0(X) \to \mathcal{J}^1(X)$. In other words, $H^1(X, \mathcal{F}) \simeq H^1(\mathcal{J}(X)^\circ)$. To prove that $H^i(X, \mathcal{F}) \simeq H^i(\mathcal{J}(X))$ for $2 \leq i \leq m$, we use again the long exact sequence. Since by assumption, $H^{i-1}(\mathcal{J}^\circ) = H^i(\mathcal{J}^\circ) = 0$, we deduce that there is an isomorphism $H^i(\mathcal{F}) \simeq H^{i-1}(\mathcal{K})$. But we have a resolution $0 \to \mathcal{K} \to \widetilde{\mathcal{J}}^\circ$ where we define the complex $\widetilde{\mathcal{J}}^\circ$ by setting $\widetilde{\mathcal{J}}^j = \mathcal{J}^{j+1}$ for $j \geq 0$ and keeping the same differentials. Moreover, we have $H^i(X, \widetilde{\mathcal{J}}^j) = 0$ for $i + j \leq (n-1), i \neq 0$. Hence by the induction assumption we see that $H^i(X, \mathcal{K}) \simeq H^i(\widetilde{\mathcal{J}}(X)^\circ) = H^{i+1}(\mathcal{J}(X)^\circ)$ if $1 \leq i \leq m-1$. Hence $H^i(X, \mathcal{F}) \simeq H^i(\mathcal{J}(X)^\circ)$ if $2 \leq i \leq m$, as was to be proved.

In order to use the above observation effectively, we need to find criteria for the vanishing of $H^i(X, \mathcal{F})$ for all $i \geq 1$. We have already seen (Remark 2.12, 3)) that this is true for injective sheaves. There are also other types of sheaves for which higher cohomologies vanish.

3.3. Proposition. *i) If $\mathcal{F}$ is a flabby sheaf, then $H^i(X, \mathcal{F}) = 0$ for all $i \geq 1$.*

ii) If X is paracompact and $\mathcal{F}$ is soft, then $H^i(X, \mathcal{F}) = 0$ for all $i \geq 1$.

Proof. Let $\mathcal{F} \to \mathcal{I}^0$ be an inclusion of $\mathcal{F}$ in an injective sheaf and let $\mathcal{G}$ be its cokernel. By Proposition 2.15, we have a long exact sequence

$$\cdots \to H^i(\mathcal{F}) \to H^i(\mathcal{I}^0) \to H^i(\mathcal{G}) \to H^{i+1}(\mathcal{F}) \to H^{i+1}(\mathcal{I}^0) \to \cdots .$$

If $i \geq 1$, then $H^i(\mathcal{I}^0) = H^{i+1}(\mathcal{I}^0) = 0$ and hence $H^{i+1}(\mathcal{F})$ is isomorphic to $H^i(\mathcal{G})$. But then both $\mathcal{F}$ and $\mathcal{I}^0$ are flabby and hence so is $\mathcal{G}$ (Corollary 1.12, i)). Thus we will be through by induction on i, if we can show that for a flabby sheaf $\mathcal{F}$, we have $H^1(X, \mathcal{F}) = 0$. But we have the exact sequence

$$0 \to \mathcal{F}(X) \to \mathcal{I}^0(X) \to \mathcal{G}(X) \to H^1(X, \mathcal{F}) \to H^1(X, \mathcal{I}^0) = 0.$$

The content of Proposition 1.8, namely that $\mathcal{I}^0(X) \to \mathcal{G}(X)$ is surjective if $\mathcal{F}$ is flabby, is just that $H^1(X, \mathcal{F}) = 0$. This completes the proof.

ii) The proof is similar to i), using this time Corollary 1.12, ii).

One can get many refinements and generalisations of Proposition 3.3. We state one of them here without proof since it is quite straightforward.

3.4. Proposition. *Let $0 \to \mathcal{F} \to \mathcal{I}^\circ$, $0 \to \mathcal{G} \to \mathcal{J}^\circ$ be two resolutions of sheaves satisfying the condition in Lemma 3.2 and $f : \mathcal{F} \to \mathcal{G}, \tilde{f} : \mathcal{I}^\circ \to \mathcal{J}^\circ$ be morphisms of complexes fitting in a commutative diagram*

$$
\begin{array}{ccc}
\mathcal{F} & \to & \mathcal{I}^\circ \\
\downarrow f & & \downarrow \tilde{f} \\
\mathcal{G} & \to & \mathcal{J}^\circ
\end{array}
$$

Then the map $\tilde{f}(X) : \mathcal{I}(X)^\circ \to \mathcal{J}(X)^\circ$ induces maps $H^i(X, \mathcal{F}) \to H^i(X, \mathcal{G})$ according to the isomorphism in Proposition 3.3, and these coincide with $H^i(f)$.

4. Singular and Sheaf Cohomologies

In [Ch. 1, Example 1.3, 4)], we defined the group of A-valued singular cochains (where A is any abelian group.) We will digress here to use this notion and define the *singular cohomology groups* of a space with values in A and to sketch briefly some of its properties with a view to relating them with sheaf cohomology, via Lemma 3.1.

4.1. Definition of singular cohomology.

We recall [Ch. 2, 6.10, 1)] the definition of the singular complex and singular cohomology of a topological space. If σ is a singular n-simplex in X, namely, a continuous map of the standard n-simplex Δ_n into X, then for each i, $0 \leq i \leq n$, we can define a singular $(n-1)$-simplex as follows. Consider the map $F_i : \Delta_{n-1} \to \Delta_n$ obtained by restricting the *linear* map $\mathbb{R}^n \to \mathbb{R}^{n+1}$ mapping (in the increasing order) the standard basis $(e_0, \ldots, e_{n-1})$ of $\mathbb{R}^n$ into that of $\mathbb{R}^{n+1}$ missing only the ith element of the latter. More explicitly, we have

$$
F_i(e_j) = e_j \text{ for } j \leq i-1, \text{ and } F_i(e_j) = e_{j+1} \text{ if } j \geq i.
$$

By composing this map with σ, we obtain a singular $(n-1)$-simplex $\sigma \circ F_i$, which may be viewed as the ith *face* of σ. If A is any abelian group, an A-valued n-cochain associates to any singular n-simplex, an element of A. For any A-valued cochain α, we define its *coboundary* $d\alpha$ to be the $(n+1)$-cochain defined by $(d\alpha)(\sigma) = \sum_{i=0}^{n+1}(-1)^i \alpha(\sigma \circ F_i)$. It is then a straightforward verification to see that $d \circ d = 0$. In fact, in this computation, terms of the form $\alpha(\sigma \circ F_i \circ F_j)$ occur twice but with opposite signs and so cancel out. Since it only involves routine checking, we will not carry out the computation here. We thus get a complex

$$
\mathcal{S}(X)^\circ : 0 \to S_X^0 \to S_X^1 \to \cdots \to S_X^n \to S_X^{n+1} \to \cdots.
$$

This is called the *singular cochain complex* of X. Its ith cohomology is called the *singular cohomology of X with values in A* and is denoted $H^i(X, A)$.

4.2. Remarks.

1) We can also define $S_i(X)$ to be the free abelian group over singular i-simplices and define differentials $S_i(X) \to S_{i-1}(X)$ by mapping any singular simplex s to $\sum (-1)^i s \circ F_i$. This defines a complex whose homologies are called *integral singular homology groups* of X. The integral singular cochain complex $S^\circ(X)$ is then the dual of the singular chain complex $S_\circ(X)$ with transposed differentials. If A is a coefficient group, then $\mathrm{Hom}(S_\circ(X), A)$ is the singular cochain complex with values in A. Similarly we can also define a singular chain complex with values in A to be the complex $S_\circ(X) \otimes A$. All these can be understood in the context of complexes in general. If $C_\circ$ is a complex of R-modules (with differentials taking C_i into C_{i-1}), then we can define complexes $(C \otimes M)^\circ$ and $\mathrm{Hom}(C, M)^\circ$ for any R-module M. However the relation between the homologies, say, of $C_\circ$ and $(C^*)^\circ$ is not just taking the duals again. If A is a field, it is indeed so.

2) The singular complex is unwieldy and not amenable to computation. But it is good for establishing the basic properties. For computational purposes, one has to use other complexes which give cohomologies isomorphic to singular cohomology. For a differential manifold the de Rham complex turns out to be one such.

4.3. Exercise. Using the singular complex, compute the singular cohomology of the space consisting of a single point.

4.4. Effect of a continuous map.

Suppose X, Y are topological spaces and $f : X \to Y$ a continuous map. Then any singular n-simplex σ in X gives rise, by composition with f, to a singular n-simplex $f\sigma$ in Y. If α is an n-cochain in Y, then it gives rise to an n-cochain $f^*\alpha$ in X by the prescription $(f^*\alpha)(\sigma) = \alpha(f\sigma)$. Clearly this gives rise to a morphism of the singular cochain complex of Y into that of X and hence a homomorphism $H^i(f) : H^i(Y, A) \to H^i(X, A)$.

4.5. Definition. Two continuous maps $f, g : X \to Y$ of topological spaces are said to be *homotopic* if there exists a continuous map $h : X \times I \to Y$ such that $h(x, 0) = f(x)$ and $h(x, 1) = g(x)$ for all $x \in X$. A space is said to be *contractible* if the identity map is homotopic to a constant map into itself.

4.6. Theorem. *If $f, g : X \to Y$ are continuous maps which are homotopic, then $H^i(f) = H^i(g)$ for all $i \geq 0$. In particular, a contractible space has the cohomology of a point.*

4.7. Remark. The above result that homotopic maps induce the same homomorphism on singular cohomology groups may be thought of as a 'principle of continuity'. Since the maps $h_t : X \to Y$ obtained by restricting h to $X \times \{t\}$, depend continuously on t, the induced maps $H^i(h_t)$ may also be expected to depend continuously on t. The objects H^i being discrete, these maps remain constant under continuous deformation. In particular, $H^i(h_0) = H^i(h_1)$.

Proof of Theorem 4.6. Let $i_0, i_1 : X \to X \times I$ be continuous inclusions given by $x \mapsto (x,0), x \mapsto (x,1)$. Since $h \circ i_0 = f$, and $h \circ i_1 = g$, we have $i_0^* \circ h^* = f^*$ and $i_1^* \circ h^* = g^*$. Hence it is enough to show that i_0 and i_1 induce the same maps on cohomology groups.

We will prove this by showing that the induced maps on the singular complex of X by the two inclusions are homotopic in the sense of Definition 2.7.

In order to give such a homotopy, we note first that for any singular n-simplex in X, we have a canonical singular 'prism', namely a continuous map $\Delta_n \times I \to X \times I$. We first decompose the 'prism' $\Delta_n \times I$ into pieces with an identification of the pieces with the standard $(n+1)$-simplex. Let us denote the 'lower' vertices $(e_i, 0)$ of $\Delta_n \times I$ by $e_0, \ldots e_n$ and the 'upper' vertices $(e_i, 1)$ by $e_0', \ldots, e_n'$. For each i with $0 \leq i \leq n$, we consider the convex closure of $e_0', \ldots, e_i', e_i, \ldots, e_n$. This piece can be identified with the standard Δ_{n+1} by the restriction to Δ_{n+1} of the linear map which takes the standard base of $\mathbb{R}^{n+2}$ into $e_0', \ldots, e_i', e_i, \ldots, e_n$.

If σ is a singular n-simplex in X, then we get a natural continuous map $\sigma \times I$ of $\Delta_n \times I$ into $X \times I$. For each $0 \leq i \leq n$, we obtain a singular $(n+1)$-simplex $P_i(\sigma)$ in $X \times I$ by composing $\sigma \times \mathrm{Id} : \Delta_n \times I \to X \times I$ with the natural (linear) map of Δ_{n+1} into $\Delta_n \times I$ given above.

For any singular $(n+1)$-cochain α in $X \times I$, we associate the n-cochain $P\alpha$ in X by setting $(P\alpha)(\sigma) = \sum(-1)^i \alpha(P_i(\sigma))$. An easy verification gives us the formula

4.8. $$dP(\alpha) + Pd(\alpha) = \alpha_1 - \alpha_0$$

where α_0 (resp. α_1) is the pull-back of α by the map $x \to (x,0)$ (resp. $(x,1)$) of X into $X \times I$. This gives the required homotopy.

One can easily show that the singular cohomologies with coefficients in an abelian group A of a point, are given by $H^0 = A$, $H^i = 0$ for $i \neq 0$. Hence any contractible space has the same cohomology. Since any vector space is contractible, $h(t,v) = tv$ giving a homotopy between the identity map and the constant map, we conclude that $\mathbb{R}^n$ has the same cohomology groups.

4.9. Barycentric subdivision.

If $i_0, i_1, \ldots, i_r$ are distinct integers between 0 and n, then we will denote the centre of gravity or the *barycentre* of the points $e_{i_0}, \ldots, e_{i_r}$ in $\mathbb{R}^{n+1}$ (where $e_0, \ldots, e_n$ is the standard basis) by $[i_0, \ldots, i_r]$. If λ is a permutation of $(0, \ldots, n)$, then the linear map $\Delta_n \to \Delta_n$ given by sending e_i onto $[\lambda(0), \ldots, \lambda(i)]$ will be denoted b_λ. The set of all b_λ, where λ runs through all permutations of $(0, \ldots, n)$, is to be viewed as splitting up Δ_n into $(n+1)!$ parts and identifying each piece with Δ_n. Now we define a map $b : S^n(X) \to S^n(X)$ by setting

$$b(\alpha)(\sigma) = \sum (\operatorname{sgn} \lambda) \alpha(\sigma \circ b_\lambda).$$

Geometrically speaking, a singular simplex σ and the collection $(\sigma \circ b_\lambda)$ represent the same object and hence σ and $b(\sigma)$ are not very different. The following result is therefore only to be expected.

4.10. Proposition. *The map $\alpha \mapsto b\alpha$ is a morphism $S^\circ(X) \to S^\circ(X)$ of complexes which is functorial in X in the sense that if $f : X \to Y$ is any continuous map, then we have a commutative diagram*

$$
\begin{array}{ccc}
S^\circ(Y) & \to & S^\circ(X) \\
\downarrow b & & \downarrow b \\
S^\circ(Y) & \to & S^\circ(X)
\end{array}
$$

Moreover b induces the identity map on the cohomology groups.

Proof. If λ is a permutation of $(0, 1, \ldots, n)$ and $0 \leq k \leq n-1$, then it is easy to verify that $b_\lambda \circ F_k = b_{\lambda \circ t_k} \circ F_k$, where t_k is the transposition $(k, k+1)$. In fact, both are maps $\Delta_{n-1} \to \Delta_n$ induced by linear maps $\mathbb{R}^n \to \mathbb{R}^{n+1}$ taking e_i, $i \leq k-1$, into $[\lambda(0), \ldots, \lambda(i)]$ and e_i, $i \geq k$, into $[\lambda(0), \ldots, \lambda(i+1)]$. If α is an $(n-1)$-cochain, then by definition we have

$$
\begin{aligned}
((b \circ d)(\alpha))(\sigma) &= \sum (\operatorname{sgn} \lambda)(d\alpha)(\sigma \circ b_\lambda) \\
&= \sum (\operatorname{sgn} \lambda) \sum_{i=0}^{n} (-1)^i \alpha(\sigma \circ b_\lambda \circ F_i).
\end{aligned}
$$

From what we have seen, the surviving terms are $(-1)^n \sum_\lambda \operatorname{sgn}(\lambda) \alpha(\sigma \circ b_\lambda \circ F_n)$. Now it is easy to verify that $b_\lambda \circ F_n = F_{\lambda(n)} \circ b_{\widetilde{\lambda}}$ where $\widetilde{\lambda}$ is the composite of λ and the cyclic permutation $(n, n-1, \ldots, \lambda(n))$, treated as a permutation of $(0, \ldots, n-1)$. Hence

$$
\begin{aligned}
((b \circ d)\alpha)(\sigma) &= \sum_\lambda (-1)^{\lambda(n)} (\operatorname{sgn} \widetilde{\lambda}) \alpha(\sigma \circ F_{\lambda(n)} \circ b_{\widetilde{\lambda}}) \\
&= ((d \circ b)\alpha)(\sigma).
\end{aligned}
$$

This proves the first assertion. The functoriality of b is obvious from the definition. To prove the last assertion, we will give a homotopy operator between the identity and b. This is done in the following lemma. It is more convenient to work with the singular chain complex $S_\circ$ than with the singular cochain complex S°. We recall (4.2) that S_i is the free abelian group on the set of singular simplices and that the differential $S_i \to S_{i-1}$ is given by $ds = \sum (-1)^i s \circ F_i$. Also our definition of b can be interpreted as the transpose of the map $S_i \to S_i$ defined by $bs = \sum \operatorname{sgn}(\lambda) s \circ b_\lambda$. It is then easy to see that it is enough to check that b and Id on $S_\circ$ are homotopic. This is accomplished in the following lemma.

4.11. Lemma. *To every singular n-simplex σ we associate an $(n-1)$-chain $k\sigma$ such that for any chain α we have $\alpha - b\alpha = k(d\alpha) + dk(\alpha)$. Here we denote by k the linear extension to all chains of the map k defined on simplices. In other words, b and Id are homotopic and in particular, b induces the identity map on singular cohomology groups.*

Proof. We will not give an explicit homotopy operator but will prove its existence by induction.

We will assume that we have constructed such an operator $k_i : S_i(X) \to S_{i+1}(X)$ for all $i \le n-1$, satisfying the equality

$$
\operatorname{Id} - b_i = k_{i-1} d + dk_i
$$

not only for the given space X, but *for all* spaces at the same time, and that the maps k_i are functorial in the obvious sense. We will need then to construct similar maps k_n. This is done as follows. Treat the standard n-simplex Δ_n as a topological space and consider $S_n(\Delta_n)$. There is a canonical singular n-simplex u_n on it, namely the identity map! We have to define in particular $k_n(u_n)$. Our choice is dictated by the requirement $u_n - b_n(u_n) = k_{n-1}(du_n) + dk_n u_n$. Since $u_n - b_n(u_n) - k_{n-1}du_n$ is already defined, such a choice is possible if we show this to be a boundary. But then Δ_n is contractible, and hence by Theorem 4.6, it is enough to check that it is a cycle. We therefore compute $d(u_n - b_n u_n - k_{n-1}du_n)$. We use now the fact that b commutes with d and the consequence of the induction assumption, namely $du_n - b_{n-1}du_n = dk_{n-1}du_n$, to conclude that it is indeed a cycle. We therefore define $k_n(u_n)$ to be v_n where $dv_n = u_n - b_n u_n - k_{n-1}du_n$.

We have however to define k_n on all singular n-simplices of all topological spaces. But then u_n is universal in the sense that given any singular n-simplex s in X, treat the map $s : \Delta_n \to X$ as a map of topological spaces and consider the image of u_n in $S_n(X)$ under the map of singular complexes induced by s. This image is the composite of the identity map of Δ_n and s, and hence is s. We are thus forced to define $k_n s$ (in view of the functoriality requirement) to be the image in $S_{n-1}(X)$ of $v_n \in S_{n-1}(\Delta_n)$ by the morphism $\Delta_n \to X$ induced by s. This proves our assertion.

Proposition 4.10 has lent substance to our claim that splitting up simplices to smaller ones does not change the homological picture. We have only carried out this systematically, with proper bookkeeping! We will now show how it can be used. If $\mathcal{U} = (U_i)_{i \in I}$ is any open covering of X, then suppose one takes only singular simplices whose images are contained in some U_i (called $\mathcal{U}$-*small* simplices) and builds up a singular complex with those simplices. The cohomology of the subcomplex should be no different from that of the full complex, because given any singular simplex, one can divide it as many times as necessary in order to make the subsimplices $\mathcal{U}$-small. This means that we have

4.12. Proposition. *i) Let $\mathcal{U} = (U_i)_{i \in I}$ be an open covering of X and $S^{\mathcal{U}}(X)$ the complex of singular cochains defined only on simplexes with image in some U_i. Then the natural surjection $S^\circ \to (S^{\mathcal{U}})^\circ$ is a morphism of complexes and induces isomorphisms on cohomology groups.*

ii) The presheaf $U \mapsto S(U)$ gives rise to a complex $\widetilde{S}^\circ$ of sheaves on X. The natural morphism $S(X)^\circ \to \widetilde{S}(X)^\circ$ induces isomorphisms on cohomology groups.

Proof. Again we will prove the statement in the set-up of singular chain complex. The first remark is that for any singular r-simplex s, there exists N such that $b^N s$ is $\mathcal{U}$-small. In fact, considering s as a continuous map $\Delta_r \to X$ we take the open covering $s^{-1}(U_i)$, $i \in I$, of Δ_r. Since the standard simplex is compact, there exists a number l (called the *Lebesgue number* of the covering) such that any subset of diameter less than l is contained in one of the sets $s^{-1}(U_i)$. It is easy to see that the diameter of all the barycentric pieces $b^N(\Delta_r)$ tends to zero as N tends to infinity. Hence for large enough N the diameter of all these pieces is less than l. This shows that $b^N s$ is a linear combination of $\mathcal{U}$-small simplices, i.e. $b^N s \in S_r^{\mathcal{U}}(X)$.

Since all the maps b^N leave the subcomplex $S_\circ^{\mathcal{U}}$ invariant, assertion i) follows from the following algebraic statement.

4.13. Lemma. *Let $K_\circ$ be a complex and $L_\circ$ a subcomplex, with K_i free and L_i generated by a subset of the basis. Let $b : K_\circ \to K_\circ$ be a morphism which* *takes $L_\circ$ into itself. Assume that there is a homotopy operator k between b and Id and that k takes L into itself. Finally assume that for every $x \in K_i$ there exists N such that $k^N x \in L_i$. Then the inclusion $L \subset K$ is a homotopy equivalence.*

Proof. It is clear that b^m is also homotopic to the identity. Indeed, if we take $k_m = (1 + b + b^2 + \cdots + b^{m-1}) \circ k$ where b^i denotes the ith iterated composite of b with itself, then $dk_m + k_m d = \sum_{i=0}^{m-1} b^i (dk + kd) = \sum_{i=0}^{m-1} b^i (1 - b) = 1 - b^m$. With this in mind, we define a homotopy operator $k_\infty : K_r \to K_{r-1}$ by setting $(k_\infty x) = (k_{N(x)} x)$, on basis elements of K_r, where $N(x)$ is the least integer N such that $b^N x \in L$. Then we compute, as above, the term $(dk_\infty + k_\infty d - \mathrm{Id})x = (dk_{N(x)} + k_{N(x)} d - \mathrm{Id})x - (k_N - k_\infty)dx$ to be $-b^{N(x)} x - \sum a_j (k_N - k_\infty)(y_j)$, where $dx = \sum a_j y_j$. Now $(k_N - k_\infty) y_j = \sum_{i=N(y_j)}^{N(x)-1} b^i k y_j$. But note that all the terms in the summation as well as $b^{N(x)} x$ belong to L. Hence we can *define* a linear map $\varphi : K \to L$ by sending x to $(\mathrm{Id} - dk_\infty - k_\infty d)x$. It is obvious that it is a morphism of complexes. If x were already in L, we have that $N(x)$ is zero by definition, and $k_\infty x = kx$ and similarly $k_\infty dx = kdx$. Hence $(\mathrm{Id} - dk_\infty - k_\infty d)x$ is simply bx. In other words, we have given a morphism $\varphi : K \to L$ such that it is homotopic to the identity on L. If we treat φ as a morphism of K into itself it is clearly homotopic to the identity with k_∞ providing a homotopy operator between it and the identity.

Proof of Proposition 4.12, ii). For every open covering $\mathcal{U}$ we have morphisms $S^\circ \to (S^{\mathcal{U}})^\circ \to (\widetilde{S})^\circ$. We have just seen that the first map is a homotopy equivalence. If $v \in H^i(\widetilde{S}^\circ)$ and c is a cocycle in $\widetilde{S}^i(X)$ representing it, we first lift c to an i-cochain c' in $S^i(X)$ by [Ch. 1, Proposition 1.14]. Since dc' maps to $dc = 0$, every $x \in X$ has an open neighbourhood such that dc' vanishes on it. Thus, although dc' itself may not be 0, there exists a covering $\mathcal{U}$ such that the image of dc' in $(S^{\mathcal{U}})^{i+1}$ is zero. In other words, the image of c' in $(S^{\mathcal{U}})^i$ is a cocycle. The class in $H^i((S^{\mathcal{U}})^\circ)$ which it defines, maps to v. Since the map $H^i(S^\circ) \to H^i((S^{\mathcal{U}})^\circ)$ is an isomorphism, this shows that the image of $H^i(S^\circ)$ in $H^i(\widetilde{S}^\circ)$ contains v. Since v is arbitrary, this map is surjective. If $u \in H^i(S^\circ)$ maps to zero in $H^i(\widetilde{S}^\circ)$, and x is a cocycle in S^i representing u, then its image in $\widetilde{S}^i$ is of the form dv. Now v can be lifted to a cochain $y \in S^{i-1}$, again by [Ch. 1, Proposition 1.14], and $x - dy$ maps to zero in $\widetilde{S}^i$. As above, this implies that there exists a covering $\mathcal{U}$ such that $x - dy$ maps to zero in $(S_{\mathcal{U}})^i$, showing that the class in $H^i(S^\circ)$ defined by x maps to zero in $H^i((S^{\mathcal{U}})^\circ)$. Hence the cohomology class of x in $H^i(S^\circ)$ is itself zero.

4.14. Theorem. *Let X be a topological space which is locally contractible. Then the singular cohomology groups of X with coefficients in an abelian group A are naturally isomorphic to the cohomology groups of the constant sheaf A.*

Proof. The first thing to notice is that we have a sheaf complex

$$A \to \widetilde{S}^0 \to \widetilde{S}^1 \to \cdots \to \widetilde{S}^i \to \widetilde{S}^{i+1} \to \cdots .$$

Secondly by 4.12, ii), the cohomology of the complex $\widetilde{S}^{\mathcal{U}}(X)^\circ$ is naturally isomorphic to that of the complex $\mathcal{S}(X)^\circ$. But by assumption, every $x \in X$ has a contractible neighbourhood U_x, which has therefore, by Theorem 4.6, the singular cohomology of a point, namely, $H^0(U_x, A) = A; H^i(U_x, A) = 0$ for $i \geq 1$. This means that the sequence

$$0 \to A \to S^0(U_x) \to \cdots \to S^i(U_x) \to S^{i+1}(U_x) \to \cdots$$

is exact and hence also

$$0 \to A \to \widetilde{S}^0(U_x) \to \cdots \to \widetilde{S}^i(U_x) \to \widetilde{S}^{i+1}(U_x) \to \cdots .$$

In particular, $A \to \widetilde{S}^\circ$ is a resolution. By Example 1.10, 2), the sheaves $\widetilde{S}^i$ are all flabby. Hence by 3.1 and 3.3, we may use this resolution for the computation of the sheaf cohomology of A. This proves our assertion and also justifies our notation $H^i(X, A)$ for sheaf cohomology.

5. Cěch and Sheaf Cohomologies

In algebraic topology there is also another way of associating cohomology spaces to a topological space, due to Cěch. Like singular cohomology, it is also defined *a priori* as having coefficients in an abelian group A. But its definition is well adapted to having coefficients in a sheaf of abelian groups. Let us first give the definition, and later show how it is related to sheaf cohomology.

Let $\mathcal{U} = (U_i)_{i \in I}$ be an open covering of a space X. Then we will construct a resolution $\mathcal{F} \to \check{\mathcal{C}}^\circ$ of any sheaf $\mathcal{F}$ on X. For every $U \subset X$ we will define in a natural way a complex $\check{\mathcal{C}}^\circ(U)$ and a natural map $\mathcal{F}(U) \to \check{\mathcal{C}}^0(U)$. We will do this for $U = X$ and simply observe that for an arbitrary U, we have only to replace $(U_i)_{i \in I}$ by the induced covering $(U \cap U_i)_{i \in I}$ of U.

For any finite sequence $\alpha : [0, 1, \ldots, r] \to I$ we denote by $|\alpha|$ the integer r and by U_α the set $\bigcap U_{\alpha(i)}$. Then we define $\check{C}^r = \prod_{|\alpha|=r} \mathcal{F}(U_\alpha)$. In order to define the differential $d : \check{C}^r \to \check{C}^{r+1}$, we define maps $\sigma_k : \check{C}^r \to \check{C}^{r+1}$ for each $0 \leq k \leq r + 1$ and set $d = \sum(-1)^k \sigma_k$. To do this we need to define maps $\pi_\alpha \circ \sigma_k : \check{C}^r \to \mathcal{F}(U_\alpha)$ for each α with $|\alpha| = r + 1$, where $\pi_\alpha : \check{C}^{r+1} \to \mathcal{F}(U_\alpha)$ is the projection. For each k, consider the (monotonic) inclusion of $[0, 1, \ldots, r]$ in $[0, 1, \ldots, r+1]$ which misses k, and compose it with α to get α_k with $|\alpha_k| = r$. Then $\pi_\alpha \circ \sigma_k$ is defined to be the composite of the projection $\check{C}^r \to \mathcal{F}(U_{\alpha_k})$ and the restriction $\mathcal{F}(U_{\alpha_k}) \to \mathcal{F}(U_\alpha)$.

It is easy to see that we have thus defined a complex. In particular, any element ζ of $\check{C}^0$ is given by sections (s_i) of $\mathcal{F}(U_i)$, $i \in I$. If s is a section of $\mathcal{F}$ then of course it gives rise, on restriction, to such a bunch (s_i) of sections. Moreover, $d : \check{C}^0 \to \check{C}^1$ can be described as follows. For $\zeta \in \check{C}^0$, $(d\zeta)_{i,j}$ is the section $s_j - s_i$ over $U_i \cap U_j$. The sheaf conditions $\mathbf{S_1}, \mathbf{S_2}$ ensure precisely that the kernel of this differential is $\mathcal{F}(X)$. It is clear that all our constructions are compatible with restrictions so that we have a complex $\check{C}^\circ$ of presheaves and a natural map $\mathcal{F} \to \check{C}^\circ$, giving an exact sequence

$$0 \to \mathcal{F} \to \check{C}^0 \to \check{C}^1.$$

5.1. Proposition. *The complex with respect to a covering* $\mathcal{U} = (U_i)_{i \in I}$ *defined above, namely*

$$\check{C}^\circ : 0 \to \mathcal{F} \to \check{C}^0 \to \check{C}^1 \to \cdots ,$$

is a resolution of $\mathcal{F}$.

Proof. We have already checked the exactness of $0 \to \mathcal{F} \to \check{C}^0 \to \check{C}^1$. It is enough to show that for any $x \in X$ and $u \in \check{C}^i(X)$ with $du = 0$, there exist a neighbourhood V of x and $v \in \check{C}^{i-1}(V)$ such that $dv = u|V$. Now x belongs to U_a for some $a \in I$. Then we will take $V = U_a$. In order to define $v \in \check{C}^{i-1}(U_a)$, we have to define $v_\alpha \in \mathcal{F}(U_a \cap U_\alpha)$ for each $\alpha : [0, i-1] \to I$. We set $v_\alpha = u_\beta$, where β is defined by $\beta(0) = a, \beta(j) = \alpha(j-1)$ for all $1 \le j \le i$. Notice that $U_a \cap U_\alpha = U_\beta$ so that our definition is meaningful. Then we have $(dv)_\alpha = \sum(-1)^k v_{\alpha_k} = \sum(-1)^k u_{\beta_k}$, where $\beta_k : [0, i] \to I$ is given by $\beta_k(0) = a, \beta_k(j) = \alpha(j-1)$ if $j \le k, \beta_k(j) = \alpha(j)$ if $j > k$. On the other hand, since $du = 0$ by assumption, we have in particular, $du_{\gamma_k} = 0$, where $\gamma_k : [0, i+1] \to I$ is given by $\gamma_k(0) = k$, and $\gamma_k(j) = \alpha(j-1)$ for $1 \le j \le i+1$. This gives the equality $u_\alpha = \sum(-1)^k u_{\beta_k}$, proving that $dv = u$.

5.2. Definition. Let $\mathcal{F}$ be a sheaf of abelian groups on a topological space X and $\mathcal{U}$ be an open covering. Then the complex $\check{C}(X)^\circ$ is called the *Čech complex* associated to the covering.

Applying 3.1 we get the following consequence.

5.3. Theorem. *If* $\mathcal{F}$ *is a sheaf on* X *and* $(U_i)_{i \in I}$ *is an open covering such that for every* $i_0, \ldots, i_r \in I$ *with* $U_{i_0} \cap \cdots \cap U_{i_r} \ne \emptyset$ *and* $j > 0$, *we have*

$H^j(U_{i_0,\ldots,i_r}, \mathcal{F}) = 0$, *then there is a canonical isomorphism of the cohomologies of the Čech complex of $\mathcal{F}$ with respect to the covering and the sheaf cohomology $H^r(X, \mathcal{F})$.*

Proof. We have seen above that $\mathcal{F} \to \check{\mathcal{C}}^\circ$ provides a resolution. In order to apply 3.1, we have only to show that $H^i(X, \check{\mathcal{C}}^j) = 0$ for all $i > 0$. But by definition, $\check{\mathcal{C}}^j$ is the sheaf $U \mapsto \prod_{|\lambda|=j} \mathcal{F}(U_\lambda)$. The components here are simply the sheaves $\widetilde{\mathcal{F}}_\lambda$ which are the extensions of $\mathcal{F}|U_\lambda$ to the whole of X, as in Lemma 1.6. From this it easily follows that, for all $i \geq 1$, we have

$$H^i(X, \check{\mathcal{C}}^j) \simeq \prod_{|\lambda|=j} H^i(X, \widetilde{\mathcal{F}}_\lambda) \simeq \prod_{|\lambda|=j} H^i(U_\lambda, \mathcal{F}) = 0.$$

One might write down explicit conditions on the covering in order to conclude that the cohomology of the Čech complex and the cohomology of the sheaf coincide in some range, using the more precise Lemma 3.2 instead of Lemma 3.1. In particular, one may conclude

5.4. Theorem. *Let $(U_i)_{i \in I}$ be an open covering of X. Assume that U_i are all simply connected and that $U_i \cap U_j$ are all connected. Then $H^1(X, \mathbb{Z})$ can be computed using the Čech complex. In other words, let $c = (c_{ij})$ be integers corresponding to $i, j \in I$ with $U_i \cap U_j \neq \emptyset$, and satisfying $c_{ij} + c_{jk} = c_{ik}$ for $i, j, k \in I$ with $U_i \cap U_j \cap U_k \neq \emptyset$. The set of such c modulo those that satisfy $c_{ij} = d_j - d_i$, where d_i are integers indexed by I, gives the first cohomology group.*

5.5. Remarks.

1) We have defined above the Čech cohomology with respect to a covering. In most cases, we may choose a covering as in Theorem 5.3 and so this suffices for practical purposes. Indeed we will show later (Ch. 6, Remark 1.16) that if M is a differential manifold, then there exists an open covering which satisfies the condition of Theorem 5.3 for the constant sheaf. However, from the theoretical angle one has to free this cohomology from dependence on the covering. That can be done by defining the cohomology for all coverings, ordering all coverings by refinement and passing to a limit.

2) When $\mathcal{F}$ is a sheaf of nonabelian groups, we have no notion of cohomology. One can of course set $H^0(X, \mathcal{F}) = \mathcal{F}(X)$, which is a group. If $(U_i)_{i \in I}$ is an open covering, one can also define the Čech cohomology $H^1(X, \mathcal{F})$ for the covering as follows. Consider the set of assignments $s_{ij} \in \mathcal{F}(U_i \cap U_j)$ for all i, j such that $U_i \cap U_j \neq \emptyset$. Assume that they satisfy the equality $s_{ij} s_{jk} = s_{ik}$ on U_{ijk}, for $i, j, k \in I$ with $U_i \cap U_j \cap U_k \neq \emptyset$. Set $(s) \sim (t)$ if there exist $\sigma_i \in \mathcal{F}_i$ such that

$\sigma_i s_{ij} = t_{ij}\sigma_j$ on U_{ij}. The quotient *set* is defined to be $H^1(X,\mathcal{F})$. It is not a group, but a set with a special point, namely that defined by $e = (e_{ij})$, e_{ij} being the unit section over U_{ij}. Again in principle, one has to pass to a limit, but if the open sets are 'good' enough, then one can make do with a single covering. If M is a compact manifold, this can be taken to be a finite covering, so that all the abelian groups occurring in the complex for this covering are free of finite rank. Hence the cohomology groups are finitely generated.

6. Differentiable Simplices; de Rham's Theorem

Consider the de Rham complex associated to a differential manifold [Ch. 2, 6.13]. We have shown in [Ch. 2, 6.14] that the complex

$$0 \to T^* \xrightarrow{d} \Lambda^2 T^* \to \cdots$$

is a resolution of the constant sheaf $\mathbb{R}$. We note that all the sheaves $\Lambda^i T^*$ are $\mathcal{A}$-modules so that they are soft sheaves by Example 1.10, 1). Hence by Lemma 3.1 and Proposition 3.3, ii), we can compute the cohomology of the sheaf $\mathbb{R}$ as the cohomology of the de Rham complex

$$0 \to \mathcal{T}^*(X) \to \Lambda^2 \mathcal{T}^*(X) \to \cdots \to \Lambda^n \mathcal{T}^*(X) \to 0$$

where $n = \dim X$. Thus we have

6.1. Theorem (de Rham). *The ith cohomology of the de Rham complex of a differential manifold is canonically isomorphic to the ith cohomology of the constant sheaf $\mathbb{R}$.*

Since we have already shown that the sheaf cohomology is isomorphic to the singular cohomology of the space, it follows that the de Rham and singular cohomologies are isomorphic. We will give below an explicit isomorphism between them. In order to do this, the first remark we will need is that in order to compute the singular cohomology of a differential manifold, one may just use 'differentiable' simplices.

6.2. Definition. Let M be a differential manifold. A singular simplex $s : \Delta_n \to M$, said to be *differentiable* if s can be extended to a neighbourhood of Δ_n in $\mathbb{R}^{n+1}$ as a differentiable map.

Following through the same constructions as in Section 4, one concludes:

a) A singular cohomology complex DS°_M based on differentiable singular simplices can be defined.

b) The assignment of $(DS)^\circ_U$ for every open set U in M and the natural restriction maps build a complex $\mathcal{D}S^\circ$ of presheaves. This in turn gives rise to a complex $\widetilde{\mathcal{D}S}^\circ$ of sheaves.

c) The map $A \to \widetilde{\mathcal{D}S}^{\circ}$ is a flabby resolution of the coefficient group A.

d) The natural morphism $\mathcal{D}S(X)^{\circ} \to \widetilde{\mathcal{D}S}(X)^{\circ}$ induces isomorphism on cohomology.

e) The natural morphism $\widetilde{\mathcal{S}}^{\circ} \to \widetilde{\mathcal{D}S}^{\circ}$ fits in a commutative diagram of complexes of presheaves:

$$
\begin{array}{ccc}
\mathcal{S}^{\circ} & \to & \widetilde{\mathcal{S}}^{\circ} \\
\downarrow & & \downarrow \\
\mathcal{D}S^{\circ} & \to & \widetilde{\mathcal{D}S}^{\circ}
\end{array}
$$

All these maps are compatible with the inclusions of A in each of the complexes. All sheaves in the above diagram are flabby.

In particular, we have

6.3. Proposition. *The surjection $\mathcal{S}(X) \to \mathcal{D}S(X)$ gives an isomorphism of cohomology groups and these are in turn isomorphic to $H^*(X, A)$.*

Going back to our aim of giving explicitly the de Rham isomorphism, we will give a morphism from the de Rham complex $\mathcal{D}R^{\circ}$ into $\mathcal{D}S^{\circ}$ rather than into $\mathcal{S}^{\circ}$. By the precise isomorphism statement in Proposition 3.4, we have only to give such a morphism which is compatible with the respective inclusions of $\mathbb{R}$ into $\mathcal{D}R$ and into $\mathcal{D}S$ in order to ensure that the induced maps on cohomologies are isomorphisms. Let then w be a differential form of degree r. In order to associate to it, an $\mathbb{R}$-valued differentiable singular cochain of dimension r, we proceed as follows. Let $s : \Delta_r \to X$ be a differentiable singular simplex. Then the pull-back of w is a differential r-form α on (a neighbourhood of) Δ_r. To produce a real number, we will 'integrate' this r-form over Δ_r and denote this by $\int_s w$. Let us now give a precise meaning to this integral. We will identify Δ_r with $D_r = \{(X_1, \ldots, X_r) \in \mathbb{R}^r : \sum_{i=1}^{r} X_i \leq 1, X_i \geq 0\}$ by the projection $(X_0, X_1, \ldots, X_n) \to (X_1, \ldots, X_n)$. Then α can be written as $f(X_0, \ldots, X_n) dX_1 \wedge \cdots \wedge dX_n$. Then we integrate f with respect to the Lebesgue measure, over D_r.

We have thus defined a linear map $\Lambda^r T_X^* \to \mathcal{D}S^r(X)$.

6.4. Theorem (Stokes). *The map $\Lambda^r T^* \to \mathcal{D}S^r$ given by integration on simplices is a morphism of the de Rham complex into the differentiable singular complex.*

Proof. What we have to show is that if w is an $(r-1)$-form on M and s a singular simplex, then $\int_s dw = \sum (-1)^k \int_{s \circ F_k} w$, or what is the same, $\int_{\Delta_r} s^*(dw) = \sum (-1)^k \int_{\Delta^{r-1}} (s \circ F_k)^*(w)$.

The map $w \mapsto s^* w$ commutes with the exterior derivative d so that it is enough to show that if α is a form of degree $r - 1$ on Δ_r, then

6.5.
$$\int_{\Delta_r} d\alpha = \sum (-1)^k \int_{\Delta_{r-1}} F_k^*(\alpha).$$

We identify Δ_r with D_r and Δ_{r-1} with D_{r-1}. Then we have to prove 6.5 with Δ_r, Δ_{r-1} replaced respectively by D_r and D_{r-1}. Since both sides of 6.5 are additive, there is no loss of generality in assuming that $\alpha = f(X_1, \ldots, X_n) dX_1 \wedge \cdots \wedge \hat{d}X_j \wedge \cdots \wedge dX_r$.

We will first compute the face operators F_k as maps from D_{r-1} to D_r. By definition, we have $F_0(X_1, \ldots, X_{r-1}) = (1 - \sum_{i=1}^{r-1} X_i, X_1, \ldots, X_{r-1})$, and $F_k(X_1, \ldots, X_{r-1}) = (X_1, \ldots, X_{k-1}, 0, X_k, \ldots, X_{r-1})$ for $k > 0$. Hence $F_0^*(dX_1) = -\sum dX_i$, and $F_0^*(dX_i) = dX_{i-1}$ for $2 \leq i \leq r$. On the other hand, if $k > 0$, we have $F_k^*(dX_i) = dX_i$ for $i \leq k - 1, F_k^*(dX_k) = 0$, and $F_k^* dX_i = dX_{i-1}$ for $i > k$.

It follows that $F_0^*(\alpha) = -(f \circ F_0)(\sum dX_i \wedge dX_1 \wedge \cdots \wedge \widehat{dX_j} \wedge \cdots \wedge dX_{r-1}) = (-1)^{j-1}(f \circ F_0) dX_1 \wedge \cdots \wedge dX_{r-1}$. For $k > 0$, we have $F_k^*(\alpha) = 0$ unless $k = j$ and $F_j^*(\alpha) = (f \circ F_j) dX_1 \wedge \cdots \wedge dX_{r-1}$. Thus we see that the the right side of 6.5 is equal to the Lebesgue integral over D_{r-1} of $(-1)^{j-1} f \circ F_0 + (-1)^j f \circ F_j$. The left side of 6.5 is the Lebesgue integral of $(-1)^{j-1} \frac{\partial f}{\partial X_j}$ over D_r. The theorem therefore reduces to proving that if f is a differentiable function of $(X_1, \ldots, X_r)$, then

$$\int_{D_r} \frac{\partial f}{\partial X_j} = \int_{D_{r-1}} \left(f\left(1 - \sum_{i=1}^{r-1} X_i, X_1, \ldots, X_{r-1}\right) - f(X_1, \ldots, 0, \ldots, X_{r-1}) \right).$$

This is just the fundamental theorem of Integral Calculus after the observation that

$$\int_{D_{r-1}} f\left(X_1, \ldots, 1 - \sum X_i, \ldots, X_{r-1}\right) = \int_{D_{r-1}} f\left(1 - \sum_i X_i, X_1, \ldots, X_{r-1}\right).$$

This proves the precise version of

6.6. Theorem (de Rham). *The map which associates to each differential form of degree r, the differentiable singular cochain obtained by integrating on simplices, induces isomorphisms of the de Rham cohomology groups with singular cohomology groups.*

If $f : X \to Y$ is a differentiable map of one differential manifold into another, then f induces a morphism of the de Rham complex of Y into that of X. We will denote this map by f^*. Obviously the de Rham map given above is compatible with the induced morphism of the singular complexes as well. In particular, we have the following fact.

6.7. Proposition *If f and g are homotopic differentiable maps $X \to Y$, then the induced maps f^*, g^* on the de Rham cohomology groups are the same.*

6.8. Exercise. Prove the above proposition directly on de Rham cohomology, without going through singular cohomology.

We have defined singular cohomology and proved some of its properties. The above considerations say that when the coefficient group is $\mathbb{R}$ or $\mathbb{C}$, and the space is question is a differential manifold M, the singular cohomology groups can be computed using the de Rham complex. In other words, one can access the topological information contained in the singular cohomology groups in terms of analytic data. But the de Rham complex is still quite big and not very computation-friendly. We will see [Ch. 9, Theorem 2.6] that when we endow M with a Riemannian metric, de Rham cohomology can be computed more efficiently by analytical means.

6.9. Computation.   *The de Rham cohomology groups of $M = S^1$ are given by*

$$H^0 = \mathbb{R}; \quad H^1 = \mathbb{R}; \qquad H^i = 0, \quad \text{for all other } i.$$

Proof. The complex is given by

$$0 \to \mathcal{A}(M) \to \mathcal{T}^*(M) \to 0.$$

Since M is connected, we have already noticed that the equation $df = 0$ for a differentiable function f implies that f is a constant, which proves the computation of H^0. Writing S^1 as a quotient of $\mathbb{R}$ by $\mathbb{Z}$, we note that dx is an invariant form on $\mathbb{R}$ and so defines a 1-form dx on S^1 which is nonzero at all points. Hence any 1-form α on S^1 can be written as $f dx$. Since we have automatically $d\alpha = 0$, the only question to settle is when it is of the form $dg = \frac{dg}{dx} dx$. In other words, we need to know which functions f are of the form dg/dx. Clearly this is true in $\mathbb{R}$ and the solution for g is unique up to a scalar factor. In particular, if α is the form dx, we get x as the solution in $\mathbb{R}$ for the above equation. But x is not invariant under $\mathbb{Z}$ and so does not go down to a function on S^1. This shows that $H^1 \neq 0$. In general the solution g on $\mathbb{R}$ is given explicitly by $\int_0^x f(t) dt$. This is invariant under translation by 1 if and only if $\int_0^1 f(t) dt = 0$. In other words, the map $\alpha \mapsto \int_M \alpha$ induces an isomorphism of the first de Rham cohomology space with $\mathbb{R}$.

6.10. Remark. The last statement above is indeed a general fact. Consider the map of the space of densities on an r-dimensional compact, connected differential manifold into $\mathbb{R}$ given by $\mu \mapsto \int \mu$. Note that all densities are cocycles in the twisted de Rham complex. Moreover, by Stokes theorem, if a density is a coboundary, the integral vanishes [Ch. 3, Theorem 3.3]. Hence we get a linear map from the de Rham cohomology space $H^r(M)$ into $\mathbb{R}$. It is obviously surjective since if we take an everywhere positive density the integral is nonzero. It is a fact that this linear form is actually an isomorphism. We do not prove it here, but it is a consequence of each of many theorems in Chapters 7 and 8.

6.11. Invariant de Rham complex.

Suppose G is a Lie group acting on a differential manifold. Then it is clear that the exterior derivative of a G-invariant form is again invariant. Hence invariant forms define a subcomplex of the de Rham complex, which we call the *invariant de Rham complex.*

Let G be compact. Then we take the positive invariant measure on G whose total measure is 1, and *average* any form ω over the group. Thus we get a new form by the formula $I(\omega) = \int g^*(\omega)$. By definition this form takes on r vector fields $X_1, \ldots, X_r$ the value $\int \omega(gX_1, \ldots, gX_r)$. It is clear that $I(\omega)$ is itself invariant under G since $x^*(I(\omega)) = x^*(\int g^*\omega dg) = \int (gx)^*(\omega)dg = I(\omega)$. The map $\omega \mapsto I(\omega)$ gives a morphism of the de Rham complex into the invariant de Rham complex. If ω is already invariant, we obviously have $I(\omega) = \omega$. Thus the above morphism is the identity on the subcomplex of invariant forms.

If G is connected in addition, we would like to show that every de Rham cohomology class is represented by a G-invariant form. This would prove that the inclusion of the invariant de Rham complex in the total de Rham complex induces isomorphism on cohomologies.

For any $g \in G$, we know by the homotopy theorem that any closed r-form ω and $g^*(\omega)$ differ by a coboundary. Since the integral can be approximated by a finite convex linear combinations of forms of the type $g_i^*(\omega)$, it follows that ω differs from these combinations by a coboundary. If we knew that in some sense the set of coboundaries is closed (and this will be shown precisely in [Ch. 9, 1.10], then it would follow that $I(\omega)$ is closed and defines the same cohomology class as ω.

But we can argue more simply as follows. Let ω be any closed r-form. Consider the singular cocycle defined by $I(\omega) - \omega$ under the de Rham morphism. It is enough to show that this is a coboundary. This would follow if it were zero on singular cycles. Accordingly, let $c = \sum a_i c_i$ be a cycle with c_i singular r-simplices and $a_i \in \mathbb{R}$. Then $\sum a_i \int_{\Delta_i} c_i^*(I(\omega) - \omega)) =$

$\sum a_i \int_{\Delta_i} \int_G (g^*(\omega) - \omega) dg$. Now using Fubini's theorem, as well as the fact that $g^*(\omega) - \omega$ is a de Rham coboundary and hence the corresponding singular cochain is also a coboundary, we see that the above integral is zero.

6.12. Theorem. *The invariant de Rham complex can be used to compute the cohomology of a differential manifold on which a compact connected Lie group G acts.*

The immediate application that comes to mind is that of a compact connected Lie group acting on itself by left translations. Thus the cohomology can be computed by using left invariant forms. In other words, consider the complex

$$0 \to \mathfrak{g}^* \to \Lambda^2 \mathfrak{g}^* \to \cdots \to \Lambda^n \mathfrak{g}^* \to 0.$$

The differentials are given by

$$d\alpha(X_1, \ldots, X_n) = \sum (-1)^{i+j} \alpha([X_i, X_j], X_1, \ldots, \hat{X}_i, \ldots, \hat{X}_j, \ldots, X_n).$$

Incidentally, we have managed to compute the cohomology of the *Lie group* in terms of this complex, which is defined purely in terms of its Lie algebra.

One can indeed go one step further. One might as well make $G \times G$ act on the Lie group G by $(g_1, g_2).x = g_1 x g_2^{-1}$. This implies that one can actually do with the *biinvariant de Rham complex*. Note that the transformation $\iota : x \mapsto x^{-1}$ acts on this complex. This action is easy to compute. In fact, it acts as $-\operatorname{Id}$ on the tangent space at 1 and so as $(-1)^r$ on the rth term of the complex. Let ω be any biinvariant form. Then $\iota^*(d\omega) = d\iota^*(\omega)$. But ι acts as $(-1)^{r+1}$ on the left and as $(-1)^r$ on the right side. Hence all the differentials in the biinvariant complex are zero, and hence we have the following computation.

6.13. Theorem. *There is a canonical isomorphism of $H^i(G, \mathbb{R})$ with the space of adjoint invariant i-forms.*

Proof. The only point is to check that the action on the right by G on left invariant i-forms is simply the ith exterior power of the adjoint action on $\Lambda^i(\mathfrak{g}^*)$. But then this is obvious.

The above principle can be used to make many computations, and we shall leave some as exercises.

Exercises

1) Assume given a differentiable map $h : M \times I \to N$, where I is the unit interval. This means that the map h can be extended to a differentiable map h' from $M \times I'$ where I' is an open set containing I. To any differential form ω in N, associate $A(\omega) = \iota_{\frac{\partial}{\partial t}}(h'^*(\omega))$.

Note that $A(\omega)$ is a form on M depending on $t \in I'$. Then define $k(\omega) = \int_0^1 A(\omega)dt$ and check that it gives a homotopy between the morphisms $\mathcal{D}R(N) \to \mathcal{D}R(M)$ given by h_0 and h_1.

2) Show that if M and N are differential manifolds, then the tensor product of the pull-back of the de Rham complexes of M and N to $M \times N$, can be used to compute the cohomology of $M \times N$ with coefficients in $\mathbb{R}$.

3) Let G be a compact connected Lie group. Show that the symmetric bilinear form $(X, Y) \mapsto \mathrm{tr}(\mathrm{ad}\, X \, \mathrm{ad}\, Y)$ on its Lie algebra is negative. If G has discrete centre, show that it is nondegenerate. (In this case, one says the group is compact semisimple.) Show that the first and second Betti numbers of a compact, semisimple group G are 0 but the third Betti number is not.

4) Let G be a compact, connected Lie group acting on a differential manifold M. Assume that for every $m \in M$, there is an element $s \in G$ such that m is an isolated fixed point and that its action on the tangent space is $-\,\mathrm{Id}$. Show that the odd Betti numbers of G are all zero.

5) Compute the Betti numbers of $SU(2)$ by finding the biinvariant forms on it.

6) Let C° be the complex over $\mathbb{Z}$ obtained by dualising a complex $C_\circ$. Show that there is a natural map of the dual of the homology group H_i of the latter complex into the cohomology group of the former. Give an example to show that in general this is not an isomorphism.

7) To any complex C° we associate another complex $C[1]^\circ$ by $C[1]^i = C^{i+1}$ and $d[1] = -d$. If $f : D^\circ \to C^\circ$ is a morphism of complexes, define another complex C_f° whose ith term is $C^i \oplus D[1]^i$ and the differential takes $(x, y), x \in C^i, y \in D[1]^i$ to $(dx + f(y), d[1]y)$. Check that this is a complex and that there is a natural exact sequence of complexes

$$0 \to C^\circ \to C_f^\circ \to D[1]^\circ \to 0.$$

If f is an inclusion, show that there is a natural morphism of C_f° into $(C/D)^\circ$ and that the induced map $H^i(C_f^\circ) \to H^i((C/D)^\circ)$ followed by the boundary homomorphism $H^i((C/D)^\circ) \to H^{i+1}(D^\circ) = H^i(D[1]^\circ)$ of the exact sequence

$$0 \to D^\circ \to C^\circ \to (C/D)^\circ \to 0$$

is the homomorphism induced by the morphism $C_f^\circ \to D[1]^\circ$.

8) Let M be a complex manifold and N a closed submanifold. Write down a sufficient condition, in terms of cohomology of sheaves, for any holomorphic function on N to be extendable to M.

9) Let $\mathcal{F}'$, $\mathcal{F}''$ be two $\mathcal{R}$-modules where $\mathcal{R}$ is a sheaf of rings on X. An extension of $\mathcal{F}''$ by $\mathcal{F}'$ is an $\mathcal{R}$-module $\mathcal{F}$ which sits in an exact sequence

$$0 \to \mathcal{F}' \to \mathcal{F} \to \mathcal{F}'' \to 0.$$

Two extensions are said to be equivalent if there is an isomorphism between the middle sheaves which is the identity on $\mathcal{F}'$ and induces the identity on $\mathcal{F}''$. Show that there is a bijection between the set of equivalence classes of extensions which locally split, and $H^1(\mathrm{Hom}(\mathcal{F}'', \mathcal{F}'))$.

10) Assume that a complex E° over $\mathbb{C}$ has only finitely many nonzero cohomologies. Define $\chi(E^\circ)$ to be $\sum(-1)^i \dim H^i(E^\circ)$. Suppose that

$$0 \to A^\circ \to B^\circ \to C^\circ \to 0$$

is an exact sequence of complexes over $\mathbb{C}$, all of which satisfy the above condition. Then show that $\chi(B^\circ) = \chi(A^\circ) + \chi(C^\circ)$.

Chapter 5

Connections on Principal and Vector Bundles; Lifting of Symbols

There are several angles from which the notion of connections can be looked at. We will start with the definition in terms of differentiation of tensor fields and later discuss other points of view. Given a vector field we would like to be able to *differentiate* any tensor field. We have already explained why the Lie derivative does not serve the purpose. A linear connection is a rule by which such a differentiation can be performed. We will deal here with the more general notion of a connection in any vector bundle E, which allows one to differentiate sections of E with respect to a vector field. A little later, we will deal with the even more general notion of principal bundles and connections on them. We will turn to linear connections in the next chapter.

1. Connections in a Vector Bundle

Recall that a differential operator of order ≤ 1 is of the form $X + m(f)$ where X is a vector field, and f is a function. Its symbol is X. The corresponding symbol sequence is

1.1.
$$0 \to \mathcal{A} \to \mathcal{D}^1 \to \mathcal{T} \to 0.$$

Although $\mathcal{D}^1$ is an $\mathcal{A}$-bimodule, the maps in this diagram are homomorphisms for either of the structures. The map $D \mapsto D(1)$ which associates to $X + m(f)$ the function f, gives a splitting of the above sequence. The corresponding splitting map of $\mathcal{T}$ to $\mathcal{D}^1$ is just the inclusion of a vector field in $\mathcal{D}^1$. Note, however, that this splitting is for the *left* $\mathcal{A}$-module structure on $\mathcal{D}^1$ and not for the right structure.

The point of the above observation is the following. Let E be a differentiable vector bundle on a differential manifold. Consider its first order symbol sequence:

1.2. $$0 \to \mathrm{Hom}(\mathcal{E}, \mathcal{A}) \to \mathcal{D}^1(\mathcal{E}, \mathcal{A}) \to \mathrm{Hom}(\mathcal{E}, \mathcal{T}) \to 0.$$

Note that by definition the middle term, namely $\mathcal{D}^1(\mathcal{E}, \mathcal{A})$, is $\mathrm{Hom}(\mathcal{E}, \mathcal{D}^1)$ where $\mathcal{D}^1$ is equipped with the *right* $\mathcal{A}$-module structure. With this understanding, the sequence 1.2 is obtained from 1.1 by applying $\mathrm{Hom}(\mathcal{E}, .)$. Therefore the left splitting of 1.1 does not lead to a splitting of 1.2.

Being an exact sequence of vector bundles, the sequence 1.2 does split. One can take local splittings and patch them together by a partition of unity for example. But, unlike the corresponding sequence in the case $\mathcal{E} = \mathcal{A}$, this sequence does not split *naturally*. As a consequence, we cannot give a meaning to the phrase: 'a first order homogeneous operator from $\mathcal{E}$ to $\mathcal{A}$'.

1.3. Definition. Let E be a differentiable vector bundle on a differential manifold M. A splitting of the first order symbol sequence

$$0 \to \mathrm{Hom}(\mathcal{E}, \mathcal{A}) \to \mathcal{D}^1(\mathcal{E}, \mathcal{A}) \to \mathrm{Hom}(\mathcal{E}, \mathcal{T}) \to 0$$

is called a *connection* on E. When E is taken to be the tangent bundle, we call it a *linear connection*.

As noted above, a connection always exists. According to the definition, a connection is a section of $\mathrm{Hom}(\mathrm{Hom}(\mathcal{E}, \mathcal{T}), \mathcal{D}^1(\mathcal{E}, \mathcal{A}))$ lying over the identity automorphism of $\mathrm{Hom}(\mathcal{E}, \mathcal{T})$. The homomorphism (which we will denote by ∇, pronounced 'nabla') of $\mathrm{Hom}(\mathcal{E}, \mathcal{T}) = \mathcal{E}^* \otimes \mathcal{T}$ into $\mathcal{D}^1(\mathcal{E}, \mathcal{A})$ can be interpreted in many ways.

For example, it is an $\mathcal{A}$-linear homomorphism $\mathcal{T} \to \mathcal{E} \otimes \mathcal{D}^1(\mathcal{E}, \mathcal{A}) = \mathcal{D}^1(\mathcal{E}, \mathcal{E})$. The image of a vector field X in this interpretation is denoted ∇_X. The symbol of ∇_X is clearly $X \otimes \mathrm{Id}_E$. Fix a section s of E. Then both operators $\mathcal{A} \to \mathcal{E}$ defined by $f \mapsto \nabla_X(fs)$ and $f \mapsto Xf \otimes s$ have the same symbol. So $f \mapsto \nabla_X(fs) - (Xf)s$ is an $\mathcal{A}$-linear homomorphism of $\mathcal{A}$ into $\mathcal{E}$. In other words, there is a section φ_s of E such that $\nabla_X(fs) - (Xf)s = f\varphi_s$ for all functions f. Taking $f = 1$, we conclude that $\varphi_s = \nabla_X(s)$.

Thus we have another formulation of our definition, namely,

1.4. Equivalent formulation. *A connection in a vector bundle E is an $\mathcal{A}$-linear map $X \mapsto \nabla_X$ from $\mathcal{T}$ to $\mathcal{D}^1(\mathcal{E}, \mathcal{E})$ satisfying (the Leibniz rule)*

$$\nabla_X(fs) = f\nabla_X(s) + (Xf)s$$

for all $f \in \mathcal{A}(M)$.

1.5. Remark. A connection gives a rule of differentiation of sections of E with respect to vector fields. If $m \in M$, then the value of the section $\nabla_X s$ at m (which is an element of the fibre E_m) depends only on the value of X at m. In order to see this we have only to check that if X vanishes at a point m of M, then $\nabla_X s$ also vanishes at m for all s. Now X can be written, at any rate locally, as $\sum f_i X_i$ where f_i are functions vanishing at m. But then $\nabla_X = \sum f_i \nabla_{X_i}$ and so, $(\nabla_X s)(m) = \sum f_i(m)(\nabla_{X_i} s)(m)$ vanishes at m for all s. Thus for any tangent vector v at $m \in M$, the expression $\nabla_v(s)$ makes sense as an element of E_m, namely $(\nabla_X s)(m)$, where X is any vector field with $X_m = v$. It is in this respect that the notion of Lie derivative of a tensor by a vector field X was found deficient. For while the derivative at a point of a section will naturally depend on the section in the neighbourhood of the point, the value at a point of the Lie derivative of a section, depends also on *the vector field* in a neighbourhood.

1.6. Definition. Given a connection ∇ on a vector bundle, we call $\nabla_X(s)$ the *covariant derivative of s with respect to the vector field X.*

1.7. Remark. Another way to interpret a connection on E is as a section of $\mathcal{T}^* \otimes \mathcal{E} \otimes \mathcal{D}^1(\mathcal{E}, \mathcal{A}) = \mathcal{D}^1(\mathcal{E}, \mathcal{T}^* \otimes \mathcal{E})$, that is to say a differential operator of order 1 from E into $T^* \otimes E$ which takes a section s into the 1-form with values in E given by $X \mapsto \nabla_X(s)$. Its symbol is a homomorphism $T^* \otimes E \to T^* \otimes E$. We may compute this as follows. Let X be any vector field and s a section of E. Contraction with X gives the operator ∇_X, which has $X \otimes \mathrm{Id}$ as its symbol. This shows that the symbol of the operator we defined above is the identity homomorphism of $T^* \otimes E$.

The image of a section s is called its *absolute derivative* and is denoted by $d_\nabla s$.

Notice that $E = \mathcal{A}$ has a natural connection, namely that for which the covariant derivative of any function f with respect to a vector field X is given by Xf. The absolute derivative is then the usual differential $f \mapsto df$.

Let ∇ be a connection on a vector bundle E. Let F be any vector bundle. Then ∇ can be used to 'lift' any first order symbol $\sigma : E \to T \otimes F$ into a first order differential operator $E \to F$. In fact, we have only to take the $\mathcal{A}$-homomorphism $\tilde{\sigma} : T^* \otimes E \to F$ associated to σ and compose it with the differential operator $d_\nabla : E \to T^* \otimes E$. Since the symbol of the latter is the identity, the symbol of the composite is σ.

Thus we have the following characterisation of a connection.

1.8. Proposition. *Let E be a vector bundle. A connection on E gives rise to a lift of every first order symbol $\sigma : E \to T \otimes F$ to a differential operator $\nabla_\sigma : E \to F$ with the property that if $f : F \to G$ is any vector bundle homomorphism, then $\nabla_{((\mathrm{Id})_T \otimes f) \circ \sigma} = f \circ \nabla_\sigma$. Moreover, if $F = T^* \otimes E$, then the natural symbol $E \to T \otimes T^* \otimes E$ is lifted to d_∇. Conversely, such a consistent lifting as above, arises from a connection.*

Proof. The last assertion follows by *defining* a connection ∇ on E by lifting the natural symbol $E \to T \otimes T^* \otimes E$ into an operator $E \to T^* \otimes E$ and checking that the given assignment of lift for any symbol $\sigma : E \to T \otimes F$ coincides with ∇_σ.

1.9. A generalisation.

We will now give a mild generalisation of the procedure of using connections to lift first order symbols to differential operators. Suppose D is a first order differential operator $F \to G$ with symbol $\sigma : F \to T \otimes G$. We have remarked [Ch. 2, 7.24] that D *does not* give rise to a natural operator from $F \otimes E$ to $G \otimes E$ since D is not $\mathcal{A}$-linear. However, if we are given a connection ∇ on E, we may use it in order to lift the symbol $\sigma \otimes (\mathrm{Id})_E : F \otimes E \to T \otimes G \otimes E$ to such an operator. Indeed, for $s \in \mathcal{F}(U)$ and $t \in \mathcal{E}(U)$ we define $\tilde{D}(s,t) = Ds \otimes t + \langle \sigma s, d_\nabla(t) \rangle$. Here $\langle \, , \, \rangle$ denotes the contraction map $(T \otimes G) \times (T^* \otimes E) \to G \otimes E$. We will now show that this $\mathbb{R}$-bilinear map, although not $\mathcal{A}$-bilinear, is nevertheless $\mathcal{A}$-*balanced*. By this we mean that for any function f over U, we have $\tilde{D}(fs,t) = \tilde{D}(s,ft)$. In fact, the left side here is by definition $D(fs) \otimes t + \langle \sigma fs, d_\nabla t \rangle$. Note that σ is $\mathcal{A}$-linear and $\langle \, , \, \rangle$ is $\mathcal{A}$-bilinear. Hence $\tilde{D}(fs,t) - f\tilde{D}(s,t) = (D(fs) - fD(s)) \otimes t$. But then this is the same as $\langle \sigma(D)s, df \rangle \otimes t$, with the same contraction notation for the pairing $(T \otimes G) \times T^* \to G$. Now we have $\tilde{D}(s,ft) - f\tilde{D}(s,t) = \langle \sigma(D)s, d_\nabla(ft) - fd_\nabla(t) \rangle = \langle \sigma(D)s, df \otimes t \rangle$. This shows that the above map is balanced as claimed and hence induces an $\mathbb{R}$-linear homomorphism $\overline{D} : \mathcal{F} \otimes_\mathcal{A} \mathcal{E} \to \mathcal{G} \otimes_\mathcal{A} \mathcal{E}$. In the course of our computation above, we have shown that $\overline{D}f(s \otimes t) - f\overline{D}(s \otimes t) = \langle \sigma s, df \rangle \otimes t = \langle (\sigma \otimes (\mathrm{Id})_E)(s \otimes t), df \rangle$ or, what is the same, $\overline{D}$ is a differential operator of first order from $F \otimes E$ to $G \otimes E$ with symbol $\sigma \otimes (\mathrm{Id})_E$.

1.10. Examples.

1) Let us start with a vector field X. Then we may try to define the analogue of the Lie derivative with respect to X, of tensors with values in E. For instance, one might try to define an operator $T^* \otimes E \to$

$T^* \otimes E$, using the usual Lie derivative $L(X)$ and a connection ∇ on E. Then the above procedure gives us an operator which takes (α, t) (where α is a 1-form and t is a section of E) to

$$L(X)(\alpha) \otimes t + \langle X \otimes \alpha, d_\nabla t \rangle = L(X)(\alpha) \otimes t + \alpha \otimes \nabla_X(t).$$

If ω is a differential 1-form with values in E given by $X \mapsto \alpha(X)t$, this means that $(\overline{L}(X)(\omega))(Y) = X\alpha(Y).t - \alpha([X,Y])t + \alpha(Y)\nabla_X t = -\alpha([X,Y])t + \nabla_X(\alpha(Y)t)$. We may also write this out as the formula:

$$(\overline{L}(X)(\omega))(Y) = -\omega([X,Y]) + \nabla_X(\omega(Y))$$

for all E-valued differential forms ω of order 1.

2) Consider the symbol σ of the exterior derivative, say from 1-forms to 2-forms. It is simply the natural homomorphism $T^* \to T \otimes \Lambda^2(T^*)$. This can be tensored with $(\mathrm{Id})_E$ to give a first order symbol of a potential operator from $T^* \otimes E$ to $\Lambda^2(T^*) \otimes E$. If E is provided with a connection ∇, we can lift it to an operator which takes any E-valued 1-form α to the E-valued 2-form which sends (X, Y) to $\nabla_X(\alpha(Y)) - \nabla_Y(\alpha(X)) - \alpha([X,Y])$.

1.11. Remark. We may thus use the same procedure to define the analogue of the exterior derivative at all stages, namely an operator $\Lambda^i(T^*) \otimes E \to \Lambda^{i+1}(T^*) \otimes E$. We will use the notation d_∇ for all these. However, we do *not* in general get an analogue of the de Rham complex this way. That is to say the composite of the operators $\Lambda^{i-1}(T^*) \otimes (E) \to \Lambda^i(T^*) \otimes E$ and $\Lambda^i(T^*) \otimes E \to \Lambda^{i+1}(T^*) \otimes E$ need not be 0. Since the composites of the symbols gives the second order symbol 0, it follows that the composite is actually a first order operator! It will turn out to be actually an $\mathcal{A}$-linear homomorphism as one can check directly. For example, when $i = 1$, this is a homomorphism $E \to \Lambda^2(T^*) \otimes E$ or, what is the same, an $\mathrm{End}(E)$-valued 2-form, namely $(d_\nabla^2)(X,Y)(s) = \nabla_X((d_\nabla s)(Y)) - \nabla_Y((d_\nabla(s)(X)) - d_\nabla s([X,Y]) = \nabla_X(\nabla_Y(s)) - \nabla_Y(\nabla_X(s)) - \nabla_{[X,Y]}(s)$. In other words, it is the $\mathrm{End}(E)$-valued 2-form $\nabla_X \circ \nabla_Y - \nabla_Y \circ \nabla_X - \nabla_{[X,Y]}$. This form is called the *curvature form of the connection*. We will return to a detailed study of this form later.

1.12. Exercises.

1) Prove that the exterior derivative of exterior r-forms with values in a vector bundle E provided with a connection ∇ is given by

$$d\alpha(X_1, \ldots, X_{r+1}) = \sum (-1)^{i+1} \nabla_{X_i} \alpha(X_1, \ldots, \hat{X}_i, \ldots, X_{r+1})$$

$$+ \sum (-1)^{i+j} \alpha([X_i, X_j], X_1, \ldots, \hat{X}_i, \ldots, \hat{X}_j, \ldots, X_{r+1}).$$

2) Compute the Lie derivative of a differential form of degree r with values in a vector bundle E.

1.13. Connections in associated bundles.

Let $\nabla^{(i)}$, $i = 1, 2$, be connections in bundles E_i. Then one can define in an obvious way a connection in $E_1 \oplus E_2$. For any vector field X one simply takes the operator $(s_1, s_2) \mapsto \nabla_X^{(1)}(s_1) + \nabla_X^{(2)}(s_2)$ as the new connection. One can also define an induced connection on $E = E_1 \otimes E_2$ as follows. Corresponding to any vector field X, and sections s_i of E_i, consider the section of $E_1 \otimes E_2$ given by

$$\nabla_X^{(1)} s_1 \otimes s_2 + s_1 \otimes \nabla_X^{(2)} s_2.$$

It is easy to check that (fs_1, s_2) and (s_1, fs_2) give rise to the same section of $E_1 \otimes E_2$. From this we conclude that this gives rise to an $\mathbb{R}$-linear homomorphism of $E_1 \otimes E_2$ into itself. We define this to be the operator ∇_X on $E_1 \otimes E_2$. One can also define a connection on the bundle $\operatorname{Hom}(E_1, E_2)$. Indeed, if X is a vector field and f a homomorphism of E_1 into E_2, then we set

$$(\nabla_X(f))(v) = \nabla_X^{(2)}(f(v)) - f(\nabla_X^{(1)}(v_1)).$$

In particular, if E_2 is taken to be trivial with the trivial connection, then one sees that a connection in a bundle gives rise to a connection in its dual.

We may take several copies of the bundles E and E^* and use any given connection on E to define a connection on $\bigotimes^r(E) \bigotimes^s(E^*)$ by iterating the procedure above. Moreover, symmetric (resp. alternating) tensors are left invariant under this extension, that is to say, a connection on E gives rise in a natural way to connections on the bundle $S^k(E)$ (resp. $\Lambda^k(E)$) as well. We will denote all these satellite connections by the same symbol ∇.

Thus the structure of a linear connection on a manifold, namely a connection on the tangent bundle T, gives rise to a connection on all the associated tensor bundles and hence allows us to differentiate tensor fields with respect to tangent vectors.

1.14. Connection as a $\mathcal{C}$-module.

A connection on E assigns to every vector field a differential operator $E \to E$. Can we also associate by iteration, natural higher order operators? If f is a function, that is to say, a 0th order operator, then of course we could define $f.s$ as the obvious product. Notice that this already gives, for every differential operator $D = X + m(f) \in \mathcal{D}^1$, a first order differential operator $E \to E$, namely $s \mapsto \nabla_X s + fs$. Moreover the resulting map $\mathcal{D}^1 \to \mathcal{D}^1(\mathcal{E}, \mathcal{E})$ is a $\mathcal{A}$-bimodule homomorphism, thanks to the Leibniz condition we have imposed on the connection. Hence this can be extended as an $\mathcal{A}$-bimodule homomorphism of $\mathcal{D}^1 \otimes \mathcal{D}^1$ into $\mathcal{D}^2(\mathcal{E}, \mathcal{E})$. In other words, if X, Y are vector fields, we do get a second order operator $E \to E$ by composing ∇_X and ∇_Y. Proceeding in this way, we obtain a homomorphism of the connection algebra $\mathcal{C}$ into $\mathcal{D}(E, E)$. This makes $\mathcal{E}$ a $\mathcal{C}$-module. Indeed, any $\mathcal{C}$-module, which is locally free as an $\mathcal{A}$-module actually occurs in the above way. So one can take a purely algebraic point of view and say that *a connection is simply the structure of a $\mathcal{C}$-module on a given locally free $\mathcal{A}$-module $\mathcal{E}$ of finite rank.*

Notice that we *cannot* associate a differential operator $E \to E$ to every differential operator in $\mathcal{D}$. The reason for this is that the above $\mathcal{C}$-module structure does not in general go down to a $\mathcal{D}$-module structure. Recall that $\mathcal{D}$ is the quotient of $\mathcal{C}$ by the two-sided ideal generated by elements of the form $R(X, Y) = \nabla_X \nabla_Y - \nabla_Y \nabla_X - [X, Y]$. Obviously these elements have to act trivially on a module if it is to become a $\mathcal{D}$-module. (Here ∇_X denotes the element in $\mathcal{C}$ corresponding to a vector field X.)

2. The Space of All Connections on a Bundle

If ∇ and ∇' are two connections on a bundle E, then by definition they are first order differential operators $E \to T^* \otimes E$ with the same symbol. Hence their difference is an $\mathcal{A}$-linear homomorphism $E \to T^* \otimes E$, that is to say, a differential form with values in $\mathrm{End}(E)$. In other words, there exists a 1-form α with values in $\mathrm{End}(E)$ such that $\nabla'_X - \nabla_X = \alpha(X)$ for all vector fields X. One might say that the space of all connections is therefore an *affine space* based on the vector space of all 1-forms with values in $\mathrm{End}(E)$. We will now make a digression to make precise the notion of an affine space, based on a vector space.

2.1. Definition. An *affine space A based on a vector space V*, or a *V-affine space*, is a set A together with a simply transitive action of the vector space V on it. If $v \in V$ and $a \in A$, then we denote the action of v on a by $v + a$ and refer to the map $a \mapsto v + a$ as a *translation*.

2.2. Remark. This only formalises the intuitive idea that the affine space is 'the same as a vector space, except that it has no origin'! The simplest example that one can think of is the following. Consider the set of all instances of time. Then it does not have any 'canonical origin'. On the other hand all periods of time do form a one-dimensional vector space, clearly acting on the former simply transitively.

2.3. Examples.

1) Any vector space acts on itself by left translations. It is obvious that this action is simply transitive and makes it possible to consider a vector space as an affine space based on itself.

2) If
$$0 \to V' \to V \to V'' \to 0$$
is an exact sequence of vector spaces, then the set of all splittings of this sequence is an affine space based on the vector space $\mathrm{Hom}(V'', V')$.

3) Let V be a vector space and f a nonzero linear form on it. Then the space $A = \{v \in V : f(v) = 1\}$ is an affine space based on the vector space $N = \ker(f)$. In fact, it is clear that N acts on A by translations and makes it an affine space.

4) In the above example, the restriction to A of the natural map $V \setminus \{0\} \to P(V)$ gives a natural bijection onto $P(V) \setminus P(N)$. Hence $P(V) \setminus P(N)$ is an affine space based on N. The affine space is thus imbedded in the projective space as an open set. Another way of putting it is that the projective space is the compactification of the affine space, the points at infinity constituting a projective space of dimension one less.

Suppose E is a vector bundle. Then a differentiable automorphism A is called a *gauge transformation*. One can transform a connection ∇ on E by a gauge transformation A, which we denote by $A_*(\nabla)$. This is by definition given by

2.4.
$$(A_*(\nabla))_X(s) = A(\nabla_X(A^{-1}s)).$$

There are now two connections on E, namely ∇ and $A_*(\nabla)$, and so their difference is a 1-form with values in $\mathrm{End}(E)$. It is indeed easy to compute this. Use the associated connection on $\mathrm{End}(E)$, and write $(\nabla_X(A))(s) = \nabla_X(As) - A(\nabla_X s)$. Substituting for s the section $A^{-1}(s)$, we see that $\nabla_X - (A_*(\nabla))_X$ takes s to $(\nabla_X(A) \circ A^{-1})(s)$. Thus this difference is the 1-form $d_\nabla(A) \circ A^{-1}$. We will write it up as a formula.

2.5. Formula. *The difference* $\nabla - A_*(\nabla)$ *between a connection* ∇ *and its transform by a gauge transformation* A, *is given by the 1-form*
$$X \mapsto \nabla_X(A) \circ A^{-1}$$
with values in $\mathrm{End}(E)$.

To describe the nature of this action, we again need to digress in order to define the notion of an affine transformation.

2.6. Definition. An *affine map* T of a V-affine space S into a V'-affine space S' consists of a map T of the set S into S', and a linear map $l(T) : V \to V'$ such that $T(v + a) = l(T)v + Ta$ for all $a \in S$ and $v \in V$. The linear map $l(T)$ is referred to as the *linear part* of the affine map.

2.7. Remark. The linear part of A is uniquely determined by the map $S \to S'$. The composite of two affine transformations is again affine and the linear part of the composite is the composite of the linear parts. An affine transformation is invertible if and only if its linear part is invertible.

2.8. Examples.

1) The translations given by an element $w \in V$ is itself an affine map $S \to S$ with linear part $(\mathrm{Id})_V$, since we have $w + (v + a) = \mathrm{Id}(v) + (w + a)$, for all $v \in V$ and $a \in A$.

2) If W is a vector space and V is a subspace of codimension 1, then any automorphism A of W which leaves V invariant, gives rise to a transformation of $P(W) \setminus P(V)$. It is a routine matter to check that this is an affine transformation.

3) Let V, V' be vector spaces. Then an affine map of the associated affine spaces is simply a linear map $V \to V'$ followed by a translation.

2.9. Exercise. What can one say about an affine map whose linear part is zero?

2.10. Definition. The group of all invertible affine transformations of an affine space S, is called the *affine group* and is denoted $GA(S)$.

In particular if V is a vector space, then $GA(V)$ makes sense. Let S be a V-affine space. Then the set of all affine maps of S into the affine space V is a vector space under addition. Moreover, it is a Lie algebra under the operation $[(T, l), (T', l')] = (l \circ T' - l' \circ T, l \circ l' - l' \circ l)$. It is denoted by $\mathfrak{g}a(S)$. If S is a vector space V regarded as an affine space, we will denote it by $\mathfrak{g}a(V)$.

2.11. Remark. We have natural exact sequences

$$1 \to V \to GA(S) \to GL(V) \to 1,$$
$$0 \to V \to \mathfrak{g}a(S) \to \mathfrak{g}l(V) \to 0.$$

When S is the affine space V, then there is a natural splitting of these two sequences.

The reason why we have interposed all this here is that if A is a gauge transformation of a vector bundle E, then its action on the space of connections on E is an affine transformation.

2.12. Proposition. *Any automorphism of a vector bundle E induces an automorphism of the space of differential 1-forms with values in $\mathrm{End}(E)$. It also acts on the affine space of connections on E as an affine transformation with the above action as its linear part.*

Proof. The action of a gauge transformation A on $T^* \otimes \mathrm{End}(E)$ is clear, namely $(\mathrm{Id})_{T^*} \otimes \mathrm{Int}(A)$, while its action on the space of connections has been described in 2.4. Now our assertion is a consequence of the following identity, which follows from the definition of the gauge action on connections:

2.13. $$A_*(\alpha(X) + \nabla_X) = A \circ \alpha(X) \circ A^{-1} + (A_*(\nabla))_X.$$

We may make things a little more explicit in the case when E is a trivial bundle. In this case there is a trivial connection on it so that any connection can be written as $\nabla_X(s) = Xs + \alpha(X)(s)$, for any vector-valued function s. Here α is an $\mathrm{End}(V)$-valued differential form, so that $\alpha(X)$ is an endomorphism of the trivial bundle. Another way of expressing the same thing is to say that $d_\nabla s = ds + \alpha(s)$. Here $\alpha(s)$ is interpreted as the differential form $X \mapsto \alpha(X)(s)$.

Thus if E is any vector bundle, and ∇ a connection on it, we may trivialise E in a sufficiently small open set, and express the connection as $d + \alpha$. If we take another trivialisation, it would differ from the above by a gauge transformation A of the trivial bundle. Thus the new expression for the connection is $d + A\alpha A^{-1} + (dA)A^{-1}$. (See formula 2.5.)

2.14. Proposition. *If E is a vector bundle which is trivialised on each set of an open covering (U_i), with transition functions (m_{ij}) on the overlaps $U_i \cap U_j$, then any connection ∇ on the whole of E is given by the operators $d + \alpha_i$ on U_i with the transition formulae*

$$\alpha_i = m_{ij}\alpha_j m_{ij}^{-1} - dm_{ij}.m_{ij}^{-1} = m_{ji}^{-1}\alpha_j m_{ji} + m_{ji}^{-1}dm_{ji}$$

on U_{ij}.

Proof. Choose the trivialisations $c_i : \pi^{-1}(U_i) \to U_i \times \mathbb{C}$. The given connection goes over to the trivial bundle as $d + \alpha_i$. The transition functions $m_{ij} = c_i \circ c_j^{-1}$ over U_{ij} are gauge transformations of the trivial bundle taking $d + \alpha_j$ to $d + \alpha_i$. According to formula 2.5, it transforms the trivial connection to $d - dm_{ij}.m_{ij}^{-1}$. Hence it transforms $d + \alpha_j$ to $d - dm_{ij}.m_{ij}^{-1} + m_{ij}\alpha_j m_{ij}^{-1}$.

So the connection $d + \alpha_i$ coincides with this transform if and only if the equation in the proposition is satisfied.

2.15. Remarks.

1) Gauge theory in Physics treats the space of all connections as a configuration space with the gauge group as the symmetry group.

2) The action of the gauge group is not faithful. In fact, any nonzero scalar acts trivially on the space of connections.

3) There is a natural group, which we call *outer gauge group*, containing the gauge group as a normal subgroup of index 2 and acting on $\mathcal{C}$ by affine transformations.

2.16. Exercise. If f is an everywhere nonzero function on M and it is considered as a gauge transformation of the trivial line bundle, then what is the linear part of its action?

3. Principal Bundles

There is another point of view, which is useful and more general, to look at connections. In order to explain it, we need to introduce the notion of *principal bundles*.

3.1. Definition. Let M be a differential manifold and G any Lie group. A *principal bundle with structure group G* consists of a differential manifold P, a fixed point free action (conventionally on the right) of G on P and a differentiable map $\pi : P \to M$ such that the fibres are simply orbits under the action of G. It is supposed to satisfy the 'local triviality axiom', namely, every point $m \in M$ admits an open neighbourhood U such that the map $\pi : \pi^{-1}(U) \to U$ can be identified with $U \times G \to U$ together with the G-action on the second factor.

3.2. Remark. The example when $P = M \times G$ and π is the projection to M and the action of G is given by $(p, s)g = (p, sg)$ is called the *trivial principal bundle*. This justifies the terminology 'locally trivial'.

3.3. Examples.

1) Take for P, the space $\mathbb{R}$ and for π the map of $\mathbb{R}$ onto $M = S^1$ given by $x \mapsto \exp(2\pi i x)$. In this case G is the discrete group $\mathbb{Z}$ acting on $\mathbb{R}$ by translations. This is a particular case of the universal covering space of a manifold M, on which the fundamental group acts by deck transformations. This gives a principal bundle with $\pi_1(M)$ as structure group.

2) Let V be a vector space of dimension n and $E = V \setminus \{0\}$. The group $\mathbb{C}^{\times}$ acts on it by scalar multiplication and the quotient is the complex projective space $P(V)$. Take for π the natural map of $E = V \setminus \{0\}$ onto $P = P(V)$. Let p be a point of E, namely a one-dimensional subspace p of V. If f is any linear form on V which does not vanish on p, then the open subspace $V \setminus \ker(f)$ of $V \setminus \{0\}$ is invariant under the action of $\mathbb{C}^{\times}$ and defines an open subspace U_f of P containing p. Now $\pi^{-1}(U_f) = V \setminus \ker(f)$ can be identified with the product $f^{-1}(1) \times \mathbb{C}^{\times}$ by the map $v \mapsto (v/f(v), f(v))$. Since $f^{-1}(1)$ is mapped diffeomorphically by π to U_f, we see that π is locally trivial and hence defines a principal $\mathbb{C}^{\times}$-bundle.

3) Consider the space of all r-frames in $\mathbb{C}^n$. An r-frame means here an ordered set of r linearly independent vectors. This can be made into a differential manifold, by identifying it with the set of all (n, r)-matrices of rank r obtained by writing out the r vectors as r columns in any basis (say the standard basis) in $\mathbb{C}^n$. This manifold is called the *Stiefel manifold*. To each such matrix, we may associate an element of the r-Grassmannian, namely the vector subspace generated by the column vectors. The group $GL(r, \mathbb{C})$ acts on the Stiefel manifold by multiplication on the right, and the quotient is clearly the Grassmannian. One can check that this gives a principal bundle with $GL(r, \mathbb{C})$ as the structure group.

Local description.

By definition, there is an open covering (U_i) of the space M and trivialisations $t_i : P|U_i \to U_i \times G$. Over the intersection $U_{ij} = U_i \cap U_j$ we then have two trivialisations of P. These give rise to the automorphism $A_{ij} = (\text{res } t_i) \circ (\text{res } t_j)^{-1}$ of the trivial bundle $U_{ij} \times G$ over U_{ij}. This is given by a diffeomorphism of the form $(x, g) \mapsto (x, m_{ij}(x)g)$. This follows from the fact that any map A of G into itself which commutes with right multiplication, is left multiplication by $A(1)$. The functions m_{ij} satisfy $m_{ij} = m_{ji}^{-1}$ and also, over $U_{ijk} = U_i \cap U_j \cap U_k$, the identity

$$m_{ij}.m_{jk} = m_{ik}.$$

The functions $m_{ij} : U_{ij} \to G$ are called *transition functions*. If we choose some other trivialisations t_i' over U_i, then we get a different set m_{ij}' of transition functions. The two trivialisations differ by an automorphism $\varphi_i = t_i \circ (t_i')^{-1}$ of the trivial bundle over U_i. Again φ is of the form $(x, g) \mapsto (x, f_i(x)g)$ for some function $f_i : U_i \to G$. It is clear that the two sets m_{ij} and m_{ij}' are related, through f_i, by the identity

$$m_{ij} f_j = f_i m_{ij}'$$

over U_{ij}.

The above data can be interpreted in terms of sheaf cohomology [Ch. 4, Remark 5.5, 2)]. The set of transition functions defines a Čech 1-cocycle with coefficients in the sheaf $\tilde{G}$ of differentiable functions on M with values in G. As such it defines an element of $H^1(M, \tilde{G})$. Note that $\tilde{G}$ is a sheaf of groups and not necessarily of abelian groups. Hence the first cohomology group is only a set with a special point. Conversely, given a 1-cocycle (m_{ij}), we may take the manifolds $U_i \times G$, piece them together by the diffeomorphisms $(x, g) \mapsto (x, m_{ij}(x)g)$ and define a manifold P. Now it is clear that this forms a principal bundle. Thus one easily concludes that the set of principal bundles is in bijection with the cohomology set $H^1(M, \tilde{G})$.

When $G = \mathbb{C}$ (resp. $\mathbb{C}^\times$), the sheaf $\tilde{G}$ can be identified with $\mathcal{A}$ (resp. $\mathcal{A}^\times$). Thus we may say that the set of line bundles is in bijection with $H^1(M, \mathcal{A}^\times)$. In this case, the cohomology $H^1(X, \mathcal{A}^\times)$ is actually a group, and the group operation is given by tensor product of line bundles.

3.4. Example. Consider the principal $\mathbb{C}^\times$-bundle given by the natural map $\pi : \mathbb{C}^{n+1} \setminus \{0\} \to \mathbb{CP}^n$. For every i, consider the open set U_i defined by $z_i \neq 0$, and the trivialisation $t_i : \pi^{-1}(U_i) \to U_i \times \mathbb{C}^\times$ given by $(z_1, \ldots, z_n) \mapsto ((z_0, \ldots, z_n), z_i)$. The map $t_i \circ t_j^{-1} : ((z_0, \ldots, z_n), \lambda) \mapsto ((z_0, \ldots, z_n), z_i/z_j \cdot \lambda)$ takes one trivialisation to the other. Hence the transition functions are given by the $\mathbb{C}^\times$-valued functions $m_{ij} = z_i/z_j$ on U_{ij}.

Let $\pi : P \to M$ be a principal G-bundle and $f : N \to M$ be a differentiable map. Then we may define the *pull-back* of P by f to obtain a principal G-bundle on N as follows. Consider the subspace of $N \times P$ given by $\{(x, \xi) : f(x) = \pi(\xi)\}$. The group G acts on $N \times P$ via its action on the second factor and this leaves the subspace invariant. If P were trivial, say $P = M \times G$, then this subspace would obviously be the same as $N \times G$. In general, if P is trivial, say over U, then the pull-back is therefore trivial over $f^{-1}(U)$. We can put together these local trivialisations to get a differentiable structure on the pull-back to make it a principal G-bundle. If m_{ij} is a set of transition functions defining P with respect to a covering (U_i) of M, the pull-back is given by the transition functions $m_{ij} \circ f$, with respect to the covering $(f^{-1}(U_i))$ of N. Our present interest in principal bundles stems from the fact that if P is a principal bundle with structure group G, then corresponding to any representation ρ of G, that is to say, a differentiable homomorphism of G into $GL(V)$, we obtain a vector bundle $E(P, \rho)$, called the *vector bundle associated to P by the representation ρ.*

In fact, suppose P is a principal G-bundle and ρ a representation of G in a vector space V. Consider the manifold $P \times V$. The group G acts on it by the prescription $(p, v)s = (ps, \rho(s)^{-1}(v))$. Take the quotient to be $E = E(P, \rho)$. There is a natural map of E into M which sends (p, v) to $\pi(p)$. Moreover,

over a locally trivial open neighbourhood U of any point $m \in M$, we may replace P by $\pi^{-1}(U) = U \times G$ and hence E by the quotient of $U \times G \times V$ by the action of G given by $(m, g, v)s = (m, gs, \rho^{-1}(s)(v))$. This quotient may be identified with $U \times V$ via the identification $(m, g, v) \mapsto (m, \rho(g)v)$. Thus E can be provided with the structure of a differential manifold by glueing these together. One has of course to check that it is Hausdorff. But if (p, v), (q, w) are two points, then it is obvious that they can be separated by open neighbourhoods if $\pi(p) \neq \pi(q)$. On the other hand, if $\pi(p) = \pi(q)$, then they lie over one local trivial neighbourhood and so can be separated again. Now one can check easily that E is a vector bundle over M.

On the other hand, if E is a complex (resp. real) vector bundle of rank k, then one can associate to it a principal bundle with structure group $GL(k, \mathbb{C})$ (resp. $GL(k, \mathbb{R})$) in such a way that the vector bundle is associated to the standard representation (namely the identity representation) is E.

We will now indicate this construction. Let E be any vector bundle. A *frame* is a linear isomorphism of a fibre over any point, say m of M, with $\mathbb{C}^k$. Now take the set of all frames at all points of M. If E is trivial, the set of frames can be identified with $M \times GL(k, \mathbb{C})$. This is clearly a manifold. One can easily check that by locally trivialising E over an open covering (U_i), one can glue all these manifolds together and get a principal bundle P with the required properties.

3.5. Remarks.

1) Let us start with a vector bundle E. Construct the corresponding principal $GL(k, \mathbb{C})$-bundle P. Any representation ρ of $GL(k, \mathbb{C})$, gives rise to the associated vector bundle $E(P, \rho)$. Thus we have devised a method of associating to a vector bundle of rank k and a representation ρ of $GL(k, \mathbb{C})$, another vector bundle which we may as well denote $\rho(E)$. For example, we have already mentioned that if ρ is the identity representation, then the associated vector bundle is E itself. If we take the r-fold tensor representation of $GL(k)$, it is easy to see that the associated vector bundle is actually $\bigotimes^r(E)$.

2) The association of a vector bundle to a representation can be carried out in greater generality. If the structure group G acts on any manifold F, we can associate a bundle over M which is locally a product with F.

3) Any representation of the fundamental group π_1 gives rise to a vector bundle, namely the vector bundle associated to the universal covering space, viewed as a principal bundle with π_1 as the structure group. Any covering space of a manifold is in fact a bundle associated to the universal covering space considered as a principal $\pi_1(M)$-bundle for its action on the coset space modulo a subgroup H.

4) If F has any additional structure which is preserved under G, then the fibres of the associated fibration over M inherit this structure and the local trivialisations respect this structure. If ρ is a representation of G, then the fibre is a vector space, and since the vector space structure is preserved under G, the associated bundle acquires the structure of a vector bundle.

5) Consider the adjoint representation of a Lie group in its Lie algebra. It preserves the Lie algebra structure. Hence the associated vector bundle, which is called the *adjoint bundle* (and is denoted $\mathrm{ad}(P)$), carries a Lie algebra structure on all its fibres.

6) If G is a Lie group and H a closed subgroup, then the natural right action of H on G gives a principal H-bundle $G \to G/H$.

Remark 4) above may be applied to the following situation. Let P be a principal G-bundle and $G \to H$ a homomorphism of Lie groups. We can then associate a bundle with H as fibre, since G acts naturally on H on the left. But the action does not respect the group structure on H so that the fibres do not come equipped with a group structure. However, if we make H act on itself on the right by translations, then the left action of G on H via the given homomorphism, commutes with the right H-action on itself. Hence the associated bundle with H as fibre comes with an action on the right by H, and it is easy to see that this makes it a principal H-bundle.

3.6. Definition. Let $f : G \to H$ be a homomorphism of Lie groups. If P is a principal G-bundle, then the above procedure gives rise to a principal H-bundle. It is called the *bundle obtained by extension of the structure group through f*. We will often drop the phrase 'through f', particularly when f is an inclusion.

3.7. Remark. Even if the map f is not injective, the phrase of extending the structure group is still used.

3.8. Example. Consider any principal bundle P with a *discrete* group G as structure group. Then fixing a point m in M and a point p over it in P, we get a homomorphism of $\pi_1(M)$ into G as follows. Lift any loop at m to a path in P with p as origin. By the monodromy theorem the end point of the lift depends only on the homotopy class of the loop. Thus we get a map f of $\pi_1(M,m)$ into G defined by mapping any homotopy class of loops at m onto the unique element g which takes p into the end point of the lift of any path in this class.

3.9. Exercise. Assume that G is a discrete group and that P, a connected topological space, is a principal G-bundle over M. Show that the associated G-bundle obtained by extending the structure group via f in the above example is isomorphic to P.

If we *start* with a principal H-bundle Q and $f : G \to H$ is a homomorphism, there is no reason to expect that there is a principal G-bundle P such that Q is obtained from P by extending the structure group through f. For example, take $G = \{1\}$ and f to be the inclusion. Then any principal G-bundle is trivial, and so also is any bundle obtained by extension of the structure group. But there is no reason why an arbitrary H-bundle should be trivial!

3.10. Definition. Let $f : G \to H$ be a homomorphism of Lie groups. If P is a principal H-bundle, then we say that the structure group can be *reduced to G through f* if there exists a principal G-bundle Q such that P is isomorphic to the bundle obtained from Q through extension of the structure group through f. If f is surjective we sometimes say that the structure group *can be lifted to G*.

One may think of reduction of the structure group as providing the bundle with some additional structure. We will presently formalise this but we will give some examples first. The claims made in these examples will be proved in 3.14.

3.11. Examples.

1) Let E be a vector bundle and P the associated principal $GL(k, \mathbb{R})$-bundle of frames. To provide a metric on all fibres, i.e., a differentiable section of the associated bundle $S^2(E^*)$, which is positive definite on each fibre, is equivalent to giving a reduction of the structure group to $O(k, \mathbb{R})$. Such a structure will be simply referred to as a *metric* on the vector bundle.

2) In the above example, we may require, instead of positive definiteness of the symmetric form on each fibre, nondegeneracy with a given signature, say, (k, l). Here it is a question of reduction of the structure group to $O(k, l)$.

3) Suppose that there is a nondegenerate alternating form on each fibre, or a differentiable section of $\Lambda^2(E^*)$ which is nondegenerate on all fibres. It is clear that this is not always possible. For one thing, E has to be of even rank $2k$, since there are no nondegenerate alternating forms on vector spaces of odd dimension. But there are other topological obstructions as well. This is the same as giving a reduction of the structure group to the symplectic subgroup $Sp(k)$.

4) If E is a real vector bundle of rank $2k$, then in order to provide it with a complex vector bundle structure, what we need is a section of $\text{End}(E)$ whose square is $-(\text{Id})_E$. This would give a reduction of the structure group from $GL(2k, \mathbb{R})$ to $GL(k, \mathbb{C})$.

3.12. Remark. There is a significant difference between Example 1) and the rest in that, while all vector bundles do admit a positive definite metric, an arbitrary vector bundle may not admit reductions to other subgroups mentioned above. As for 1), we note that if we trivialise the bundle locally and provide the trivial bundle with a metric structure, then these can be glued by the use of a differentiable partition of unity, thanks to the fact that a convex combination of metrics is also a metric. This last assertion is not valid for forms of other signatures. For example in the case of a bundle of rank 2, the existence of a nondegenerate form of signature $(1, 1)$ would imply that the fibres E_m can be written as a direct sum of two subspaces, namely the two isotropic subspaces for the form. The tangent bundle of S^2 does not admit such a form with signature $(1, 1)$. In Examples 3) and 4), we need at least that the rank of the bundle is even.

3.13. Examples.

1) Let G be a Lie group and H a closed subgroup. Consider the principal H-bundle $G \to G/H$. If we now extend the structure group to G, then the new bundle is given by $G \times G$ modulo the action of H on it by $(x, y)h = (xh, h^{-1}y)$. But this quotient can be identified with $G/H \times G$ by means of the map $G \times G \to G/H \times G$ given by $(x, y) \mapsto (xH, xy)$. In other words, the H-bundle $G \to G/H$ which could well be nontrivial, becomes trivial on extension of the structure group from H to G! We may also say that the trivial G-bundle has a reduction to H which is nontrivial. On the other hand the trivial H-bundle can also be obtained by reduction from it. Thus two different reductions of the same G-bundle to a subgroup H, may yield nonisomorphic H-bundles.

2) Suppose

$$0 \to E' \to E \to E'' \to 0$$

is an exact sequence of vector bundles with $rk(E') = k$ and $rk(E'') = l$. Instead of considering all frames, namely isomorphisms of E_m, $m \in M$, with a fixed vector space, say $\mathbb{R}^{k+l}$, we may restrict ourselves only to those frames which take E'_m into $\mathbb{R}^k$. We get a principal bundle all right but the structure group is now the set of all linear automorphisms of $\mathbb{R}^{k+l}$ which leave the subspace $\mathbb{R}^k$ invariant. In other words, the structure group has been reduced in this case to the so-called 'parabolic' group, $\{g \in GL(k+l, \mathbb{R}) : g_{ij} = 0, \text{ for } i > k, j \leq k\}$.

3.14. Proposition. *There is a one-one correspondence between reductions of the structure group of P from G to a subgroup H and sections of the associated bundle with G/H (with the obvious G-action) as typical fibre.*

Proof. In fact, in this case, the associated bundle has a very simple description, namely $P/H \to M$. Indeed, the map of $P \times G/H$ to P/H taking (p, gH) to the H-orbit of pg takes $(ps, s^{-1}gH)$ to the same orbit and goes down to a map of the associated bundle as we have defined, onto P/H. The map which takes the H-orbit of p to (p, H) provides the inverse. It is obvious then that the map $P \to P/H$ is itself a principal H-bundle. Now any section $s : M \to P/H$ can be used to pull back this H-bundle to an H-bundle on M. It is easy to check that the G-bundle obtained from it by extension of the structure group is simply P. Conversely, if Q is any reduction of P to an H-bundle, then we have an isomorphism $(Q \times G)/H \to P$ of principal G-bundles. Of course, this yields an isomorphism of associated bundles with typical fibre G/H. But this is given on the left side by passing to a further quotient of $(Q \times G)/H$ by the right action of H. This is easily identified with the trivial bundle $M \times (G/H)$. Now the canonical section of this bundle gives a section of P/H.

3.15. Remark. The claim in Examples 3.11 is equivalent, after the above proposition, to giving sections of associated bundles with fibres $GL(n)/O(n)$, $GL(n)/Sp(n)$, etc. These reductions in the case of the tangent bundle will be studied in Chapter 7.

In order to write down things in an analytical fashion, we often trivialise a bundle on a sufficiently small open set. The approach of principal bundles allows us to do something similar. This is somewhat neater, and the glueing data gets replaced by 'invariance under G'. It is rather like taking the set of *all* trivialisations, instead of taking one trivialisation, and getting analytic expressions in terms of each of them, with appropriate transformation formulae for a change of trivialisations. This is in some sense closer to the classical use of 'quantities' in physics.

Since the principal bundle consists of 'all frames' (assuming it came from a vector bundle E), it follows that when E is pulled back to P itself, it admits a frame at every point, that is to say that E pulls back to a trivial bundle. We may rephrase this by saying that when the principal bundle P is pulled back to the manifold P by π, then it is trivial. Although this is obvious, in view of its importance we will enunciate it as a proposition.

3.16. Proposition. *The pull-back of a principal G-bundle $\pi : P \to M$ by π is canonically trivial.*

Proof. This assertion is almost a tautology. Note that the pull-back is given by the subspace $\{(p_1, p_2) : \pi(p_1) = \pi(p_2)\}$ of $P \times P$. The trivial principal bundle $P \times G$ on the manifold P is mapped isomorphically to the pull-back by the map $(p, g) \mapsto (p, pg)$.

Let ρ be a representation of G and $E = E(P, \rho)$ the associated vector bundle. If σ is a section of E, then its pull back to the manifold P gives a section of $\pi^*(E)$. But then this is associated to $\pi^*(P)$, which is canonically trivial. Hence the pull-back of σ to P may be considered a section of the trivial vector bundle or, what is the same, a vector-valued function, on P. Conversely, let f be a V-valued function on P satisfying $f(pg) = \rho(g)^{-1} f(p)$ for all $p \in P$ and $g \in G$. Then the section of the trivial bundle taking p to $(p, f(p))$ of $P \times V$ induces a section σ_f of $P/G = M$ to $(P \times V)/G = E$. We may sum this up as follows.

3.17. Proposition. *Let P be a principal G-bundle, ρ a representation of G in a vector space V and $E = E(P, \rho)$ the associated vector bundle. Then there is a canonical isomorphism between the vector space of sections of E and the vector space of V-valued functions f on P that satisfy $f(pg) = \rho(g)^{-1} f(p)$ for all $p \in P$ and $g \in G$.*

In the same way the following statement may also be verified.

3.18. Proposition. *Under the same assumptions as above, there is a canonical bijection between the space of differential r-forms on M with values in the vector bundle $E(P, \rho)$ and the space of differential r-forms ω on P with values in V such that*

i) $\omega(v_1, \ldots, v_r) = 0$ if at least one of the vectors is vertical, i.e. mapped by π to 0;

ii) If $w_1, \ldots, w_r$ are tangent vectors at pg obtained as translates by g of tangent vectors $v_1, \ldots, v_r$ at p, then $\omega(w_1, \ldots, w_r) = \rho(g)^{-1} \omega(v_1, \ldots, v_r)$.

4. Connections on Principal Bundles

Let S be any manifold on which a Lie group G acts (on the right) simply transitively. Then the tangent bundle of S is of course trivial, since G can be identified with S on choosing a point s_0 in S, by means of the map $g \mapsto s_0 g$. It is even *canonically* trivial. Indeed, we have an isomorphism of the trivial vector bundle $S \times \mathfrak{g}$ over S with the tangent bundle given by $(s, X) \mapsto v$, where v is the image of X by the differential of the map $g \to sg$ at 1.

In particular, there is a linear projection of the tangent bundle of S into $\mathfrak{g}$. This may be considered as a $\mathfrak{g}$-valued 1-form on S.

4.1. Definition. Let G be a Lie group and $\mathfrak{g}$ its Lie algebra. Let S be any differential manifold on which G acts simply transitively on the right. Then there is a canonical 1-form on S with values in $\mathfrak{g}$, which associates to any tangent vector v at $s \in S$, the unique vector X at 1 which is mapped by the differential of $g \mapsto sg$ to v. This form is called the *Maurer-Cartan form* on S. In particular, if $S = G$ with G acting on itself by right translations, it is called the *Maurer-Cartan form* on G. We will denote this generally by m_G.

4.2. Definition. Let G be a Lie group and P a principal G-bundle. Then a *connection form* on P is a 1-form on the manifold P with values in the Lie algebra $\mathfrak{g}$ whose restriction to the fibres (which are all submanifolds of P on which G acts simply transitively) are Maurer-Cartan forms and which is equivariant for the action of G on 1-forms on P on the one hand and the right adjoint action on $\mathfrak{g}$ on the other.

4.3. Remark. Any two connection forms on P differ by a 1-form on P with values in $\mathfrak{g}$ which vanishes on vertical vector fields and is equivariant for the action of G on P on the one hand and the adjoint action on $\mathfrak{g}$ on the other. In other words, they differ by a 1-form on the *base M* with values in the associated bundle $\mathrm{ad}(P)$ with $\mathfrak{g}$ as fibre (considered as a $\mathfrak{g}$-valued form on P). Thus the space of all connections on P is an affine space based on the vector space of $\mathrm{ad}(P)$-valued 1-forms on M.

We will show that a connection form γ on a principal bundle P with structure group G, gives rise to covariant differentiation on any vector bundle associated to a representation ρ of G in a vector space V. We will denote the induced representation of the Lie algebra $\mathfrak{g}$ also by ρ. As we have seen above, sections of the associated vector bundle may be identified with functions f on P which satisfy $f(sg) = \rho(g)^{-1}(f(s))$ for all $s \in P$ and $g \in G$. We will first show the following.

4.4. Lemma. *If v is any vertical tangent vector at $p \in P$, i.e. $\pi : P \to M$ maps v to 0, and f is a ρ-equivariant function with values in V, then* $vf + \rho(\gamma(v))f(p) = 0.$

Proof. Note that the map $g \mapsto pg$ induces an isomorphism of G with the fibre through p, and the Lie algebra $\mathfrak{g}$ with the space of vertical vectors at p. Therefore the assertion is equivalent to the following. If $f : G \to V$ is given by $f(g) = \rho(g)^{-1}x$, and $X \in \mathfrak{g}$, then $Xf = -\rho(X)x$, which is obvious.

This means that if v is now any tangent vector, and we consider the map $f \mapsto vf + \rho(\gamma(v))(f)$, then this value depends only on v modulo the vertical tangent space at p or, what is the same, only on the image by π of v in the tangent space at $m = \pi(p)$. Next we will investigate the dependence of this value on the choice of the point p over m. Suppose then that $q = pg$ is another point of P over m. If w is the g-translate of v, then $wf = v(h)$, where h is the function defined by $h(x) = f(xg)$. Since $f(xg) = \rho(g)^{-1}f(x)$ we have $wf = \rho(g)^{-1}vf$. On the other hand, $\rho(\gamma(w))(h) = \rho(g)^{-1}(\gamma(v)g)(\rho(g)^{-1})f = \rho(g)^{-1}(\gamma(v)f)$. In other words, for any tangent vector v at a point $m \in M$, we may define $\nabla_v(\sigma_f)$ by taking *any* lift of v at *any* point $p \in P$ over m and setting its value to be $vf + \rho(\gamma(v))(f)(p)$. We will skip the routine check that this does define a covariant differentiation. In sum we have the following theorem.

4.5. Theorem. *Any connection form on a principal G-bundle, gives rise canonically to a covariant differentiation in every associated vector bundle.*

In view of this theorem, the datum of a connection form on a principal bundle is often said to be a *connection* on it.

4.6. Remark. Let $\pi : P \to X$ be a principal G-bundle and $f : Y \to X$ a differential map. Clearly there is a G-equivariant differentiable map $\tilde{f}$ of the pull-back $f^*(P)$ into P which induces f on the base spaces. Therefore, if γ is a connection form on P, we may take its inverse image $\tilde{f}^*(\gamma)$ on $f^*(P)$. Since the fibre of $f^*(P)$ at $y \in Y$ is canonically the fibre at $f(y)$ of P, the restriction of $\tilde{f}^*(\gamma)$ to this fibre is the Maurer-Cartan form. Also the equivariance of $\tilde{f}^*(\gamma)$ with reference to the action of G on $f^*(P)$ and that on $\mathfrak{g}$ follows from the G-equivariance of $\tilde{f}$. Hence it is also a connection form. It is called the *pull-back connection* on $f^*(P)$.

4.7. Local expression.

Consider the trivial principal bundle $M \times G$. The connection form on it corresponding to the trivial connection is $p_2^*(m_G)$, where m_G is the Maurer-Cartan form on G. Any connection form is given by $\mathrm{Ad}(g)^{-1} \circ p_1^* \omega + p_2^*(m_G)$, where ω is any 1-form on M with values in $\mathfrak{g}$. The form ω which determines the connection is obtained by restricting the connection form to $M \times \{1\}$. In other words, if s is the section $x \mapsto (x, 1)$ then ω is the pull-back of the connection form by s.

Let P be a principal G-bundle and γ a connection on it. Let (U_i) be an open covering and s_i sections over U_i given by trivialisations t_i of the bundle over U_i. If l_{ij} are transition functions given by the above trivialisations, then we have $s_j(x) = s_i(x).l_{ij}(x)$ over U_{ij}. Note that the action map $P \times G \to P$

pulls γ back to $\mathrm{Ad}(g)^{-1} \circ p_1^*(\gamma) + p_2^*(m_G)$. Here the action by $\mathrm{Ad}(g)^{-1}$ is by the action on $\mathfrak{g}$. The pull-backs ω_i, ω_j of γ by s_i and s_j respectively, are therefore related by the formula

$$\omega_j = \mathrm{Ad}(l_{ij}^{-1}) \circ (\omega_i) + l_{ij}^*(m_G).$$

Compare this with the local expression for connections in vector bundles given in 2.14.

4.8. Remarks.

1) In order to define differentiation of a section along a vector at m, we took an *arbitrary lift* at a point p over m in the above procedure. However, note that the differential γ_p, which is a linear map of T_p into $\mathfrak{g}$, maps the vertical subspace isomorphically to $\mathfrak{g}$, and hence that its kernel (which we will call the *horizontal space* at p) is supplementary to the vertical space. So it is also mapped by π isomorphically to $T_m(M)$. It makes sense then to lift the given vector v in $T_m(M)$ to a vector in the *horizontal subspace* at p. Thus, given the connection form, there is a canonical lift of any vector at m to any point in P over it. Moreover, by the invariance property of γ we also conclude that if p and $q = pg$ are two points over m, the horizontal lift at q is obtained by translating the horizontal lift at p by g. From this it is also clear that any differentiable vector field on M admits a unique horizontal lift, that it is differentiable and that it is invariant under the action of G. Suppose then X is a vector field on M and hX its horizontal lift to a vector field on P. Our definition of covariant differentiation amounts to saying that $\nabla_X(\sigma_f)$ corresponds to the equivariant function $(hX)(f)$.

2) If F is a differential manifold on which G acts differentiably, consider the associated bundle with F as fibre. If (ξ, v) is a point of $P \times F$, then we have a tangent space decomposition $T_{(\xi,v)} = T_\xi \oplus T_v$. It contains the horizontal space H_ξ. The differential of the map $P \times F \to P \times_G F$ maps H_ξ injectively since it is mapped isomorphically when composed with further projection to M. Thus at any point of the associated bundle, we have a subspace of the tangent space mapping isomorphically to the tangent space in M. In other words, it is a space supplementary to the vertical space and so may still be called the *horizontal space.*

3) If ∇ is a connection on a vector bundle, then we can use the horizontal space in the dual bundle in order to define the covariant derivative. In fact, consider any section s of E. It can be considered as a function f on E^* which is linear on fibres. Given any tangent vector v at a point $m \in M$, we may lift it uniquely to a horizontal tangent vector t at any point ξ on the fibre over m in E^*. Then tf is a scalar, and the association $\xi \mapsto tf$ is again a linear map of the fibre and so gives an element of the dual of E_m^*, namely E_m. This is the value of $\nabla_v s$ at m.

4.9. Exercise. Prove the statement in 3) above.

4.10. Extension of the structure group and connection forms.

If $f : G \to H$ is a homomorphism of Lie groups, then it induces a homomorphism $\mathfrak{f} : \mathfrak{g} \to \mathfrak{h}$ of Lie algebras. Let γ be a connection form on a principal G-bundle P. We can then define a connection form on the principal H-bundle Q obtained by extending the structure group through f. In fact, consider on $P \times H$ the $\mathfrak{h}$-valued 1-form given by $p_1^*(\omega) + p_2^*(m_H)$, where ω is the 1-form obtained by composing $p_1^*(\gamma)$ with $\mathfrak{f}$, and m_H is the Maurer-Cartan form on H. This is of course a connection form on the trivial H-bundle $P \times H$ over P.

Both forms $p_1^*(\omega)$ and $p_2^*(m_H)$ are equivariant for the (diagonal) action of G and the adjoint action of G on $\mathfrak{h}$ via f. Hence so is the sum. If we evaluate $p_1^*(\omega)$ on a vector v tangential to an orbit of G, and identify v with an element of $\mathfrak{g}$, we get $p_1^*(\omega)(v) = \omega(pv) = f(\gamma(v)) = f(m_G(v)) = f(v)$. The action of G on H being via $f(g)^{-1}$, the term $p_2^*(m_H)(v)$ gives the value $-f(v)$. Therefore $p_1^*(\omega) + p_2^*(m_H)$ goes down to an $\mathfrak{h}$-valued 1-form on the quotient Q of $P \times H$ by G. It is easily seen now that it is a connection form on the H-bundle Q. We refer to it as the *connection form on Q obtained by extension of the structure group through f*.

If E is a vector bundle, then we have seen that a connection on it gives rise to a connection form on the principal $GL(n)$-bundle. Conversely, if P is the principal $GL(n)$-bundle associated to a vector bundle V, then a connection form on P gives rise to a connection on V, by Theorem 4.5.

In particular, if we start with a vector bundle with a connection, the frame bundle gets a connection form, and hence all the vector bundles associated to it via representations of $GL(n)$ inherit connections. It is a routine matter to check that the satellite connections that we defined on E^*, $S^i(E)$, etc. in 1.13 can all be identified with the connections so obtained.

Let P be a principal G-bundle and $f : G \to H$ a homomorphism of Lie groups. Giving a connection form on the H-bundle obtained by extension of the structure group, is entirely equivalent to giving an $\mathfrak{h}$-valued form on P which is G-equivariant and restricts to $\mathfrak{f} \circ \mu_G$ on fibres. We may refer to this form as an *H-connection form on P*, although its structure group is only G. Now we have the following exact sequence of vector spaces:

$$\mathfrak{g} \to \mathfrak{h} \to \mathfrak{h}/\operatorname{im}\mathfrak{g} \to 0.$$

This is actually a sequence of G-modules. If γ is an H-connection on the G-bundle P, then we may compose it with the natural map of $\mathfrak{h}$ to $\operatorname{coker}(\mathfrak{f}) = \mathfrak{h}/\operatorname{im}\mathfrak{g}$ to get a 1-form which

    a) is G-equivariant, and

    b) *vanishes* on vertical vectors.

In other words, it gives a 1-form with values in the vector bundle associated to G for the representation of G on $\mathfrak{h}/\mathfrak{g}$.

4.11. Definition. Let P be a principal G-bundle and $f : G \to H$ a homomorphism of Lie groups. Let γ be an H-connection on P. Then the 1-form with values in the associated bundle with $\operatorname{coker}(\mathfrak{f})$ as typical fibre, is called *the second fundamental form.*

4.12. Example. Let

$$0 \to E' \to E \to E'' \to 0$$

be an exact sequence of vector bundles, with $\operatorname{rk}(E') = l$ and $\operatorname{rk}(E'') = k$. We have already seen in Example 3.13, 2), that this amounts to giving a reduction of the structure group to the *parabolic subgroup* G of $H = GL(k + l)$, consisting of linear maps which leave the subspace spanned by the first l basis vectors invariant.

Now a connection on E which leaves E' invariant (in the sense that for any vector field X, $\nabla_X(s)$ is a section of E' whenever s is one) may be considered as a G-connection on the principal bundle of frames of E. But if we are only given a connection on E, that is to say, an H-connection, then we may restrict ∇_X to sections of E' and project to E'' to get sections of E''. Let f be a function, p the projection $E \to E''$ and s a section of E'. Then we have $p(\nabla_X(fs)) = p(Xf.s + f\nabla_X s) = fp(\nabla_X s)$. Thus the map $s \mapsto p(\nabla_X s)$ is an $\mathcal{A}$-linear map from E' into E''. In other words, we obtain in this situation, a differential 1-form on M with values in the vector bundle $\operatorname{Hom}(E'', E')$.

The quotient $\mathfrak{h}/\mathfrak{g}$ can be identified with the space of (k, l) matrices. The associated bundle can then be identified with the bundle $\operatorname{Hom}(E', E'')$, and the second fundamental form with the above 1-form.

5. Curvature

5.1. Definition. Let E be a vector bundle and ∇ a connection on it. The exterior 2-form $R_\nabla(X, Y)$ with values in $\operatorname{End}(E)$ given by $(X, Y) \mapsto \nabla_X \circ \nabla_Y - \nabla_Y \circ \nabla_X - \nabla_{[X,Y]}$ is called the *curvature form* of the connection. We will drop the subscript ∇ in R_∇ when we judge that the connection in question is obvious from the context.

5.2. Remarks.

1) We have already remarked in 1.14 that a connection endows the $\mathcal{A}$-module $\mathcal{E}$ with the structure of a $\mathcal{C}$-module. The map $(X, Y) \mapsto R(X, Y)$ of $\mathcal{T}(M) \times \mathcal{T}(M)$ into the ring $\mathcal{C}(M)$ is $\mathcal{A}(M)$-bilinear and the elements $R(X, Y)$ commute with $m(f), f \in \mathcal{A}(M)$. Elements of $\mathcal{C}(M)$ which commute with all elements $m(f)$ constitute an $\mathcal{A}(M)$-algebra. The map $(X, Y) \mapsto R(X, Y)$ is thus an alternating 2-form on $\mathcal{T}(M)$ with values in this algebra and gives rise to an $\mathrm{End}(E)$-valued 2-form on every $\mathcal{C}$-module $\mathcal{E}$.

2) We will see in [Ch. 7, Theorem 4.7] that if the manifold is provided with a Riemannian metric (for example, if it is a submanifold of $\mathbb{R}^n$), then one can define a natural connection on its tangent bundle. The associated curvature form measures the intrinsic curvature of the space. This is the origin of the terminology 'curvature'.

We would like now to find a local expression for the curvature form of a connection in terms of its local expression. This only amounts to computing $R(X, Y)$ in the case of a connection on the trivial bundle given by $d_\nabla = d + \alpha$. Before doing this, we will explain the notation we will adopt in giving this formula.

Let us suppose that there exists an alternating bilinear product b on a vector bundle E with values in another bundle F. If α is a 1-form with values in E, then one denotes by $b(\alpha, \alpha)$ the exterior 2-form with values in F which takes (X, Y) into $b(\alpha(X), \alpha(Y))$. In particular if α is a 1-form with values in $\mathrm{End}(E)$ (or for that matter, any Lie algebra) then one may use the Lie algebra bracket in $\mathrm{End}(E)$ to provide a meaning to $[\alpha, \alpha]$. On the other hand one may also define $\omega \wedge \omega$, using the composition in $\mathrm{End}(E)$. These lead to the same result. But if we use the Lie algebra bracket and use it to define the wedge product, then the resulting form would be twice the form $[\alpha, \alpha]$.

5.3. Proposition. *Let ∇, ∇' be two connections on a vector bundle V and R, R' their curvature forms. Then we have*

$$R' = R + d_\nabla \alpha + [\alpha, \alpha],$$

where α is the $\mathrm{End}(V)$-valued 1-form defined by $X \mapsto \nabla'_X = \nabla_X + \alpha(X)$.

Proof. In fact, let s be a section of V. Then

$$
\begin{aligned}
R'(X,Y)(s) &= (\nabla_X + \alpha(X))(\nabla_Y s + \alpha(Y)(s)) \\
&\quad - (\nabla_Y + \alpha(Y))(\nabla_X s + \alpha(X)(s)) \\
&\quad - (\nabla_{[X,Y]}(s) + \alpha([X,Y])(s)) \\
&= \nabla_X \nabla_Y(s) + \alpha(X)(\nabla_Y s) + \nabla_X(\alpha(Y)(s)) + \alpha(X)(\alpha(Y)(s)) \\
&\quad - \nabla_Y \nabla_X(s) - \alpha(Y)(\nabla_X s) - \nabla_Y(\alpha(X)(s)) \\
&\quad - \alpha(Y)(\alpha(X)(s)) - \nabla_{[X,Y]}(s) - \alpha([X,Y])(s) \\
&= R(X,Y)s + \alpha(X)(\nabla_Y s) + \nabla_X(\alpha(Y)(s)) - \alpha(Y)(\nabla_X s) \\
&\quad - \nabla_Y(\alpha(X)(s)) + [\alpha(X),\alpha(Y)](s) - \alpha([X,Y])(s) \\
&= R(X,Y)s + [\nabla_X, \alpha(Y)](s) - [\nabla_Y, \alpha(X)](s) \\
&\quad - \alpha([X,Y])(s) + [\alpha(X),\alpha(Y)](s) \\
&= R(X,Y)s + (d_\nabla \alpha + [\alpha,\alpha])(X,Y)(s),
\end{aligned}
$$

proving our assertion.

The above theorem says that

a) If the connection form on the principal bundle of frames of V is given by γ, then the curvature form is given by $d\gamma + [\gamma, \gamma]$.

b) If the bundle E is trivialised on an open covering $(U)_i$ and is given by transition functions m_{ij} on U_{ij}, and the connection is given by the forms α_i on the open sets (U_i), then the curvature form is given by $d\alpha_i + [\alpha_i, \alpha_i]$ on (U_i).

5.4. Remark. We may now adopt the principal bundle point of view. Let γ be a connection form on P. Then the curvature form is *defined* to be the $\mathfrak{g}$-valued 2-form $d\gamma + [\gamma, \gamma]$ on the principal bundle. This can be viewed as a 2-form on M with values in the adjoint bundle. If ρ is any representation of G in a vector space V and we take the associated connection on $E(P, \rho)$, then its curvature form is obtained by composing the curvature form of the G-connection with the induced map $\mathfrak{g} \to \mathrm{End}(V)$ of Lie algebras.

5.5. Curvature of associated bundles.

Let ∇ be a connection in a principal bundle P. If $f : G \to H$ is a homomorphism and Q is the bundle obtained by extension of the structure group, then we have seen that Q comes equipped with a connection as well. What is its curvature form? The homomorphism f gives rise to a homomorphism $\mathfrak{f}$ of $\mathfrak{g}$ to $\mathfrak{h}$. If we consider the adjoint action of G on $\mathfrak{g}$, and the composite of f and the adjoint action of H on $\mathfrak{h}$, then the homomorphism $\mathfrak{f}$ respects this action of G. Hence we have an induced homomorphism of vector bundles from $\mathrm{ad}(P)$ into $\mathrm{ad}(Q)$. The curvature form of ∇ is a 2-form with values in $\mathrm{ad}(P)$. Composing with the above homomorphism we get an $\mathrm{ad}(Q)$-valued 2-form. It follows from Remark 5.4 that it is actually the curvature form of the extended H-connection on Q.

Flat connections.

5.6. Definition. A connection on a vector bundle E (or on a principal G-bundle) is said to be *flat* or *integrable* if the curvature form is identically zero.

5.7. Remarks.

1) Sometimes this is defined to be a *locally flat* connection.

2) It follows from the definition that a flat connection form γ on a principal bundle satisfies
$$d\gamma + [\gamma, \gamma] = 0.$$
If $X_i, i = 1, 2$, are horizontal vector fields, then we have $\gamma(X_i) = 0$, so that the above equation implies that $\gamma([X_1, X_2]) = 0$. Hence $[X_1, X_2]$ is also horizontal. In other words, the subbundle of $T(P)$ given by the horizontal subspaces at each point, is integrable. Conversely assume that the bundle of horizontal subspaces is integrable. The curvature R of the connection is the 2-form on P given by $d\gamma + [\gamma, \gamma]$. Let us evaluate it on two vector fields X_1, X_2 on P. It is zero whenever one of the arguments is vertical. Hence we can conclude that γ is flat if we show that

$$R(X_1, X_2) = X_1(\gamma(X_2)) - X_2(\gamma(X_1)) - \gamma([X_1, X_2]) - [\gamma(X_1, \gamma(X_2))]$$

vanishes whenever both X_1 and X_2 are horizontal. But then $\gamma(X_i)$ are both zero, and under our assumption, $[X_1, X_2]$ is also horizontal so that $\gamma([X_1, X_2])$ is also zero. Hence the curvature form is identically zero. Thus flatness of a connection is equivalent to the integrability of the horizontal bundle.

Moreover, Frobenius theorem [Ch. 2, Theorem 4.2] ensures that through any point ξ in P, there exists a locally closed submanifold of P which is integral for the horizontal subbundle. The projection map $\pi : P \to M$ induces an isomorphism of the tangent space at p to the integral manifold, with the tangent space of M at $\pi(\xi)$. Hence π restricts (in a possibly smaller neighbourhood) to this submanifold as a diffeomorphism onto an open submanifold of M. In other words, there exists a local section at $\pi(\xi)$ which trivialises the bundle, and pulls back the connection form to 0. Thus flatness of the connection is equivalent to the local trivialisability of the bundle *as well as* the connection. Any local section s of P over an open set U trivialises the connection if and only if its differential maps the tangent spaces

$T_m(M)$ of M into horizontal spaces at the image $s(m)$, for all $m \in U$. If s is one such section, then sections of the form $t : m \mapsto s(m)g$ for some $g \in G$ also have the same property. By the uniqueness of integral manifolds, if two local sections which trivialise the connection coincide at a point, then they coincide in a neighbourhood.

Thus we have

5.8. Proposition. *If P is a principal G-bundle provided with a flat connection ∇, then every point $m \in M$ has a neighbourhood U and a trivialisation of both P and ∇ over U. If U is connected, then any two trivialisations differ by translation by an element of G.*

If we take for P a trivial bundle, the above conclusion can be restated as follows.

5.9. Lemma. *Let α be a 1-form on a differential manifold M with values in $\mathrm{End}(V)$ (or more generally with values in the Lie algebra $\mathfrak{g}$ of a Lie group G), satisfying*

$$d\alpha + [\alpha, \alpha] = 0.$$

Then every point on M has a neighbourhood U and a function φ on U with values in $GL(V)$ (or G) such that α is the pull-back of the Maurer-Cartan form m_G on G by φ. Moreover, if φ and ψ are two such functions on the same domain, then there exists $g \in G$ such that $\psi(m) = \varphi(m)g$ for all m in the domain.

5.10. Corollary. *If a principal G-bundle has a flat connection ∇, then it is given by locally constant transition functions with respect to an open covering.*

Proof. Let (U_i) be a covering with trivialisations of P as well as the connection on these open sets. On the intersections, the two trivialisations differ by locally constant transition functions, by the uniqueness statement above.

Representations of the fundamental group.

5.11. Proposition. *If a vector bundle is given by a representation of the fundamental group, then the induced connection on it is flat. The same is true of the principal G-bundle obtained by extension through a homomorphism $f : \pi_1 \to G$.*

Proof. Note that the universal covering space, considered as a principal bundle with the fundamental group (with the discrete topology) as structure group, admits the connection form 0. Therefore the connection obtained by extension of the structure group is also flat.

We have the following converse.

5.12. Theorem. *If E is a vector bundle on M, and ∇ a flat connection on it, then there is an isomorphism of E with the vector bundle E_ρ associated to a representation ρ of the fundamental group of M which takes ∇ to the natural connection on E_ρ. A similar statement is also true for a principal bundle with G as structure group.*

Proof. In view of Corollary 5.10, our assertion follows from the next proposition.

5.13. Proposition. *If a principal bundle on a connected differential manifold is defined by transition functions which are constant, then it is associated to the universal covering space, via a homomorphism of the fundamental group into the structure group.*

Proof. If we provide G with the discrete topology, the transition functions are still continuous. Hence we may construct a principal bundle Q with discrete structure group. Note that each connected component is now a covering space of X. Now using the theory of covering spaces, we get a homomorphism of the fundamental group into G and a map of the universal covering into Q inducing an isomorphism of the bundle obtained by extension of the structure group to G, with Q.

We have seen that associated with any connection on a vector bundle E, there is an 'exterior derivative' as a differential operator of order 1 from $\Lambda^{i-1}(T^*) \otimes E$ into $\Lambda^i(T^*) \otimes E$. If P is a principal G-bundle with a connection form γ, then for any representation ρ in a vector space V, we can define exterior derivatives (with respect to the connection) of forms with values in the associated vector bundle. If α is an E_ρ-valued r-form, it corresponds to an equivariant r-form β on P (vanishing whenever one of the vectors is vertical). Then by definition, we have

$$d_\nabla \beta(X_1, \ldots, X_{r+1})$$
$$= \sum (-1)^{i+1} \nabla_{X_i}(\beta(X_1, \ldots, \hat{X}_i, \ldots, X_{r+1}))$$
$$+ \sum (-1)^{i+j} \beta([X_i, X_j], X_1, \ldots, X_i, \ldots, X_j, \ldots, X_{r+1})$$
$$= \sum (-1)^{i+1} X_i \beta(X_1, \ldots, \hat{X}_i, \ldots, X_{r+1})$$
$$+ \sum (-1)^{i+1} \rho(\gamma(X_i)) \beta(X_1, \ldots, \hat{X}_i, \ldots, X_{r+1})$$
$$+ \sum (-1)^{i+j} \beta([X_i, X_j], X_1, \ldots, X_i, \ldots, X_j, \ldots, X_{r+1})$$
$$= d\beta(X_1, \ldots, X_{r+1}) + \sum (-1)^{i+1} \rho(\gamma(X_i)) \beta(X_1, \ldots, \hat{X}_i, \ldots, X_{r+1}).$$

The second term in the last line is simply $(\gamma \wedge \beta)(X_1, \ldots, X_{r+1})$ where for the wedge product, we use the bilinear product $\mathfrak{g} \otimes V \to V$, giving the Lie algebra action on V. Finally we conclude that the form $d_\nabla \alpha$ corresponds to the form on P given by

5.14.
$$d\beta + \gamma \wedge \beta.$$

We have also seen that the composite of two successive exterior derivatives is not in general zero. However if the curvature is zero, then it is indeed so. In fact, in order to check that the composite of successive exterior derivatives is zero, it is enough to verify it locally. But then we may assume that E is trivial and that the connection is also the trivial one. Thus we get a de Rham complex in this case:

$$\cdots \to \Lambda^{i-1}(\mathcal{T}^*) \otimes \mathcal{E} \to \Lambda^{i}(\mathcal{T}^*) \otimes \mathcal{E} \to \Lambda^{i+1}(\mathcal{T}^*) \otimes \mathcal{E} \to \cdots.$$

What can one say about the kernel in the first stage, namely $\mathcal{E} \to \mathcal{T}^* \otimes \mathcal{E}$? Indeed, if $\mathcal{L}$ is the kernel (which is a sheaf of $\mathbb{C}$-vector spaces), then we can compute it explicitly on open sets where E and ∇ are trivial. In that case, i.e. when $\mathcal{E} = \mathcal{A}^r$ and ∇ is the usual connection, the kernel is $\mathbb{C}^r$. In other words, this $\mathbb{C}$-sheaf is locally free of rank r, or, what is the same, a *local system* of r-dimensional vector spaces. Moreover, the natural map $\mathcal{L} \otimes_\mathbb{C} \mathcal{A}$ into $\mathcal{E}$ is an isomorphism. Thus the associated de Rham complex is the same as the de Rham complex associated to the local system $\mathcal{L}$.

5.15. Remark. The above considerations show that the existence of a flat connection on a vector bundle imposes severe *topological restrictions* on the bundle. In fact, suppose the manifold is simply connected. Then any vector bundle which admits a flat connection would have to be trivial, since the fundamental group is trivial. This suggests that in general, the curvature form may carry some information about the topological nature of the vector bundle. Indeed it is true and we will now investigate this question.

6. Chern-Weil Theory

The way we will extract topological information from the curvature is via de Rham's theorem. In other words, we will use the curvature form to obtain some closed forms on M and therefore, thanks to the de Rham theorem, cohomology classes. These may be considered as *topological invariants* of the vector bundle in question.

The first step on the road is

6.1. Bianchi's identity. *The curvature form R of a connection ∇ in a principal G-bundle P satisfies $d_\nabla R = 0$. Here R is a 2-form with values in the adjoint bundle and d_∇ is the exterior derivative of R with respect to the induced connection on $\mathrm{ad}(P)$. This means that we have, for any vector fields X, Y, Z, the identity*

$$\sum (\nabla_X(R(Y,Z)) - R([X,Y],Z))$$

where the summation is taken over cyclic permutations of X, Y, Z.

Proof. This is a routine verification following the definition of R. In fact, if γ is the connection form, substituting for R the $\mathfrak{g}$-valued 2-form $d\gamma + [\gamma, \gamma]$ on P, and applying d_∇ to it, we get $d(d\gamma + [\gamma, \gamma]) + \gamma \wedge (d\gamma + [\gamma, \gamma]) = d[\gamma, \gamma] + \gamma \wedge d\gamma$, since $d^2 = 0$ and $[\gamma, [\gamma, \gamma]] = 0$ thanks to the Jacobi identity. Recall that by our definition, $[\gamma, \gamma](X, Y) = [\gamma(X), \gamma(Y)]$, and $(\gamma \wedge d\gamma)(X, Y, Z) = \sum [\gamma(X), d\gamma(Y, Z)] = \sum [\gamma(X), Y\gamma Z - Z\gamma Y - \gamma[Y, Z]]$, where the sum extends over cyclic permutations. Substituting $d[\gamma, \gamma](X, Y, Z) = \sum X[\gamma, \gamma](Y, Z) - \sum [\gamma, \gamma]([X, Y], Z) = \sum X[\gamma Y, \gamma Z] - \sum [\gamma[X, Y], \gamma Z]$, we get the result.

In particular, if E is a complex line bundle, then the bundle $\mathrm{End}(E)$ and the induced connection on it being trivial, d_∇ on $\mathrm{End}(E)$ is the same as the usual exterior derivative. Hence the 2-form R with coefficients in $\mathbb{C}$ is closed and defines a cohomology class in $H^2(M, \mathbb{C})$. Moreover, if we start with a different connection, then the two connections differ by a 1-form α and their curvatures differ by $d\alpha$ by Proposition 5.3. In other words, the *class* in the de Rham cohomology $H^2(M, \mathbb{C})$ defined by the curvature form, is *independent* of the connection. It is obvious that this cohomological invariant of the bundle is functorial in the sense that the invariant of the pull-back of a bundle is the pull-back of the invariant. It is called the *geometric Chern class* of the line bundle.

6.2. Computation.

We will make a computation of the Chern class of a specific line bundle, namely the *Hopf bundle*. Consider the principal $\mathbb{C}^\times$-bundle $E = \mathbb{C}^{n+1} \setminus \{0\}$ over $\mathbb{CP}^n$. Consider the $\mathbb{C}$-valued 1-form γ on E given by $\frac{\sum_0^n \bar{z}_j dz_j}{\sum_0^n \bar{z}_j z_j}$. It is clear that multiplication by any nonzero scalar leaves this form invariant. Also if $(a_0, \ldots, a_n)$ is any point in E, the identification of $\mathbb{C}^\times$ with a fibre by the map $\lambda \mapsto (a_0\lambda, \ldots, a_n\lambda)$ pulls γ to dz/z, which is the Maurer-Cartan form on $\mathbb{C}^\times$. Hence γ is a connection form on E. Its curvature form is $d\gamma$ since the Lie algebra is abelian, and is given by $\sum_j d(\frac{\bar{z}_j}{\sum \bar{z}_j z_j}) \wedge dz_j$. It is a 2-form on $\mathbb{CP}^n$ pulled up to E.

Now we will assume that $n = 1$ and integrate this form on $\mathbb{CP}^1$. It is enough to perform the integration on $\mathbb{C}^\times$ since its complement in $\mathbb{P}^1$ consists only of two points. Take the section of the bundle over $\mathbb{C}^\times$ which takes z to $(z, 1)$. The pull-back of the curvature form is then $d\frac{\bar{z}}{1+|z|^2} \wedge dz = \frac{\partial}{\partial \bar{z}}(\frac{\bar{z}}{1+|z|^2})d\bar{z} \wedge dz = \frac{1}{(1+|z|^2)^2}d\bar{z} \wedge dz$. We will use the orientation given by $dx \wedge dy$ in order to compute its integral. Recall that integrating top degree forms gives an isomorphism of the top de Rham cohomology with $\mathbb{R}$ or $\mathbb{C}$.

We will use polar coordinates $z = re^{i\theta}$ on $\mathbb{C} \setminus \{0\}$ and convert the above form to $\frac{2ir}{(1+r^2)^2}dr \wedge d\theta$. Since $dr \wedge d\theta$ and $dx \wedge dy$ give the same orientation, its integral is $\int_0^\infty \int_0^{2\pi} \frac{2ir}{(1+r^2)^2}drd\theta = 2\pi i$. In particular, the Chern class is nonzero.

6.3. Exercise. Compute the first geometric Chern class of the tangent bundle of $\mathbb{P}^1$.

6.4. Remark. Since the Chern class of the trivial bundle is zero, it follows that the Hopf bundle is not trivial. Even this very simple case illustrates the power of studying the topological invariant of a bundle.

6.5. Integral Chern class.

Consider any line bundle on M. Suppose it is given by transition functions (m_{ij}) with respect to a locally trivial covering (U_i). We would like to identify the first Chern class defined above in terms of (m_{ij}). The curvature form is a closed 2-form and gives the Chern class as a de Rham cohomology class in $H^2(M, \mathbb{C})$. In order to identify this in the sheaf cohomology with coefficients in the constant sheaf $\mathbb{C}$, we will recall the explicit description of the de Rham isomorphism. How does a closed 2-form give rise to a class in sheaf cohomology? We will denote the sheaf of closed i-forms by $\mathcal{Z}^i$ here. Consider the following two exact sequences of sheaves (here $\mathcal{A}$ is the sheaf of differentiable complex-valued functions):

$$0 \to \mathcal{Z}^1 \to \mathcal{T}^* \to \mathcal{Z}^2 \to 0,$$

$$0 \to \mathbb{C} \to \mathcal{A} \to \mathcal{Z}^1 \to 0.$$

Since $\mathcal{T}^*$ is a soft sheaf, all its positive cohomology spaces vanish [Ch. 4, Proposition 3.3, ii)]. Hence we get an exact sequence

$$\mathcal{T}^*(M) \to \mathcal{Z}^2(M) \to H^1(M, \mathcal{Z}^1) \to 0.$$

This identifies the space of closed forms modulo the exact forms, namely the second de Rham cohomology, with $H^1(M, \mathcal{Z}^1)$. The second sequence identifies it in turn with $H^2(M, \mathbb{C})$ since $\mathcal{A}$ is soft. We wish to understand how to associate to a closed 2-form R, a *Čech* 1-cocycle with values in $\mathcal{Z}^1$. In fact, if we can lift R locally to sections of $\mathcal{T}^*(U_i)$, then the differences on U_{ij} of the lifted sections over U_i and U_j give a family of sections of $\mathcal{Z}^1$ over U_{ij}, namely the required Čech cocycle. In other words, we write R as $d\alpha_i$ on open sets U_i and take the differences $\alpha_j - \alpha_i$ on U_{ij}. Actually, by our convention for boundary homomorphism in [Ch. 4, Remark 2.16], the required class is its negative.

Now if (m_{ij}) are transition functions for the line bundle in question, the local forms dm_{ij}/m_{ij} are closed. A choice of forms α_i on U_i such that $\alpha_j - \alpha_i = (dm_{ij})/m_{ij}$ is just a connection on the line bundle! Its curvature form is given by the 2-forms $d\alpha_i$, which coincide on the intersections and therefore give a global 2-form. This is the 2-form of which we set out to find the cohomology class, under the de Rham isomorphism. By construction, we have already written it locally as $d\alpha_i$. The difference of α_j and α_i is actually dm_{ij}/m_{ij}, which is the negative of the Cĕch 1-cocycle with values in $\mathcal{Z}^1$ that we needed to identify at the first stage.

Therefore we conclude that the first Chern class of the line bundle is simply the image of this class in $H^2(\mathbb{C})$ by the connecting homomorphism of the second sequence above. This can again be restated as follows. Take the image of the element of $H^1(\mathcal{A}^\times)$ in $H^1(\mathcal{Z}_1)$ by the homomorphism $f \mapsto df/f$. Then take the image under the connecting homomorphism of the second sequence above. But now consider the diagram

$$
\begin{array}{ccccccccc}
0 & \to & 2\pi i\mathbb{Z} & \to & \mathcal{A} & \to & \mathcal{A}^\times & \to & 0 \\
  &   & \downarrow   &   & \downarrow &   & \downarrow &   &   \\
0 & \to & \mathbb{C}   & \to & \mathcal{A} & \to & \mathcal{Z}^1 & \to & 0
\end{array}
$$

All the maps are the obvious ones; the top right horizontal arrow is the exponential map and the right arrow downwards is the map taking f to df/f. The Cĕch 1-cocyle in $\mathcal{Z}^1$ given by dm_{ij}/m_{ij} is the image of the Cĕch 1-cocycle in $\mathcal{A}^\times$ given by m_{ij} by the right vertical map. In other words, we take the class in $H^1(\mathcal{A}^\times)$ given by the line bundle. Instead of taking its image in $H^1(\mathcal{Z}^1)$ and then go by the connecting homomorphism of the lower sequence to $H^2(\mathbb{C})$, we may as well take its image by the connecting homomorphism of the top sequence to $H^2(2\pi i\mathbb{Z})$ and then go down by the left vertical map. Thus we have proved the following statement.

6.6. Theorem. *The (geometric) Chern class of any connection in a line bundle represents the negative of the image in $H^2(M, \mathbb{C})$ of the element of $H^2(M, 2\pi i\mathbb{Z})$ obtained as the image under the connecting homomorphism of the exponential sequence, of the element in $H^1(M, \mathcal{A}^\times)$ corresponding to the line bundle.*

6.7. Remarks.

1) The Chern class of a line bundle is essentially an integral class. One might have actually defined it as an element of $H^2(M, \mathbb{Z})$ using the exponential sequence as in the theorem above and dividing it by $-2\pi i$. One may think of it as the *topological* definition. This class is clearly functorial for pull-backs. The above theorem then states that the complex cohomology class so defined is the same (up to multiplication by $2\pi i$) as the curvature class. If E is any vector bundle, we can define the first Chern class as the integral Chern class of $\det(E)$. Again we have just shown that it is the same (up to a factor of $2\pi i$) as the class defined by the first Chern form of a connection.

2) The topological Chern class of a line bundle is therefore obtained by taking the image of the boundary homomorphism of the sequence

$$0 \to \mathbb{Z} \to \mathcal{O} \to \mathcal{O}^\times \to 0$$

where the surjection is given by $f \mapsto e^{2\pi i f}$.

Can we also get a similar class for arbitrary principal bundles? Let E be a principal G-bundle with a connection. Since R has values in the adjoint bundle, in order to obtain an ordinary form, we need a homomorphism of $\mathrm{ad}(P)$ into the trivial bundle. Suppose we have a linear map $t : \mathfrak{g} \to \mathbb{C}$, which is equivariant for the adjoint action of G on $\mathfrak{g}$ on the one hand, and the trivial action of G on $\mathbb{C}$ on the other. Then the induced homomorphism of the associated bundle $\mathrm{ad}(E)$ (namely, the associated vector bundle with $\mathfrak{g}$ as fibres) into the trivial bundle respects the induced connection on $\mathrm{ad}(E)$ and the connection on $\mathcal{A}$ induced by the trivial representation, namely the trivial connection. The image of the $\mathrm{ad}(P)$-valued d_∇-closed curvature form R of P is therefore a closed $\mathbb{C}$-valued 2-form and defines a class in $H^2(M, \mathbb{C})$.

The map t is also equivariant for the induced $\mathfrak{g}$ actions, namely the bracket action on $\mathfrak{g}$ and the trivial action on $\mathbb{C}$. So the map is zero on $[\mathfrak{g}, \mathfrak{g}]$. If we take any other connection on E, then its curvature form differs from R by an additive term of the type $d\alpha + [\alpha, \alpha]$. Applying t to it, we get $d(t \circ \alpha)$. Thus the class we defined is independent of the connection and depends only on E and t.

6.8. Remarks.

1) The supposed equivariant map of $t : \mathfrak{g} \to \mathbb{C}$ is actually a homomorphism of Lie algebras. If it integrates to a group homomorphism $G \to \mathbb{C}^\times$, then we can induce a connection on the line bundle (obtained by extension of the structure group). Its curvature form is obviously $t \circ R$. So in this case, the class defined above is simply the Chern class of the associated line bundle.

2) The following is a special case of the above. Let E be a complex vector bundle with a connection. For any vector space V, the trace mapping from $\mathrm{End}(V)$ into $\mathbb{C}$ is a $GL(V)$-homomorphism (taking the adjoint representation on $\mathfrak{g} = \mathrm{End}(V)$ and the trivial representation on $\mathbb{C}$). (This only means that $\mathrm{tr}(TAT^{-1}) = \mathrm{tr}(A)$ for all $T \in GL(n)$.) Hence it gives rise to an induced homomorphism $\mathrm{End}(E) \to \mathcal{A}_{\mathbb{C}}$. Thus the image of the curvature form is a 2-form with values in $\mathbb{C}$. This is called the *first Chern form* of the connection. It is closed, as we have observed above, and gives a class in $H^2(M, \mathbb{C})$. This class is independent of the connection and is called the *first Chern class* of the vector bundle E and is denoted by $c_1(E)$. This is clearly compatible with the definition we gave for line bundles. In this case, the trace homomorphism we started with actually integrates into the group homomorphism $\det : GL(r, \mathbb{C}) \to \mathbb{C}^{\times}$. So we conclude that $c_1(E) = c_1(\det(E))$ where $\det(E)$ is the line bundle associated to the bundle E by $\det : GL(n) \to \mathbb{C}^*$, or what is the same $\Lambda^n(E)$.

3) Why have we defined the first Chern class for $GL(k, \mathbb{C})$-bundles and not for $GL(k, \mathbb{R})$-bundles? In fact, formally one can take the curvature form of a connection and apply the trace and obtain a cohomology class in $H^2(M, \mathbb{R})$ independent of the connection. But then we have already remarked that any real vector bundle E admits a positive definite metric along the fibres. If we take an orthogonal connection in E, its curvature form takes values in the Lie algebra of the orthogonal group, namely the space of skew-symmetric matrices. In particular, the traces of all $R(X, Y)$ are zero. So we conclude that the cohomology class thus defined is 0 for all bundles E! This shows the slippery nature of building formal theories without examples.

Notice again that in the case of a real vector bundle, if we define the class using curvature, we will have obtained a class in $H^2(M, \mathbb{R})$. The corresponding element of $H^2(M, \mathbb{C})$ is then the Chern class of its complexification and is therefore in the image of $H^2(M, 2\pi i \mathbb{Z})$. It is no wonder that it is zero always!

6.9. Remark. While the above discussion gave rise to an interesting invariant, namely, the first Chern class of a vector bundle, we cannot obtain any more classes by that simple procedure, for there are not many $GL(V)$-homomorphisms from $\mathrm{End}(V)$ into the trivial representation space $\mathbb{C}$. In fact, up to a scalar factor, the trace map is the only such homomorphism! However we can modify the construction by starting out with R^r instead of R itself. Notice that R belongs to $\Lambda^{(+)} \otimes \mathrm{Sym}(\mathrm{End}\, E)$, where $\Lambda^{(+)}$ denotes the (commutative) subalgebra of elements of even degree in the exterior algebra $\Lambda(T^*)$. Hence it makes sense to talk of R^r as an element of $\Lambda^{2r}(T^*) \otimes S^r(\mathrm{End}(E))$. It is the $2r$-form with values in $S^r(\mathrm{End}(E))$ given by

$$R^r(X_1, \ldots, X_r) = \sum \epsilon_\sigma \prod R(X_{\sigma(2i)}, X_{\sigma(2i+1)}),$$

where we take the summation over permutations which satisfy $\sigma(2i) < \sigma(2i+1)$ for all $i \leq r$. Then we have the same statement as the Bianchi identity for R^r, that is to say it is a $2r$-form with values in $S^r(\text{End}(E))$ which is closed for the exterior derivative associated to the induced connection on $S^r(\text{End}(E))$. This follows from the derivation property of the exterior derivative.

Thus we start with R^r as an element of $\Lambda^{2r}(T^*) \otimes S^r(\text{End } E)$, and take its image in any $\Lambda^{2r}(T^*)$ by means of a homomorphism $S^r(\text{End } E) \to \mathcal{A}$ given rise to by a $GL(V)$-homomorphism of $S^r(\text{End } V)$ into the trivial representation space $\mathbb{C}$. This then gives a closed, complex-valued $2r$-form as before.

Are there such homomorphisms? Yes, indeed! Actually there are many and we can in fact write them all out. We will first define such a homomorphism for each $r \leq \dim(V) = k$. The map which associates to each endomorphism A of V, the trace of its lift, $\Lambda^r(A) : \Lambda^r(V) \to \Lambda^r(V)$, is the homomorphism we have in mind. Up to sign, this map is the same as that of associating to an element of $\text{End}(V)$, the coefficient of t^{k-r} in its characteristic polynomial. Another way of saying it is that it associates to any matrix the rth elementary symmetric function of its eigenvalues. When $r = 1$, this is the same as the Chern form we defined earlier.

6.10. Theorem. *For every $r \leq \dim(V) = k$, the coefficient of t^{k-r} of the characteristic polynomial $\det(t + A)$ of elements of $\text{End}(V)$ gives rise to an adjoint invariant polynomial of degree k. It may be considered as giving a $GL(V)$-invariant linear map $S^r(\text{End}(V)) \to \mathbb{C}$. Let E be any vector bundle of rank k and ∇ a connection on it. By taking the image of R^r under such a linear map, we obtain a $\mathbb{C}$-valued closed form of degree $2r$ and thus a cohomology class $H^{2r}(M, \mathbb{C})$. The cohomology class so obtained is independent of the connection.*

Proof. In fact, consider the manifold $\mathbb{R} \times M$ and the bundle $p_2^* E$ on it. The projection to the first factor may be considered as a real-valued function on the product, and we denote it by t. Let ∇_1, ∇_2 be two connections on E. Consider the connection $\nabla = t\nabla_1 + (1 - t)\nabla_2$ on $p_2^*(E)$, where we have denoted by ∇_i the pulled back connections as well. Consider the Chern forms of this connection, on $\mathbb{R} \times M$. By the functoriality of the construction and the obvious fact that the pull-backs of ∇ to $\{0\} \times M$ and $\{1\} \times M$ are ∇_1 and ∇_2 respectively, we conclude that the Chern form on the product also pulls back to the Chern forms of the two connections on these two submanifolds. Now our assertion follows from [Ch. 4, Proposition 6.7].

6.11. Definition. The *rth geometric Chern form* of a complex vector bundle of rank k is the $2r$-form obtained by substituting the curvature form in the $(k-r)$th coefficient of the characteristic polynomial $t \mapsto \det(tI + A)$. The cohomology class in $H^{2r}(M, \mathbb{C})$ defined by it, is called its rth *geometric Chern class* and is denoted $c_r(E)$.

6.12. Proposition. *If E is a direct sum of two vector bundles E_1 and E_2, then the rth Chern class of E is given by*

$$c_r(E) = \sum_{i+j=r} c_i(E_1).c_j(E_2)$$

where the multiplication is taken in the cohomology algebra.

Proof. We take connections in each of the E_i's and take the direct sum connection in E and will compute the Chern form of this connection. The curvature form in E associates to any two vectors X, Y the endomorphism $R(X,Y)$ which is the direct sum of $R_1(X,Y)$ and $R_2(X,Y)$. The $2r$-form R^r is therefore given by the sum of the terms $R_1^i.R_2^j$ with $i+j=r$, where we have used the inclusion of $S^{2i}(\operatorname{End} E_1) \otimes S^{2j}(\operatorname{End} E_2)$ in $S^{2r}(\operatorname{End} E)$ and the exterior multiplication $\Lambda^{2i}(T^*) \otimes \Lambda^{2j}(T^*) \to \Lambda^{2r}(T^*)$. Applying the linear trace map, we get our assertion.

This also shows that the rth Chern class is in general nontrivial. The simplest thing to do is to take M to be the r-fold product $(\mathbb{P}^1)^r$ and E to be the direct sum of the pull-backs of the Hopf bundle from all the factors and use the above theorem.

6.13. Remarks.

1) We have seen that the first Chern class is essentially an integral class. It can also be proved that the rth Chern class comes from a class in $H^{2r}(M, (2\pi i)^r \mathbb{Z})$. From the topological point of view therefore it makes sense to define the *topological Chern class* as the class obtained by dividing our class by $(-2\pi i)^r$. These are therefore in the image of $H^{2r}(M, \mathbb{Z})$ in $H^{2r}(\mathbb{C})$. Such classes can indeed be defined as integral classes from a purely topological standpoint and proved to be functorial and having property 6.12. The general case can actually be reduced to the case of line bundles.

2) We cannot define any more invariants in the case of vector bundles by this procedure. It is easy to see that any adjoint invariant map $\operatorname{End}(V) \to \mathbb{C}$ is generated by the above, namely the coefficients of the characteristic polynomial. Hence any cohomological invariant we can define by the above procedure, is a polynomial in the Chern classes c_r.

We can apply much the same ideas for any principal G-bundle P. The curvature form R of any connection on it, is a 2-form with values in the vector bundle $\operatorname{ad}(P)$. The $2r$-form R^r has values in $S^r(\operatorname{ad} P)$. If we have an adjoint invariant homogeneous polynomial from $\mathfrak{g}$ into $\mathbb{C}$ of degree r, then we can define a cohomology invariant in $H^{2r}(M, \mathbb{C})$ exactly as above.

We will now illustrate this for the case of orthogonal bundles. Let E be a principal $SO(n, \mathbb{C})$-bundle. The curvature of a connection on E, considered as a 2-form on the total space E, takes values in the Lie algebra of $SO(n, \mathbb{C})$. We recall that if the quadratic form is taken in the standard form, namely $q(z_1, \ldots, z_n) = \sum z_i^2$, the Lie algebra of $SO(n)$ consists of skew-symmetric matrices. We then look for equivariant maps $S^r(\mathfrak{so}(n, \mathbb{C})) \to \mathbb{C}$, where the action of G is the adjoint action on the first and trivial on the second factor. The obvious maps to take are the coefficients of the characteristic polynomial again, since they are after all invariant under the adjoint action even under $GL(n)$. However, the coefficients of the characteristic polynomial of a skew-symmetric matrix are zero whenever the parity of the degree is different from that of n. This is because the lift of a skew-symmetric endomorphism of a vector space to $\Lambda^i(V)$ is again skew-symmetric for odd i and consequently is traceless. Thus we get, for every even number $2k \leq n$, a cohomology class in $H^{4k}(M, \mathbb{C})$. These are of course not very different from what we defined above, treating the orthogonal bundle just as a vector bundle. In fact, if we ignore the fact that E is an orthogonal bundle, but treat it as a vector bundle (that is to say, consider the $GL(n, \mathbb{C})$-bundle obtained by extension of the structure group), then these are just $2k$th Chern classes.

6.14. Definition. The class $(-1)^k c_{2k}(E)$ is called the *kth Pontrjagin class* of the special orthogonal bundle.

In the case when $n = 2r$, we have therefore defined r Pontrjagin classes. But we can also define one more class by our procedure. In other words, we will give one more invariant polynomial called *Pfaffian* from the space of skew-symmetric matrices. The space of skew-symmetric matrices can be identified with $\Lambda^2(V)$, compatibly with the adjoint action of the orthogonal group on the former and the second exterior of the natural representation on the latter. By taking the rth power of an element of $\mathfrak{so}(2r) = \Lambda^2(V)$, we get an element of $\Lambda^{2r}(V) = \mathbb{C}$. Thus we get a polynomial map of the Lie algebra into $\det(V) = \Lambda^{2r}(V)$ of degree r, which is equivariant for the adjoint action of $SO(n)$ on the first factor and the trivial action on the second. This polynomial is called the *Pfaffian*. Thus we get an invariant in $H^{2r}(M, \mathbb{C})$, using this invariant polynomial. It is called the *Euler-Poincaré class*. In fact, one can show that all $SO(2n)$-invariant polynomials are generated by the even coefficients of the characteristic polynomial and the Pfaffian. These however are not independent. In fact, the square of the Pfaffian is the determinant, namely the constant term of the characteristic polynomial.

Again this class can be shown to be $(2\pi)^r$ times an integral class. From the topological viewpoint the integral class is defined to be the *Euler class.*

6.15. Remarks.

1) If an oriented vector bundle of rank $2n$ admits a section nonzero everywhere, then one can introduce a metric along the fibres such that the section has unit length at all points. In other words, the structure group can be reduced to $SO(2n - 1)$. Its Lie algebra consists of elements of the Lie algebra of $SO(2n)$ which are zero on the first vector of the standard basis. But the Pfaffian of a skew-symmetric matrix which is not invertible is of course zero. Hence the substitution in the Pfaffian of the curvature form of an $SO(2n)$-connection obtained by extending one on the reduced bundle, is zero. We thus conclude that the Euler class of a bundle admitting an everywhere nonzero section, is zero. If one could compute the Euler class of the tangent bundle of a differential manifold and show that it is nonzero, it would follow that the manifold does not admit a vector field nonzero everywhere. This question was at the root of the development of the characteristic classes from the topological point of view.

2) If we start with a complex vector bundle of rank n, we can reduce its structure group to the unitary group by introducing a Hermitian structure along its fibres. Disregarding its complex structure, we may treat it as an $SO(2n)$-bundle and compute its Euler class. The result is that this class is just $(-1)^n$ times $c_n(E)$. This is only a matter of computing the Pfaffian of a skew-symmetric matrix which is associated to a skew-Hermitian matrix A, namely

$$\begin{pmatrix} \mathrm{Re}(A) & -\mathrm{Im}(A) \\ \mathrm{Im}(A) & \mathrm{Re}(A) \end{pmatrix}.$$

The Pfaffian of the latter is computed (for example by diagonalising A) to be $\det(iA) = i^n \det(A)$. Now note that to get the topological Euler class we have to multiply the geometric Euler class by $(1/2\pi)^n$ while the Chern class is obtained by multiplying by $(-1/2\pi i)^n$.

3) We can define characteristic classes in somewhat greater generality. So far we have been taking maps from $S^r(\mathfrak{g})$ into $\mathbb{C}$. Suppose G is not connected. Then we can take a (possibly nontrivial) representation ρ of G which is trivial on the connected component G° of G containing 1. Let V be such a representation, and we take a map $S^r(\mathfrak{g})$ into V, equivariant for the adjoint action of G on the first factor and the action ρ on V. Then for any principal G-bundle and a connection on it, we can take the image of the rth power R^r of the curvature R as above and obtain a $2r$-form with values in the vector bundle associated to ρ. But this bundle has G/G° as structure group. The latter being discrete, the associated connection is flat. The $2r$-form that we obtained is then closed for the exterior derivative associated to the flat connection. By de Rham's theorem this defines an invariant in $H^{2r}(M, L(\rho))$ where $L(\rho)$ is the local system defined by the associated bundle $\rho(P)$ and the flat connection on it.

6.16. Example. Consider an orthogonal bundle P, i.e. a bundle with structure group $G = O(n)$. This group has two connected components, and the connected component of 1, namely $SO(n)$, is of index 2. Take $n = 2r$ as above, but notice that the Pfaffian is actually a map $S^r \mathfrak{g} \to \Lambda^{2r}(\mathbb{C}^{2r}) = \mathbb{C}$. But it is equivariant for $O(2r)$ only if we take the determinant character on the second factor and the adjoint on the first. Hence the Euler class takes values in the cohomology of the local system given by the determinant character of $O(2r)$.

6.17. Remarks.

1) If E is a real orthogonal bundle, the same construction as above leads to Pontrjagin classes in $H^{2i}(M, \mathbb{R})$. In fact, the odd Chern classes of its complexification vanish and the even ones give the Pontrjagin classes (up to sign).

But the Euler class cannot be accessed directly for a vector bundle by our construction. This is because the Pfaffian which is an orthogonal invariant does not extend to an invariant polynomial on all matrices. However, we can introduce a metric along the fibres, that is to say, reduce the structure group to the orthogonal group $O(n, \mathbb{R})$. When $n = 2k$ is even, the Euler class is clearly defined for a real $SO(2k)$-bundle as a cohomology class in $H^k(M, \mathbb{R})$. Since all metrics form a convex set, one can easily check that the class so defined does not depend on the metric. Thus the Euler class is defined for any (oriented) real vector bundle.

2) If we take the tangent bundle of a compact oriented manifold M of even dimension $2r$, the topological Euler class is an element of $H^{2r}(M, \mathbb{R})$. Since $H^{2r}(M, \mathbb{Z}) = \mathbb{Z}$, it is an integer. It can be proved to be the *Euler-Poincaré characteristic*, namely, $\sum(-1)^i b_i$ where $b_i = \dim H^i(M, \mathbb{R})$. This is known as the *Gauss-Bonnet theorem*. See [**10**].

7. Holonomy Group; Ambrose-Singer Theorem

We will indicate a geometric understanding of connections and curvature in principal bundles. Let us begin by recalling that flatness of a connection on a simply connected manifold implies that the bundle is trivial. All connections on bundles over the interval are flat, as there are no nonzero 2-forms and so the curvature form is always 0. Hence any bundle with a connection admits a trivialisation. Suppose P is a principal bundle over any manifold M, and ∇ is a connection on it. If $q : [0,1] \to M$ is a differentiable or piecewise differentiable path in M, then the pull-back of P to the curve is trivial. Indeed, given a point p on the fibre of P over $q(0)$, there is a canonical section of the pull-back $q^*(P) \subset I \times P$, which takes the value $(0,p)$ at the point 0. This is the same as saying that there is a lifted path $\tilde{q} : I \to P$ with origin p.

7.1. Definition. A path in P is said to be *horizontal* if its differential at any point of $I = [0,1]$ is a horizontal vector.

Since the pull-back of the connection to I is also trivial, it follows that the section $\tilde{q}$ is horizontal. This amounts to saying that any path in M admits a horizontal lift. The lift is unique if the origin is prescribed. This may be referred to as a *moving frame*, since any point of the principal bundle can be thought of as a frame. Moreover, if we have a differentiable family of paths, in the sense that we have a differentiable map of $f : T \times I$ into M (where T is a differential manifold), then the map $T \times I \to P$ obtained by horizontally lifting the paths $f|\{t\} \times I \to M$ is differentiable.

Even if the path in M is a loop, there is no guarantee that its horizontal lift would be a loop. (One knows that this is not the case even for a covering space.) There is a unique element of the structure group which takes the initial point to the end point of the lifted path. This is called the *holonomy* of the loop at a given initial point p of P. Unlike the case of a covering space, we *do not* claim that the end point of the lift depends only on the homotopy class of the loop. The end point of the horizontal lift is the translate of p by an element of G. By taking all possible loops at a point $m \in M$, and lifting them with initial point at p, we get a family of elements of G. The constant loop gives $1 \in G$, and the inverse of a loop has the inverse as its holonomy. Let q_1, q_2 be two loops and q their composite. If both of them are piecewise differentiable, so is q. Let $\tilde{q}_i$ be the horizontal lifts of q_i with origin p. If s is the holonomy of q_1, i.e. $\tilde{q}_1(1) = p.s$, then the horizontal lift of q_2 with origin ps is given by $\tilde{q}_2 s$. So the composite of $\tilde{q}_1$ and $\tilde{q}_2.s$ is piecewise differentiable, horizontal, lifts q and has origin p. Hence if the holonomy of the loop q is u, then $pu = \tilde{q}_2(1).s = pt.s$ where t is the holonomy of q_2. In other words, $u = ts$. Thus all the holonomy elements at p form a subgroup of G.

7.2. Definition. The group formed by the holonomy of all loops at a point $m \in M$ obtained by lifting them with initial point $p \in P$ is called the *holonomy group* at p. The subgroup formed by taking only loops homotopic to the identity is called the *restricted holonomy group*.

7.3. Proposition. *The restricted holonomy group at $p \in P$ is an arcwise connected subgroup of G.*

Proof. In fact, let q be a loop at $m = \pi(p)$ which is homotopically trivial, i.e. there is a family of loops q_t at m parametrised by I, with $q_0 = q$, and q_1 being the constant loop at m. Their lifts with p as origin have end points pg_t, say. For $t = 1$ it is the constant loop at p. Then $t \mapsto g_t$ is a path connecting 1 and the holonomy of q at p.

7.4. Remark. It can be shown that any arcwise connected subgroup of a Lie group is a Lie subgroup.

The dependence of the holonomy group, on the point p is explained in the following proposition.

7.5. Proposition. *If M is connected, then the holonomy groups at any two points in P are conjugate.*

Proof. Take any point $p' \in P$, at first, on the fibre through p. In other words, let $p' = pg$ for some $g \in G$. If q is any loop at $m = \pi(p) = \pi(p')$, then its horizontal lift $\tilde{q}'$ with origin at p' is simply the right g-translate of the horizontal lift $\tilde{q}$ at p. From this we conclude that if the holonomies of γ at p, p' are h, h' respectively, then $\tilde{q}'(1) = \tilde{q}(1).g = ph.g = p'g^{-1}hg$. This shows that $h' = g^{-1}hg$ and hence the holonomy groups at p and p' are conjugate.

If p' is not necessarily on the same fibre, let $m' = \pi(p')$. Connect m and m' by a path δ and take its horizontal lift $\tilde{\delta}$ with origin at p. Its end point lies over m' and so is of the form $p'g$. We have just shown that the holonomies at p' and $p'g$ are conjugate. So we may as well replace p' by $p'g$. In other words we will assume that the horizontal lift of δ has p' as its end point. Then to any loop q at m, we associate the loop q' at m' by composing δ^{-1} first and then q and δ again. Its horizontal lift is obtained by composing $\tilde{\delta}^{-1}$ first, the horizontal lift of q and then $\tilde{\delta}$. This shows that the holonomy subgroups at p and p' are identical.

7.6. Theorem. *Let M be a connected differential manifold and P a principal G-bundle on it with a connection form α. Then the structure group can be reduced to the holonomy group G_p at any point $p \in P$. The connection can also be reduced to the reduced bundle in the sense that there is a G_p-connection in the reduced principal bundle such that the given connection is extended from it.*

Proof. Let p be any point of P. Consider the union of all piecewise smooth, horizontal curves with initial point p. Let us call it H_p. Since any point of M can be connected by a path to πp and the path can be lifted to a horizontal one starting at p, the map $\pi : H_p \to M$ is surjective.

The holonomy group at p leaves H_p invariant. For if $p' \in H_p$, then there is by definition a horizontal path γ joining p and p'. If $g \in G_p$, then p and pg can be connected by a horizontal lift of a loop q at $\pi(p)$. By translating the curve γ by g we see that pg can be connected by a horizontal curve to $p'g$. In particular it follows that p and $p'g$ can also be so connected. In other words, $p'g$ belongs to H_p, proving our assertion.

Conversely, if p_1 and $p_2 = p_1 g$ are two points in H_p on the same fibre, then p can be connected by horizontal paths C_1, C_2 to them. By translating the curve C_1 by g we get a horizontal path connecting pg to p_2. Composing the path C_2 with the inverse of this path, we get a horizontal path joining p and pg. This implies that it is a horizontal lift of a loop at $\pi(p)$ and hence that $g \in G_p$.

Now we wish to show that H_p is a submanifold of P. Let $p' \in H_p$ and m' be its image in M. Connecting p and p' by a horizontal path, we can identify G_p with $G_{p'}$. Consider a coordinate neighbourhood U of m'. For every point x of U let S_x be the line segment (in the coordinate system) joining x and m (which is 0 in the coordinate system). Thus we get a differentiable family of paths in M, parametrised by U. Hence there is also a differentiable map of $U \times I$ to P, its restrictions to $\{v\} \times I$ for all $v \in U$ being horizontal paths with initial point p'. The restriction to $U \times \{1\}$ gives rise to a differentiable section s of P over U which has its image in H_p. This gives an isomorphism of $\pi^{-1}(U)$ with the trivial bundle $U \times G$, namely $\xi \mapsto (\pi(\xi), g(\xi))$ where $g(\xi)$ is determined by the equation $\xi g(\xi) = s(\pi\xi)$. The restriction of $g(\xi)$ to $\pi^{-1}(U) \cap H_p$ goes into G'_p.

From our remarks above it gives an isomorphism of H_p onto $U \times G_p$. From this it follows that H_p is a differential manifold, that it is a principal G_p-bundle and that the bundle obtained by extension of the structure group to G can be identified with P.

The relationship between holonomy and curvature is given by the following theorem, due to Ambrose and Singer.

7.7. Theorem. *The Lie algebra of the restricted holonomy group at a point p in a principal G-bundle is the subalgebra of $\mathfrak{g}$ generated by the values $R(v, w)$ where v, w are tangent vectors at points of the reduced bundle as given above.*

Proof. As we have seen above, the structure group of the bundle and the connection may be reduced to the holonomy group. Therefore we may assume, without loss of generality, that the holonomy group is itself G and show that the Lie subalgebra $\mathfrak{g}'$ generated by $R(v, w)$ is actually $\mathfrak{g}$. From the G-equivariance of R we see that $\mathfrak{g}'$ is invariant under the adjoint action of G. Hence it is also invariant under the adjoint action of $\mathfrak{g}$.

The vertical subbundle of the tangent bundle of P admits a subbundle E, namely that corresponding to the Lie algebra $\mathfrak{g}'$. Consider the subbundle of $T(P)$ which is the direct sum of E and the subbundle of horizontal subspaces. We claim that it is an integrable subbundle. If we take two horizontal vector fields X_1, X_2, the vertical projection of their bracket is given by $R(X_1, X_2)$ and so it is a section of E. On the other hand, if we take two sections of E and bracket them, it is again a section of E since $\mathfrak{g}'$ is a Lie subalgebra. Finally, the bracket of an invariant vertical vector field and an invariant horizontal vector field is always zero. These assertions prove our claim. This subbundle is also invariant under the action of G. Therefore any integral manifold for this integrable subbundle has the property that it is invariant under G and the tangent space at any point contains the horizontal subspace at that point. This shows that its dimension is the same as that of P. In particular, we have $\mathfrak{g}' = \mathfrak{g}$.

Exercises

1) Let E be a vector bundle and s a nonzero section. Show that there exists a connection ∇ such that $\nabla_X s = 0$ for all vector fields, if s is nonzero at all points. Is it also true if s has zeros? What does this mean for a trivial bundle?

2) Let M be a connected manifold. Show that the action of the gauge group factors to an effective action of its quotient by the group of nonzero scalars.

3) Give a faithful representation of the group $GA(V)$ in a vector space of dimension $n + 1$ where $\dim(V) = n$.

4) Let γ be a connection form on the trivial $U(1)$-bundle on any manifold M. Show that there exists a $U(1)$-invariant differentiable map $M \times U(1) \to S^N$, for some N, such that γ is the pull-back of $\sum \bar{z}_i dz_i$.

5) Show that given an open covering (U_i) of a differential manifold, there exists a family of differentiable functions f_i which satisfy $\sum |f_i|^2 = 1$ and $\sum \bar{f}_i f_i = 0$. Hence conclude that if M is a compact manifold, and E a $U(1)$-bundle on it, then any connection on E is induced by a differentiable map into the complex projective space from the above connection on the Hopf bundle on it.

6) Let C be the connection algebra and B the subspace given by $X \in C$ such that $Xf = fX$ for all elements f in A. Show that B is a subalgebra which is invariant under brackets by elements of C.

7) Let G be any connected Lie group. Show that a unique connection can be defined on its tangent bundle with the property that $\nabla_X Y = 0$ for all left invariant vector fields X and Y. Compute its curvature form.

8) Let a connected Lie group G act transitively on a manifold M, and let I be the isotropy subgroup at a point m. Find the condition in terms of I, for the G-action to lift to a principal H-bundle.

Chapter 6

Linear Connections

We take up here the special features when we deal with a connection in the *tangent bundle* of a differential manifold. These are called *linear connections.*

1. Linear Connections

Let ∇ be a linear connection on a differential manifold M. Let $\pi : T \to M$ be the tangent fibration. Then there is a canonical vector field on the manifold T defined as follows. Take any point t of T. It is by definition a tangent vector v of M at the point $m = \pi(t) \in M$. This vector can be lifted horizontally to a tangent vector of the manifold T at the point t. The vector field we have in mind associates to the point t, this horizontal lift.

1.1. Definition. The vector field defined above on the manifold T is called the *geodesic vector field* associated to the linear connection.

If $(U, (x_i))$ is a local coordinate system in M, there is a natural local coordinate system $(\pi^{-1}(U), (x_i, y_j))$ in T, consisting of the functions $x_i = x_i \circ \pi, y_j = dx_j$. Here dx_j are 1-forms on U, considered as functions on $\pi^{-1}(U)$. A point of $\pi^{-1}(U)$ with coordinates (x_i, y_j) is then the tangent vector at the point (x_i) of U given by $\sum y_j \frac{\partial}{\partial x_j}$. We will now compute the local expression for the geodesic vector field in $\pi^{-1}(U)$. Let ∇, the given linear connection, be given by the law:

$$\nabla_{\frac{\partial}{\partial x_i}} \left(\frac{\partial}{\partial x_j} \right) = \sum \Gamma_{ij}^k \frac{\partial}{\partial x_k}.$$

The companion connection acting on sections of T^*, namely differential 1-forms, is therefore given by

$$\left\langle \frac{\partial}{\partial x_k}, \nabla_{\frac{\partial}{\partial x_i}}(dx_j) \right\rangle = -\left\langle \nabla_{\frac{\partial}{\partial x_i}}\left(\frac{\partial}{\partial x_k}\right), dx_j \right\rangle = -\Gamma^j_{ik}.$$

In other words,

$$\nabla_{\frac{\partial}{\partial x_i}}(dx_j) = -\sum \Gamma^j_{ik} dx_k.$$

Thus we have $\nabla_{\sum y_i \frac{\partial}{\partial x_i}}(dx_j) = -\sum_{i,k} y_i \Gamma^j_{ik} dx_k$. Interpreting dx_k as the functions y_k on the tangent bundle, the right side may also be written as $-\sum_{i,k} y_i y_k \Gamma^j_{ik}$. By [Ch. 5, Remark 4.8, 3)] the covariant derivative of dx_j with respect to the vector field $\sum y_i \frac{\partial}{\partial x_i}$ (considered as a function on T) is obtained by lifting the vector field horizontally and letting it act on the function y_j. So if the lift is expressed in local coordinates upstairs in $\pi^{-1}(U)$ as $\sum y_i \frac{\partial}{\partial x_i} + \sum \varphi_k(x,y) \frac{\partial}{\partial y_k}$, then it acts on the function y_j and yields $\varphi_j(x,y)$. Thus we have $\varphi_j(x,y) = -\sum_{i,k} y_i y_k \Gamma^j_{ik}$. This then gives the local expression for the geodesic vector field to be

1.2.
$$\sum_i y_i \frac{\partial}{\partial x_i} - \sum_{i,j,k} y_i y_k \Gamma^j_{ik} \frac{\partial}{\partial y_j}.$$

In particular, the geodesic vector field is differentiable.

1.3. Definition. Given a linear connection, we may consider the flow of the geodesic vector field. The image in M of any curve in T which is integral for the geodesic flow is called a *geodesic* of the linear connection.

1.4. Remarks.

1) If we take *any* path $c : I \to M$, then for every $t \in I$, the image of $\frac{d}{dt}$ by the differential of c, gives a tangent vector $c'(t)$ at $c(t)$, which is a point of the manifold T. In other words, we get a lift of c to the manifold T. This is called the *canonical lift* of a curve. If c is a geodesic, and γ is the integral curve of the geodesic vector field with $c'(0)$ as its origin, then $c = \pi \circ \gamma$, by definition. Now $\gamma'(t)$ is horizontal, and the map π maps it on the vector $c'(t)$. But since γ is integral to the geodesic vector field G, it follows that the vector $\gamma'(t)$ is the horizontal lift of the vector $\gamma(t)$ at the point $\pi(\gamma(t)) = c(t)$. Thus $c'(t)$ is the vector corresponding to $\gamma(t)$. In other words, the canonical lift of $c(t)$ is the same as the horizontal lift. This clearly characterises geodesics. This observation allows us to write down a differential equation that a geodesic must satisfy. See 1.5 below.

2) Let c be a geodesic of a linear connection. In order to translate a vector v at $c(0)$ parallelly along c, we have to take the horizontal lift of c with v as origin. In particular, take v to be $c'(0)$. Then the vector at $c(t)$ obtained by parallelly translating $c'(0)$ along c itself, is given by $\gamma(t)$, where γ is the horizontal lift of c starting from $c'(0)$. But this is the canonical lift as we have seen above, and so we conclude that the translated vector at $c(t)$ is simply $c'(t)$. We may express this by saying that the tangent to $c(0)$ remains tangent to c when translated along c. For this reason we say that geodesics are *autoparallel*. In this respect it is like a straight line in the Euclidean space.

3) To any Riemannian manifold, which will be defined in Chapter 7, one can canonically associate a linear connection as we will see in [Ch. 7, Proposition 4.8]. The geodesics for this connection turn out [Ch. 7, 5.7] to be precisely curves which minimise distances between close enough points. This is the origin of the terminology.

1.5. Proposition. *A differentiable path $c(t)$ where t varies in an open interval containing 0, is a geodesic if and only if it satisfies the following differential equation:*

$$\frac{d^2 c_i}{dt^2} + \sum \Gamma^i_{jk}(c(t)) \frac{dc_j}{dt} \frac{dc_k}{dt} = 0.$$

Here $c(t) = (c_1(t), \ldots, c_n(t))$ with respect to a system $(U, (x_i))$ of local coordinates of M.

Proof. We will use the associated local coordinates $(\pi^{-1}(U), (x_i, y_j))$ on T. Then the canonical lift of c is given locally by the functions $(c_i(t), \frac{dc_j}{dt})$. For it to be integral for the geodesic vector field, its tangent at any point should be given by the value of the geodesic vector field at that point. This means that $\sum \frac{dc_i}{dt} \frac{\partial}{\partial x_i} + \sum (\frac{d^2 c_i}{dt^2}) \frac{\partial}{\partial y_i}$ is the same as the geodesic vector at the point $(c_i(t), \frac{\partial c_j}{\partial t})$. Equating the coefficients of $\frac{\partial}{\partial y_i}$ in this and in 1.2, we get the system of differential equations

$$\frac{d^2 c_i}{dt^2} + \sum \Gamma^i_{jk}(c(t)) \frac{dc_j}{dt} \frac{dc_k}{dt} = 0$$

for the functions $c_i(t)$.

1.6. Remark. Notice that the above equation only depends on the functions $\Gamma^i_{jk} + \Gamma^i_{kj}$. Its significance will be explained presently.

By the definition of geodesics, we see that, given any point $m \in M$ and a vector v at m, there exists a *unique* geodesic c starting at m with $c'(0) = v$, namely the image of the integral curve of the geodesic vector field starting at $c'(0)$. We denote the unique geodesic with m as origin and v as the initial tangent vector by c_v. Clearly we have $c_{\lambda v}(t) = c_v(\lambda t)$.

1.7. Exercise. If c and d are two geodesics whose images coincide, then there exist constants $a \neq 0$ and b such that $d(at + b) = c(t)$ for all small t.

1.8. Exponential mapping.

We wish to construct a map $T(M) \to M \times M$ as follows. To any tangent vector v at a point $m \in M$, associate the pair $(m, c_v(1))$. This may not make sense, since the geodesic vector field may not give a global flow. Consequently $c_v(t)$ may only be defined for small values of $|t|$. If we restrict ourselves to a neighbourhood U of M imbedded in $T(M)$ as the zero section, then it is well defined and differentiable. This is called the *exponential map* associated to the linear connection.

The exponential map fits into the diagram:

$$
\begin{array}{ccc}
M & \stackrel{\mathrm{Id}}{\to} & M \\
\Big\downarrow {\scriptstyle\text{zero}} & & \Big\downarrow {\scriptstyle\text{diagonal}} \\
U \subset T(M) & \stackrel{\mathrm{exp}}{\to} & M \times M
\end{array}
$$

If we restrict the exponential map to a neighbourhood of 0 in the one-dimensional subspace $\mathbb{C}v$ of the tangent space at a point $m \in M$, then we get the map $tv \mapsto (m, c_{tv}(1) = c_v(t))$. Hence the differential at the zero vector at m of the exponential map takes v to $(0, v) \in T_m \oplus T_m$. On the other hand, from the above diagram we conclude that the image of this differential contains all vectors of the form (v, v) in $T_m(M) \oplus T_m(M)$. Hence the differential of the exponential map is actually surjective, and for dimensional reasons, an isomorphism.

1.9. Proposition. *The exponential map takes a neighbourhood of M in $T(M)$ diffeomorphically onto an open set in $M \times M$. Also by restricting to one fibre, the exponential map $v \mapsto c_v(1)$ at a point $m \in M$ is a diffeomorphism of an open neighbourhood of 0 onto an open neighbourhood of m in M.*

Proof. This is now simply the inverse function theorem.

1.10. Corollary. *For every neighbourhood W of m, there exists a neighbourhood $V \subset W$ with the property that any two points in V can be joined by a geodesic contained in W.*

Proof. Given W as above, we choose an open subset U of $T(M)$ containing the zero vector at m such that $\exp |U$ is a diffeomorphism onto an open set contained in $W \times W$. The image of U is an open subset containing (m, m) and so contains an open set of the form $V \times V$, where V is an open set in M containing m. For any $x \in V$, the exponential image of $\pi^{-1}(x) \cap U$ is contained in $\{x\} \times W$ and contains $\{x\} \times V$. Hence if $y \in V$, then x and y can be connected by a geodesic in W.

We will use the exponential local diffeomorphism to identify an open neighbourhood U of m with a neighbourhood V of 0 in $T_m(M)$. The tangent bundle is in particular trivialised and all the tangent spaces can be identified with $T_m(M)$. Then the equation of geodesics in V has a nice expression. Take the local expression for the connection as $d + \alpha$ where α is a 1-form with values in $\mathrm{End}(T_m(M))$. It fact we will consider it as a function with values in the set of bilinear maps $T_m(M) \times T_m(M) \to T_m(M)$. For every $m \in M$, let β_m denote the form obtained by symmetrising the corresponding bilinear map. Then the equation is

1.11. $$c''(t) + \beta_{c(t)}(c'(t), c'(t)) = 0.$$

(Here c', c'' denote the first and second derivatives of c.)

1.12. Definition. Take any linear isomorphism of $T_m(M)$ with $\mathbb{R}^n$. Then composition of the inverse of the exponential map at m with the isomorphism of $T_m(M)$ with $\mathbb{R}^n$ gives a coordinate system in a neighbourhood of m, called a *normal coordinate system.*

1.13. Remark. Consider a normal coordinate system. On the one hand, the curves $t \mapsto tv$ are geodesics in this system, and on the other hand, this path satisfies the differential equation 1.11. Note that since the map is linear, the term $\frac{d^2}{dt^2} c(t)$ is zero and the equation reduces to $\sum_{jk} \Gamma^i_{jk}(tv) x_j x_k = 0$, where (x_i) are the coordinates of v. Letting t tend to zero, we get $\sum \Gamma^i_{jk}(0) x_j x_k = 0$. This vanishes for all values of x_j, x_k and so the coefficients $\Gamma^i_{jk}(0) + \Gamma^i_{kj}(0)$ of this quadratic polynomial are all zero. Thus we have the following

1.14. Theorem. *The functions $\Gamma^k_{ij} + \Gamma^k_{ji}$, where Γ^k_{ij} are defined by $\nabla_{\frac{\partial}{\partial x_i}}(\frac{\partial}{\partial x_j}) = \sum_k \Gamma^k_{ij} \frac{\partial}{\partial x_k}$ in normal coordinates around p, are all zero at p.*

We have the following improvement of Corollary 1.10.

1.15. Proposition. *For any $m \in M$, there exists an open nneighbourhood U such that any two points in U can be connected by a unique geodesic contained in U.*

Proof. We will again use normal coordinates (U, x) at $m \in M$, and use the Euclidean distance from m in these coordinates. For any x, y in this neighbourhood we will denote $\|x\|^2$ by $N(x)$ and the inner product by $b(x, y)$.

We have seen that there is an open ball V of radius r such that any two points in V can be connected by a geodesic in U. If c is any geodesic inside U that has $c(0)$ and $c(1)$ in V, but is not contained in V, then there exists t_0 with $0 < t_0 < 1$ such that $N(c(t))$ attains the maximum at t_0. This implies that $\frac{dN(c(t))}{dt}|_{t_0} = 0$ and also that $\frac{d^2 N(c(t)}{dt^2}|_{t_0} \leq 0$.

We now compute $\frac{d^2 N(c(t))}{dt^2}$ to be

$$2\frac{d}{dt} b(c(t), c'(t)) = 2N(c'(t)) + 2b(c(t), c''(t)).$$

If c is a geodesic, then by 1.11, this becomes

$$2N(c'(t)) - 2b(c(t), \beta_{c(t)}(c'(t), c'(t))).$$

Now consider the function which takes any v to the quadratic form $Q_v : w \mapsto N(w) - b(v, \beta_v(w, w))$. But $\beta_0 = 0$ by Theorem 1.14 and so Q_0 is positive definite. Hence so is Q_v for all v in a neighbourhood of 0. We may cut down our neighbourhood U to an open ball such that Q_v is positive definite for all v in it. Taking $w = c'(t)$ in the above computation, we see that $\frac{d^2 N(c(t)}{dt^2}$ is positive at t_0. This contradicts our assumption.

An open set such as guaranteed by Proposition 1.15 above is called a *convex neighbourhood.*

1.16. Remark. We make two comments. Firstly, there is a fundamental system of convex neighbourhoods. Secondly, an open set which satisfies the above condition is contractible. If we are interested only in this, we could have taken any suitable exponential neighbourhood. But if U and V are two convex open sets, then clearly their intersection is also convex. Hence we conclude that there is an open covering of M such that the intersection of any finite set of the members of the covering is contractible. In order to compute the cohomology of a constant sheaf, we may use the above type of covering, thanks to [Ch. 4, Remark 5.5, 1)]. From this one can conclude

1.17. Theorem *The integral singular cohomology groups of a compact differential manifold are finitely generated.*

2. Lifting of Symbols and Torsion

2.1. Torsion of a linear connection.

Let ∇ be a linear connection on M. Then we may try to lift some natural first order symbols, using the procedure we have explained in [Ch. 5, Proposition 1.8]. Thus we may try and lift the first order symbol $T^* \otimes T^* \to \Lambda^2(T^*)$ given by the wedge product. (According to our theory this can be lifted to a differential operator if we are given a connection in T^*, but a connection in T gives a canonical dual connection in T^*.) Thus we get a first order operator from T^* into $\Lambda^2(T^*)$ with this as its symbol. Notice that the exterior derivative which is defined without reference to any connection also has the same symbol.

Let us compute the lift which uses the linear connection. According to our definition, the lift is the composite of the operator $\nabla : T^* \to T^* \otimes T^*$ and the wedging map $T^* \otimes T^* \to \Lambda^2(T^*)$ given by the symbol. In other words, it associates, to a 1-form ω, the 2-form given by $(X, Y) \mapsto (\nabla_X(\omega))(Y) - (\nabla_Y(\omega))(X) = X\omega(Y) - \omega(\nabla_X Y) - Y\omega(X) + \omega(\nabla_Y X) = d\omega(X, Y) + \omega([X, Y]) - \omega(\nabla_X Y) + \omega(\nabla_Y X)$. Thus this coincides with the exterior derivative if and only if $\nabla_X Y - \nabla_Y X - [X, Y]$ vanishes for all vector fields X, Y. Since the symbols of d and the above operator coincide, it follows that the form $(X, Y) \mapsto \nabla_X Y = \nabla_Y X - [X, Y]$ is an $\mathcal{A}$-bilinear form with values in T.

2.2. Definition. Let ∇ be a linear connection on a differential manifold M. The exterior 2-form with values in T given by $(X, Y) \mapsto \nabla_X Y - \nabla_Y X - [X, Y]$ is called the *torsion* of the connection and is denoted by $\tau(\nabla)$.

It is often convenient to use a linear connection whose torsion tensor vanishes identically. Suppose ∇ is a linear connection with torsion τ. Any other linear connection differs from ∇ by a 1-form with values in $\mathrm{End}(T)$. If $\nabla'_X = \nabla_X + \alpha(X)$, then the torsion form τ' of ∇' is given by $\tau'(X, Y) = \nabla'_X(Y) - \nabla'_Y(X) - [X, Y] = \nabla_X(Y) + \alpha(X)(Y) - \nabla_Y(X) - \alpha(Y)(X) - [X, Y] = \tau(X, Y) + \beta(X, Y) - \beta(Y, X)$ where β is the bilinear form defined by $\beta(X, Y) = \alpha(X)(Y)$.

If β is a bilinear map into T such that $\beta(X, Y) - \beta(Y, X) = -\tau(X, Y)$, then τ' is identically zero. For example, we can take $\beta(X, Y) = -\frac{1}{2}\tau(X, Y)$. Also ∇ and ∇' have the same torsion form if and only if their difference β is symmetric. Thus we have proved the following assertion.

2.3. Proposition. *There exists a torsion free connection. Any two torsion free connections differ by a symmetric 2-form with values in T.*

Often we require a torsion free connection which leaves some structure on the manifold invariant, but this is a more delicate question and will be dealt with in Chapter 7. We will only illustrate this by a simple example here.

A linear connection gives a connection on all the tensor bundles as well. In particular, it gives a connection on $\mathcal{K} = \Lambda^n(T^*)$. Since the sheaf $\mathcal{S}$ of densities is the tensor product of $\mathcal{K}$ with the local system OR, there is also a natural connection on $\mathcal{S}$. We have seen in [Ch. 3, Remark 2.10] that the sheaf of densities on a differential manifold is actually trivial, although noncanonically. Let m be an everywhere positive density, thus trivialising the sheaf $\mathcal{S}$. We will now see if there is a torsion free linear connection which leaves m invariant.

Let ∇ be any torsion free linear connection. Define a differential 1-form ω by setting $\nabla_X(m) = \omega(X)m$. By Proposition 2.3, any other torsion free linear connection ∇' is given by $\nabla'_X = \nabla_X + \alpha(X)$, where the bilinear map $\beta : T \times T \to T$ given by $(X, Y) \mapsto \alpha(X)Y$ is symmetric. Now $\nabla'_X - \nabla_X$, acting on $\mathcal{S}$, associates to X, the trace of $Y \mapsto \beta(X, Y)$. Given any linear form ω on T we can find a symmetric bilinear map $\alpha : T \times T \to T$ such that $\omega(X) = \mathrm{tr}(Y \to \beta(X, Y))$, for all X. For example, the map $(X, Y) \mapsto 1/(n+1)(\omega(X)Y + \omega(Y)X)$ has this property. This proves the following.

2.4. Proposition. *There exists a torsion free linear connection which leaves a given everywhere positive density invariant.*

2.5. Exercises.

1) Determine all linear connections with the above property.

2) Find the lift using a linear connection of the symbol of the Lie derivative with respect to a vector field.

The computation in 2.1 above can be used to get the following general result. Let ∇ be a connection on a vector bundle E. Then we have defined in [Ch. 5, 1.10, 1.12] an exterior derivative map $\Lambda^{i-1}(T^*) \otimes E \to \Lambda^i(T^*) \otimes E$. On the other hand, using a torsion free connection and the satellite connections ∇ on $\Lambda^{i-1}(T^*)$ and the connection on E, we may lift the symbol of the above, into a differential operator $\Lambda^{i-1}(T^*) \otimes E \to \Lambda^i(T^*) \otimes E$. These two coincide if the torsion of the linear connection is zero. If α is a k-form with values in E, then the exterior derivative $d\alpha$ which uses the connection on E coincides with $(X_1, \ldots, X_{k+1}) \mapsto \sum (-1)^{i+1} (\nabla_{X_i}\alpha)(X_1, \ldots, \hat{X}_i, \ldots, X_{k+1})$.

In particular we may take for E the bundle $\mathrm{End}(T)$ and for α the curvature form R. Then the (second) Bianchi identity [Ch. 5, 6.1] says that $d_\nabla R = 0$. Thus for linear connections, the identity can also be stated as follows.

2.6. Identity (Bianchi). *If ∇ is a torsion free linear connection and R its curvature form, then*

$$\sum \nabla_X R(Y, Z) = 0$$

where the sum is taken over cyclic permutations of the vector fields X, Y, Z.

2.7. Remark. Let ζ be the canonical 1-form with values in the tangent bundle T given by $\zeta(X) = X$ for all vector fields X. Then $d_\nabla \zeta$ is the torsion tensor. In fact, by definition, $d_\nabla \zeta(X, Y) = \nabla_X \zeta(Y) - \nabla_Y \zeta(X) - \zeta([X, Y])$, which is just the definition of the torsion tensor.

2.8. Bianchi identity. *The curvature form R of a torsion free linear connection satisfies the following identity:*

$$R(X, Y)Z + R(Y, Z)X + R(Z, X)Y = 0$$

for any three vector fields X, Y, Z.

Proof. This is just a question of plugging in the value $R(X, Y)Z = \nabla_X \nabla_Y Z - \nabla_Y \nabla_X Z - \nabla_{[X,Y]} Z$ and summing it cyclically. Using the vanishing of the torsion tensor, the left side of the above identity reduces to the cyclic sum

$$\sum \nabla_X([Y, Z]) + \nabla_Y([Z, X]) - \nabla_{[X,Y]} Z$$

which in turn simplifies to

$$[X, [Y, Z]] + [Y, [Z, X]] + [Z, [X, Y]].$$

This is zero, thanks to the Jacobi identity [Ch. 2, 2.1]. One can also give a slick proof by observing that the expression $\sum R(X, Y)Z$ is the exterior product of R and the 1-form ζ with values in T given by $\zeta(X) = X$ for all vector fields X. To give a meaning to the exterior product, we use the obvious pairing $\operatorname{End} T \otimes T \to T$. On the other hand, $d_\nabla \zeta$ is simply the torsion tensor T which has been assumed to be zero. Hence our assertion follows from the Bianchi identity in [Ch. 5, 6.1].

2.9. Remark. This is known as the *first* Bianchi identity. But since the second one involving derivatives, stated in [Ch. 5, 6.1], is valid for any connection on a principal bundle, it was stated earlier.

2.10. Lifting of a symbol and its adjoint.

Let $s : E \to T \otimes F$ be a first order symbol. Its adjoint has been defined to be the negative of the natural transpose map $\mathrm{adj}(s) : F^* \to T \otimes E^*$. If we choose connections ∇_E and ∇_F on the vector bundles E and F, then we can lift these into differential operators $D_s : E \to F$ and $D_{\mathrm{adj}(s)} : F^* \to E^*$ respectively. It is natural to ask how $D_{\mathrm{adj}(s)}$ is related to the adjoint operator $\mathrm{adj}(D_s)$. Firstly, the adjoint of D_s is actually an operator from $F^* \otimes \mathcal{S}$ to $E^* \otimes \mathcal{S}$. But then $\mathcal{S}$ can be trivialised. Let us in fact fix an everywhere positive density (trivialising $\mathcal{S}$) and a torsion free linear connection which leaves it invariant. We would like in this case to compare $D_{\mathrm{adj}(s)}$ and $\mathrm{adj}(D_s)$. By definition $D_s = \tilde{s} \circ d_{\nabla_E}$. Hence its adjoint is given by $\mathrm{adj}(d_{\nabla_E}) \circ \mathrm{adj}(\tilde{s})$. So we need only determine $\mathrm{adj}(d_{\nabla_E}) : T \otimes E^* \otimes \mathcal{S} \to E^* \otimes \mathcal{S}$.

We may apply evaluation $e(X)$ on a vector field X and then compute the adjoint, namely $\mathrm{adj}(\nabla_X)$. Then the adjoint of ∇ itself is obtained by the equation $\mathrm{adj}(\nabla_X)(s, \mu) = (\mathrm{adj}\, d_\nabla)(X, s, \mu)$. As for $\mathrm{adj}(\nabla_X)$ we have the following easy computation.

2.11. Proposition. *The adjoint of ∇_X, where ∇ is a connection on a vector bundle E and X a vector field, is $-L(X)$ acting on $E^* \otimes \mathcal{S}$. Here $L(X)$ is the Lie derivative of densities with values in E^* with respect to the dual connection on E^*.*

Proof. Since the statement is local, there is no harm in assuming that E is trivial. We will then compare the adjoint of the given connection and the trivial connection ∇_{tr} on E. We will identify $\mathcal{S}$ with $\mathcal{K}$ and make the computation. Let $\nabla = \nabla_{tr} + \alpha$, as usual. Then $\mathrm{adj}\,\nabla_X = \mathrm{adj}(\nabla_{tr})_X + \alpha(X)^t$. The assertion for the trivial connection is actually the definition of the adjoint of a vector field. On the other hand, the Lie derivative with respect to the dual connection and that with respect to the trivial connection differ by $-\alpha^t$, so that we have $-L_\nabla(X) = -L_{tr}(X) + \alpha(X)^t$. This proves the proposition.

We have seen that if we take a torsion free linear connection and use the induced connection on S, then the Lie derivative using the connection is the same as the canonical Lie derivative. Hence in this case, we conclude that the adjoint of ∇ is obtained by lifting the adjoint symbol, using for the lifting, the dual connection on E^* and a torsion free linear connection. From this we get in general the following result.

2.12. Theorem. *Let $s : E \to T \otimes F$ be a first order symbol and D_s the $E \to F$ which lifts s using a connection ∇ on E. Lift the adjoint symbol $\mathrm{adj}(s) : F^* \otimes \mathcal{S} \to T \otimes E^* \otimes \mathcal{S}$ using a connection on F and the connection on $\mathcal{S}$ induced by a torsion free linear connection on the manifold. If the* *symbol s is invariant under the connections chosen, then the lift of* adj(s) *is the same as the adjoint of D_s.*

Proof. We can write D_s as $\tilde{s} \circ d_{\nabla_E}$. Hence adj$(D_s) = $ adj$(d_{\nabla_E}) \circ ((s)^t \otimes I_\mathcal{S})$. We just observed that adj(d_{∇_E}) is the lift of the adjoint of the symbol of ∇_E, namely $-I_{T \otimes E^* \otimes \mathcal{S}}$. The composite symbol is $-(s)^t \otimes I_\mathcal{S} = $ adj(s). Under our assumption then, the lift of the composite symbol adj(s) is adj(D_s).

2.13. Linear connection on a submanifold.

Let M be a differential manifold and N a submanifold. Then we have the following exact sequence of bundles on N:

$$0 \to T_N \to T_M|N \to \mathrm{Nor}(N, M) \to 0.$$

We know that this sequence always splits. Assume given a specific splitting $t : T_M|N \to T_N$. If ∇ is a linear connection on M, then it induces a connection $i^*\nabla$ on the restriction $T_M|N$. Using the splitting we can define a connection on T_N. In fact, if X, Y are vector fields on N, we define $\nabla_X^N Y = t((i^*\nabla)_X Y)$, where i is the inclusion of N in M.

2.14. Exercise.
If the linear connection on M is torsion free, then find the condition on the splitting which will ensure that the induced linear connection on N is also torsion free.

2.15. Cartan connections.

We conclude this chapter by giving the definition of a related notion called Cartan connection. Suppose G is a (connected) Lie group and H a closed subgroup. Consider the principal H-bundle $G \to G/H$. If P is any principal H-bundle on a differential manifold M, then we may consider the vector bundle associated to P by the isotropy representation of H on $\mathfrak{g}/\mathfrak{h}$. The set-up for a Cartan connection is an isomorphism of this associated bundle with the tangent bundle of M.

2.16. Examples.

1) The structure group of the tangent bundle can be extended to the affine group $G = GA(n, \mathbb{R})$, using the splitting $H = GL(n, \mathbb{R}) \to GA(n, \mathbb{R})$. The isotropy representation of $GL(n, \mathbb{R})$ in $\mathfrak{g}/\mathfrak{h}$ is simply the natural representation of $GL(n, \mathbb{R})$ in $\mathbb{R}^n$. Take for P the bundle of frames on the tangent bundle. Clearly the bundle associated to the isotropy representation is the tangent bundle.

2) Let V be an $(n + 2)$-dimensional real vector space with a quadratic form q of signature $(n + 1, 1)$. The group $SO(q)$ is not connected but has two connected components. Let G be the connected component containing 1. Then there is a natural G-action on the quadric $Q = $

$\{v \in V : q(v) = 0\}$ in $\mathbb{P}(V)$. It is easy to check that G acts transitively on Q. Let H be the isotropy subgroup at a point v_0 of Q. Then the isotropy action of H is an n-dimensional representation. There is a natural homomorphism χ of H into $\mathbb{R}^{\times}$, defined by $hv_0 = \chi(h)v_0$. The kernel acts on $v_0^{\perp}/\mathbb{R}v_0$ and maps isomorphically onto $SO(q')$ where q' is the induced positive definite metric on $v^{\perp}/\mathbb{R}v_0$. Now it is easy to see that the isotropy representation in $\mathfrak{g}/\mathfrak{h}$ of H can be identified with this space and the action satisfies $q(hv) = (\chi(h))^2 q(v)$. Given a metric on $T(M)$ we can get an H-bundle to which it is associated via the isotropy representation. A G-connection is then a Cartan connection.

2.17. Definition. A *Cartan connection* is a connection on the G-bundle such that there is an isomorphism α of the tangent bundle with the isotropy bundle. This isomorphism considered as an 1-form is called the *soldering form*.

The curvature of the Cartan connection is a 2-form with values in the adjoint bundle with $\mathfrak{g}$ as fibres, and so gives also a 2-form with values in the isotropy bundle. This is called the *torsion tensor* of the Cartan connection.

Let us consider Example 1) above. In this case we take the soldering form to be the identity. The Cartan connection in this case is called an *affine connection*. Since $GL(n, \mathbb{R})$ is a *quotient* of $GA(n)$, the affine connection has as its image a linear connection. Therefore the curvature of the affine connection has also two components, namely the curvature of the linear connection, and a 2-form with values in the tangent bundle. The latter is easily computed to be the absolute derivative of the 1-form α. Thus we have $d\alpha(X, Y) = \nabla_X \alpha(Y) - \nabla_Y(\alpha)(X) - \alpha([X, Y])$. Noting that α is the identity form, we conclude that the torsion form of the affine connection, considered as a Cartan connection, is the torsion of the linear connection.

Exercises

1) Let M be a manifold provided with a linear connection. Write down the condition that the horizontal lift of a vector field X on M to the principal tangent bundle leaves the connection form invariant, in terms of the Lie and covariant derivatives with respect to X.

2) Show that the set of all vector fields on M which leave a given linear connection invariant is closed under brackets and forms a Lie algebra of dimension $\leq n^2 + n$, where n is the dimension of M.

3) Consider the principal bundle of frames of the tangent bundle of a differential manifold. Construct a canonical differential 1-form on it with values in $\mathbb{R}^n$. Show that it is equivariant under the action of $GL(n)$ and is zero on vertical vectors. Given a connection form, compute its exterior derivative with respect to the connection.

4) Suppose given a diffeomorphism A of M of finite order. Show that there exists a linear connection on M invariant under it. If $m \in M$ is an isolated fixed point, then show that it is also infinitesimally isolated, that is to say, the differential of its action at $T_m(M)$ does not have 1 as an eigenvalue.

5) Suppose G is a connected Lie group and H a closed subgroup. Assume given a vector subspace $\mathfrak{m}$ of $\mathfrak{g}$ which is invariant under the isotropy action of H and is supplementary to $\mathfrak{h}$. Then show that there is a canonical connection in the principal H-bundle $G \to G/H$. Compute its curvature form in terms of the Lie algebra structure.

6) Show in the above example that the bundle associated to the principal H-bundle $G \to G/H$ for the isotropic action of H is the tangent bundle of G/H. Under the same hypothesis as above, compute the torsion tensor field of the linear connection given by the canonical connection.

7) A linear connection on a differential manifold M is said to be locally symmetric if for every $m \in M$ there is an involution with m as an isolated point that leaves the connection invariant. Show that a linear connection is locally symmetric if and only if it is torsion free and the covariant derivative of the curvature form is zero.

8) Let M be a differential manifold and let the tangent bundle be the direct sum of two subbundles E and F. Assume that there exists a J-structure on E. Then find the condition for the existence of a linear connection which preserves the decomposition as well as the J-structure on E.

Chapter 7

Manifolds with Additional Structures

We are interested here in a general study of differential manifolds which are endowed with some additional structures. Many of these structures involve, in the first place at least, a reduction of the structure group of the tangent bundle from $GL(n, \mathbb{R})$ to some standard subgroup.

1. Reduction of the Structure Group

1.1. Orientation.

We have already encountered a few examples of such structures. We start with the case when the subgroup is $GL(n, \mathbb{R})^+$, consisting of linear transformations with positive determinant. A reduction to this subgroup of the structure group is equivalent to providing the differential manifold with an orientation. We know that such a reduction is not in general possible. In fact, we have seen in [Ch. 5, Proposition 3.14] that a reduction is equivalent to giving a section of the bundle $P/GL(n, \mathbb{R})^+$, where P is the principal bundle of frames of T. This has fibre type $GL(n, \mathbb{R})/GL(n, \mathbb{R})^+$. The determinant homomorphism into $\mathbb{R}^\times$ induces an isomorphism of this quotient with $\mathbb{R}^\times/\mathbb{R}^+$. Thus $P/GL(n, \mathbb{R})^+ \to M$ is a 2-sheeted covering space of M. If the total space of this covering is connected, it admits no section and so there is no such reduction. If it is not, it consists of two connected components both of which map isomorphically onto M. Thus in the latter case, there are two possible sections, that is to say, two orientations, said to be *opposites* of each other.

1.2. Riemannian structure.

To give a symmetric 2-form g, which is positive definite at every point, is equivalent to a reduction of the structure group of the tangent bundle to $O(n)$. This datum is called a *Riemannian structure* on M. We have already remarked [Ch. 5, Example 3.1, 1) and Remark 3.12] in the general context of arbitrary vector bundles that such a structure always exists. Indeed, locally we may choose a coordinate system $(x_1, \ldots, x_n)$ and take the form $\sum dx_i^2$. This gives a Riemannian structure locally. If g_i are Riemannian structures on open sets U_i of a covering, and φ_i a partition of unity with respect to this covering, then $\sum \varphi_i g_i$ is a global Riemannian metric, since a convex combination of positive definite forms is again one. If the manifold is oriented, then a Riemannian structure actually gives a reduction of the structure group to $SO(n)$.

If we replace the condition of positive definiteness of the symmetric tensor by nondegeneracy, then, over a connected set its signature does not change, and so we may as well say that the form has a fixed signature (p, q) at all points of M. This is equivalent to a reduction of the structure group to $O(p, q)$ for a fixed pair (p, q) of positive integers with $p + q = n$. We have remarked that such a structure (if $p \neq 0, n$) need not exist in general. If it does, the structure is called a *pseudo-Riemannian structure with signature of type* (p, q).

1.3. Definition. A manifold provided with a Riemannian (resp. pseudo-Riemannian) structure is called a *Riemannian* (resp. a *pseudo-Riemannian*) *manifold.*

1.4. Examples.

1) The tangent space at any point of any real vector space V of finite dimension is identified with the vector space V itself. Hence any positive definite symmetric bilinear form on V gives rise to a Riemannian structure on it. Usually on $\mathbb{R}^n$, we take the Euclidean inner product $g(x, y) = \sum x_i y_i$. On the other hand, if we take the bilinear form on $\mathbb{R}^4$ given by $g(x, y) = -x_1^2 + x_2^2 + x_3^2 + x_4^2$, we get a pseudo-Riemannian manifold with signature $(3, 1)$. This is the Lorentz metric on $\mathbb{R}^4$, basic to the study of special relativity.

2) There is a natural metric on the unit sphere S^n. The tangent space at a point $x \in \mathbb{R}^{n+1}$ with $\sum x_i^2 = 1$ consists of the vectors $v = (v_i)$ with $\sum x_i v_i = 0$. If v is such a vector we define $g(v) = \sum v_i^2$. It is clear that this gives a Riemannian structure on S^n. Whenever we refer to S^n as a Riemannian manifold, we have in mind this structure.

1.5. Exercise. Carry out the construction of the second example in the context of a vector space V provided with a symmetric nondegenerate bilinear form β of signature (p, q), $p \neq 0$. Show that the subset $W = \{v \in V : \beta(v) = 1\}$ is a closed submanifold of V. What is the signature of the induced pseudo-Riemannian structure on the submanifold W?

1.6. Remarks.

1) Obviously the second example above is derived from the first. This is a general procedure. If M is a Riemannian manifold, and N is a submanifold, then there is a natural Riemannian structure on the submanifold. In fact, for any $x \in N$, we may restrict the metric on the tangent space $T_x(M)$ to the subspace $T_x(N)$ and obtain in this way a Riemannian structure on N. The Riemannian manifold thus obtained will be called a *Riemannian submanifold*. In fact, for this construction, it is enough if N is taken to be a manifold together with a differentiable map into M such that the differentiable map is injective at all points of N. The structure obtained in this way is said to be *induced* from that of M.

2) Notice, however, that a similar construction *does not* work in the generality of pseudo-Riemannian manifolds. For, while the restriction of a positive definite form to any vector subspace remains positive definite, the restriction of a nondegenerate form need not remain nondegenerate. However, if M is provided with a pseudo-Riemannian structure of signature (p, q), and N is a submanifold of dimension q such that the restriction of the bilinear form on $T_x(M)$ to $T_x(N)$ is positive definite for any point x of N, then N gets a natural Riemannian structure. Although this is quite an artificial situation, it does occur in the following case. Consider the closed subspace H of $\mathbb{R}^{n+1}$ given by the equation $\sum_{i=1}^{n} x_i^2 - x_{n+1}^2 = -1$. It is easily verified (using Ch. 1, Example 3.2, 4)) that it is actually a differential submanifold of $\mathbb{R}^{n+1}$. It is not connected since x_{n+1} is never zero. Indeed it has two connected components given by $x_{n+1} > 0$ and $x_{n+1} < 0$. Consider the restriction of the projection $(x_1, \ldots, x_{n+1}) \mapsto (x_1, \ldots, x_n)$ of $\mathbb{R}^{n+1}$ to $\mathbb{R}^n$ to each of these. Clearly it is bijective and the inverse is given by $(x_1, \ldots, x_n) \mapsto \left(x_1, \ldots, x_n, \sqrt{\sum_{i=1}^{n} x_i^2 + 1}\right)$. It is also differentiable and has the map $(v_i) \mapsto \left(v_1, \ldots, v_n, \frac{\sum a_i v_i}{\sqrt{\sum a_i^2 + 1}}\right)$ as its differential at a point $(a_i) \in \mathbb{R}^n$. We now provide $\mathbb{R}^{n+1}$ with the *pseudo-Riemannian* structure given by the quadratic form $\sum_{i=1}^{n} x_i^2 - x_{n+1}^2$. This gives the quadratic form $(v_i) \mapsto \left(\sum v_i\right)^2 - \frac{\left(\sum a_i v_i\right)^2}{\sum a_i^2 + 1}$ on the tangent space at $(a_i) \in \mathbb{R}^n$. It is always positive whenever at least one v_i is nonzero. In other words, the induced quadratic form on the tangent space at any point of H is positive definite. Hence we get a Riemannian structure on H. This space is called the *hyperbolic space* of dimension n. This example will come up again later in 6.13.

3) Clearly the Riemannian structure on $\mathbb{R}^n$ is invariant under translations. Since the translatory action is transitive, it is determined by the metric on the tangent space at 0. If T is any linear transformation, then T takes the Riemannian metric to another metric, also invariant under translations. The new metric has, on the tangent space at 0, the transform by T of the standard metric, namely $T'T$. In particular, the Euclidean metric is invariant under the natural action of the orthogonal group. Putting these together we conclude that the Riemannian metric is invariant under the action of the Euclidean motion group, namely the group generated by orthogonal transformations and translations.

Also the sphere is invariant under the above action of $O(n)$. The induced Riemannian structure on S^n is clearly invariant under this action.

4) If M and N are Riemannian manifolds, then one can provide $M \times N$ with a natural Riemannian structure, called the *product of M and N*. In fact, the tangent space at any point (m, n) of $M \times N$ is canonically the direct sum of $T_m(M)$ and $T_n(N)$. Since both of these spaces come with natural metrics, we can provide the direct sum with a metric structure with respect to which the two subspaces are orthogonal.

5) Consider a compact Lie group G acting on a manifold M. Let g be a Riemannian structure on M. Then we may *average* the metric over the group G to obtain a G-invariant Riemannian structure. In other words, if v, w are tangent vectors at a point $m \in M$, then we define $\tilde{g}(v, w)$ to be $\int g(xv, xw)dx$ where the integration is taken over G with respect to the invariant measure dx on G, and xv, xw are the vectors at xm obtained as translates of v, w by the action of $x \in G$. It is easy to verify that the Riemannian structure is G-invariant in the obvious sense.

6) We may construct on any Lie group G a metric invariant under left translations. As in the case of the Euclidean space, this metric depends only on the metric on the tangent space at one point, say 1. If we need this to be right invariant as well, this metric ought to be invariant under the action of G on the tangent space at 1, given by left translating by g and right translating by g^{-1}, in other words, the adjoint representation of G in its Lie algebra $\mathfrak{g}$. If G is compact, such a metric exists in view of 5) above, and thus we can find a *biinvariant* Riemannian metric in this case.

1.7. Almost symplectic structure.

Let ω be an alternating 2-form on T. We assume again that this form is nondegenerate. This is called an *almost symplectic structure* on M.

1.8. Definition. An *almost symplectic manifold* is a differential manifold together with a nondegenerate (exterior) 2-form.

The existence of a nondegenerate alternating bilinear form on a vector space implies that the dimension of the vector space is even. In particular, it follows that an almost symplectic manifold is even-dimensional. However this alone will not ensure that the manifold admits an almost symplectic structure. For example the symplectic form on a 2-dimensional manifold gives an everywhere nonzero section of $\Lambda^2(T^*)$. This implies that the differential manifold is oriented.

1.9. Exercise. Show that an almost symplectic manifold M of dimension $2m$ has an everywhere nonzero $2m$-form. Deduce that M is oriented.

1.10. Remark. This datum is equivalent to a reduction of the structure group of the tangent bundle to the symplectic group $Sp(n, \mathbb{R})$. For, given any nondegenerate, alternating bilinear form ω on a real vector space V, one can find a basis (e_i, f_i) such that $\omega(\sum a_i e_i + \sum b_j f_j, \sum a'_i e_i + \sum b'_i f_i) = \sum a_i b'_i - \sum b_i a'_i$. In fact, choose any nonzero vector (if $V \neq (0)$), say e_1. Since ω is nondegenerate, there is at least one vector v such that $\omega(e_1, v) \neq 0$. Dividing v by a nonzero scalar, we may as well assume that $\omega(e_1, v) = 1$. Define f_1 to be such a vector. Clearly e_1 and f_1 span a two-dimensional subspace V'. Take the subspace $\{w \in V : \omega(e_1, w) = 0 \text{ and } \omega(f_1, w) = 0\}$. If $\dim V$ is greater than 2, this space is nonzero and supplementary to V'. The restriction of ω to it remains nondegenerate. Thus we find a basis as above by an induction on the dimension.

The group $GL(2n, \mathbb{R})$ acts on the space of all nondegenerate alternating bilinear forms by the prescription $g(\omega)(v, w) = \omega(g^{-1}v, g^{-1}w)$. Our remark above shows that this action is transitive, and the isotropy group is by definition $Sp(n, \mathbb{R})$. Thus an almost symplectic structure is simply a section of $P/Sp(n, \mathbb{R})$ where P is the principal tangent bundle of frames.

1.11. Examples.

1) Consider $\mathbb{R}^{2n}$. The tangent space at any point can be canonically identified with $\mathbb{R}^{2n}$ itself. Hence the choice of any nondegenerate alternating bilinear form on the *vector space* $\mathbb{R}^{2n}$, for example, $\omega(x, y) =$

$\sum_{i=1}^{n}(x_i y_{n+i} - x_{n+i} y_i)$, gives rise to a symplectic structure on the manifold $\mathbb{R}^{2n}$. This differential form has the coordinate expression $\sum_{i=1}^{n} dx_i \wedge dx_{n+i}$.

2) Let N be any manifold. Consider its cotangent bundle as a manifold M. Let π be the bundle projection $M \to N$. Any point $m \in M$ consists, by definition, of a point $a \in N$ and a 1-differential α at a. Then we have $\pi(m) = a$. If $v \in T_m(M)$ is a tangent vector at m, then the differential of π at m maps v to a tangent vector at the point a. The scalar $\langle d\pi(v), \alpha \rangle$ depends linearly on v and so defines a tautological 1-form β on M.

It is easy to compute it in local coordinates. Let (x_i) be a coordinate system on an open set U on N. Then $\frac{\partial}{\partial x_i}$ may be considered a function on $\pi^{-1}(U)$ which is linear on fibres. A coordinate system for M in $\pi^{-1}(U) = U \times \mathbb{R}^n$ is given by $(x_i \circ \pi, \frac{\partial}{\partial x_j})$. Let us denote this system of coordinates by (q_i, p_i). We will now compute $(d\pi)(\frac{\partial}{\partial q_i})$. Its value on the coordinate function x_j is given by $\frac{\partial}{\partial q_i}(x_j \circ \pi) = \frac{\partial}{\partial q_i}(q_j) = \delta_{ij}$. Hence $d\pi(\frac{\partial}{\partial q_i}) = x_i$. Also $d\pi(\frac{\partial}{\partial p_i})(x_j) = \frac{\partial}{\partial p_i}(x_j \circ \pi) = \frac{\partial}{\partial p_i}(q_j) = 0$. If v is the vector $\sum a_i \frac{\partial}{\partial q_i} + \sum b_j \frac{\partial}{\partial p_j}$ at the point (q_i, p_j), then $d\pi(v) = \sum a_i \frac{\partial}{\partial x_i}$. On the other hand the point (q_i, p_j) represents the differential $\sum p_j \frac{\partial}{\partial x_j}$. Hence $\beta(v) = \langle \sum a_i \frac{\partial}{\partial x_i}, \sum p_j dx_j \rangle = \sum a_i p_i$. In other words, we have $\beta = \sum p_j dq_j$. Incidentally this also shows that the form β on M that we defined is differentiable. The exterior derivative of this form is the 2-form $\sum dp_j \wedge dq_j$, which is clearly nondegenerate. Thus there is a canonical almost symplectic structure on $M = T^*(N)$.

1.12. Remarks.

1) Notice that in the examples we have given above, the symplectic form is actually a *closed* 2-form. In that case, the manifold is said to be a *symplectic manifold*. We will return to this question later (4.11, 4.12).

2) In the Hamiltonian formulation of classical mechanics, the cotangent bundle with the symplectic form is the basic manifold on which the dynamical equations are written. In this context, if M is the configuration space, the manifold $T^*(M)$ is called the *phase space*.

1.13. Almost complex structure.

1.14. Definition. Let M be a differential manifold. An automorphism J of the tangent bundle (that is to say, a gauge transformation of the tangent bundle) satisfying $J^2 = -I$ is called an *almost complex structure* on M. A manifold provided with such a structure will be called an *almost complex manifold*.

As in the case of a symplectic manifold, an almost complex manifold is necessarily even-dimensional. This follows from the fact that any transformation J of a (finite-dimensional, real) vector space V with $J^2 = -\operatorname{Id}$ is semisimple and has eigenvalues $\pm i$, the multiplicities of these two eigenvalues being equal. The vector space $V \otimes_{\mathbb{R}} \mathbb{C}$ is a direct sum of the eigenspaces corresponding to i and $-i$ (which we denote respectively by $V^{1,0}$ and $V^{0,1}$). The complex conjugation taking $v + iw \in V \otimes \mathbb{C}$, with $v, w \in V$, to $v - iw$, interchanges the two eigenspaces. Hence $\dim V \otimes \mathbb{C} = 2 \dim V^{1,0} = 2 \dim V^{0,1}$. Taking V to be the tangent space T_m at a point m, we deduce that the dimension of M, which is the same as the dimension of T_m for any $m \in M$, is even. The *complex dimension* of an almost complex manifold is defined to be half the dimension of the differential manifold.

An almost complex structure is simply a reduction of the structure group of the tangent bundle from $GL(2n, \mathbb{R})$ to the subgroup $GL(n, \mathbb{C})$. Here $GL(n, \mathbb{C})$ is imbedded in $GL(2n, \mathbb{R})$ by the map

$$(a_{ij}) \mapsto \begin{pmatrix} \operatorname{Re}(a_{ij}) & -\operatorname{Im}(a_{ij}) \\ \operatorname{Im}(a_{ij}) & \operatorname{Re}(a_{ij}) \end{pmatrix}.$$

It is clear that $GL(2n, \mathbb{R})$ acts by inner automorphisms on the set of all transformations J of $\mathbb{R}^{2n}$ satisfying $J^2 = -\operatorname{Id}$. We have shown above that with respect to a suitable basis of V, the matrix of J takes the form $\begin{pmatrix} 0_n & -I_n \\ I_n & 0_n \end{pmatrix}$. Hence the above action is transitive. Identifying $\mathbb{R}^{2n}$ with $\mathbb{C}^n$ we get one such transformation, namely multiplication by i. The isotropy group at this point is the set of real linear transformations of $\mathbb{R}^{2n}$, commuting with multiplication by i, that is to say, the set of complex linear transformations. Thus we may identify $GL(2n, \mathbb{R})/GL(n, \mathbb{C})$ with the set $\{J \in GL(2n, \mathbb{R}) : J^2 = -I\}$. From this we conclude that an almost complex structure on a differential manifold is simply a section over M of the principal bundle of frames of the tangent bundle, modulo the action of $GL(n, \mathbb{C})$. This amounts to a reduction to $GL(n, \mathbb{C})$ of its structure group.

We note that if V is as above, then V^* also has a transformation J with $J^2 = -\operatorname{Id}$. In particular, it is a direct sum of two subspaces $V^*_{1,0}$ and $V^*_{0,1}$ as above. These may also be characterised as the spaces of (complex) linear forms on $V \otimes \mathbb{C}$ that vanish on $V^{0,1}$ and $V^{1,0}$ respectively. If we take the exterior powers $\Lambda^r(V^* \otimes \mathbb{C}) = \Lambda^r(V^*) \otimes \mathbb{C}$, then they break up into a direct sum of subspaces $\Lambda^p(V^*_{1,0}) \otimes \Lambda^q(V^*_{0,1})$ with $p + q = r$. These subspaces may be denoted $\Lambda^{p,q}(V^*)$. If we regard elements of Λ^r as alternating r-forms on $V \otimes \mathbb{C}$, then $\Lambda^{p,q}(V^*)$ consists of those forms which vanish whenever more than p of the vectors belong to $V^{1,0}$ or more than q of the vectors belong to

$V^{0,1}$. It is easy to check that the complex conjugation in $V \otimes \mathbb{C}$ induces an isomorphism of $\Lambda^{p,q}(V^*)$ with $\Lambda^{q,p}(V^*)$ (as real vector spaces). Thus in an almost complex manifold it makes sense to talk of a differential form of *type* (p, q). It is a section of the bundle $\Lambda^r(T^*) \otimes \mathbb{C}$ which belongs to $\Lambda^{p,q}(T_m^*)$ for all $m \in M$. The *real vector bundles* $\Lambda^{p,q}(T^*)$ and $\Lambda^{q,p}(T^*)$ are isomorphic, thanks to complex conjugation.

1.15. Remark. Any complex manifold has a natural almost complex structure. For if $(U, z_1, \ldots, z_n)$ is a complex coordinate system, then the real tangent space at any point in U can be identified with $\mathbb{C}^n$. Hence there is an operator J on the tangent space, obtained by transporting the operation of multiplication by the complex scalar i. The key point is that if we take any other coordinate system, the two differ by a *holomorphic* map $(f_i(z))$ of a domain in $\mathbb{C}^n$ with another. The induced map on the tangent space considered as $\mathbb{R}^{2n}$ is given by a matrix of the type

$$\begin{pmatrix} \dfrac{\partial \operatorname{Re}(f_i)}{\partial x_j} & \dfrac{\partial \operatorname{Re}(f_i)}{\partial y_j} \\ \dfrac{\partial \operatorname{Im}(f_i)}{\partial x_j} & \dfrac{\partial \operatorname{Im}(f_i)}{\partial y_j} \end{pmatrix}$$

Here x_j, y_j are respectively the real and imaginary parts of z_j, and the matrix has been written with respect to the basis $\frac{\partial}{\partial x_j}, \frac{\partial}{\partial y_j}$ of the tangent space. In this basis the transformation J we have defined is given by the matrix

$$\begin{pmatrix} 0 & -1 \\ 1 & 0 \end{pmatrix}.$$

The condition that these two matrices commute is equivalent to the conditions:

$$\frac{\partial \operatorname{Re}(f_i)}{\partial x_j} = \frac{\partial \operatorname{Im}(f_i)}{\partial y_j},$$
$$\frac{\partial \operatorname{Im}(f_i)}{\partial x_j} = -\frac{\partial \operatorname{Re}(f_i)}{\partial y_j}.$$

These are the Cauchy-Riemann equations satisfied by holomorphic functions. Since our transition functions are holomorphic, it follows that the almost complex structure we have defined using a coordinate system, is indeed canonical.

1.16. Almost Hermitian structures.

1.17. Definition. A differential manifold which has a Riemannian structure g as well as an almost complex structure J such that $g(Jv, w) + g(v, Jw) = 0$ for all pairs of tangent vectors v, w at any point, is called an *almost Hermitian manifold*.

Notice first that if we consider the real tangent space as a complex vector space, using J, then g is the real part of a unique Hermitian form on it. This is again a statement on vector spaces. Let V be a finite-dimensional complex vector space and $g : V \times V \to \mathbb{R}$ a symmetric $\mathbb{R}$-bilinear form satisfying $g(iv, w) + g(v, iw) = 0$ for all $v, w \in V$. Then the $\mathbb{R}$-bilinear form ω given by $(v, w) \mapsto g(v, iw)$ is actually *alternating*. In fact, setting $v = w$ in the above equality, we get $g(v, iv) = -g(iv, v)$, while symmetry implies that $g(v, iv) = g(iv, v)$. Moreover, ω also satisfies $\omega(v, iw) + \omega(iv, w) = 0$. Now the $\mathbb{R}$-bilinear map $V \times V \to \mathbb{C}$ given by

$$B : (v, w) \mapsto g(v, w) + i\omega(v, w)$$

is directly verified to be a Hermitian form, namely $\mathbb{C}$-linear in v, antilinear in w and Hermitian symmetric, i.e. $B(v, w) = \overline{B(w, v)}$.

1.18. Exercise. Prove that any alternating form ω which satisfies $\omega(iv, w) + \omega(v, iw) = 0$ is the imaginary part of a unique Hermitian form on it.

Now the data given by the pair (J, g) as above is equivalent to giving a reduction of the structure group from $GL(2n, \mathbb{R})$ to $U(n)$. In fact, the almost complex structure gives a reduction to the subgroup $GL(n, \mathbb{C})$. To every Hermitian form H one can associate the matrix $H(e_i, e_j)$ where (e_i) is the standard basis of $\mathbb{C}^n$. This gives a bijection between Hermitian forms and matrices A that satisfy $A = \overline{A}'$. The group $GL(n, \mathbb{C})$ acts on all Hermitian matrices by the prescription $(g, H) \mapsto gH\overline{(g^t)^{-1}}$. This action is transitive on the set of positive definite Hermitian matrices, and the isotropy group at the identity is the unitary group $U(n)$. Thus a Hermitian structure on the reduced complex vector bundle gives a further reduction to $U(n)$. On the other hand, we have remarked that the form $\omega(v, w) = g(v, Jw)$ at every point is actually an alternating form. Moreover, this form is also nondegenerate. For, if $\omega(v, w) = 0$ for all w, i.e., $g(v, Jw) = 0$ for all w, we conclude, since g is nondegenerate, that $w = 0$. Thus an almost Hermitian structure induces an almost symplectic structure as well.

1.19. Exercises.

1) Show that any almost symplectic manifold admits an almost complex structure.

2) Show that any almost complex manifold admits an almost Hermitian structure and hence an almost symplectic structure also.

2. Torsion Free G-Connections

Since all the structures we have discussed above are simply reductions of the structure group of the tangent bundle to a subgroup G of $GL(n, \mathbb{R})$, it is clear that given such a reduction, there exists a connection on the reduced principal G-bundle. A connection on the reduced bundle gives rise to a $GL(n, \mathbb{R})$-connection on the tangent bundle, namely, a linear connection. In each of these cases, they may be characterised as linear connections which leave a suitable tensor invariant. For example, in the case of a Riemannian structure, an $O(n)$-connection is simply a linear connection which satisfies $\nabla_X(g) = 0$ for all vector fields X. This can be expanded to the equation

$$Xg(Y, Z) - g(\nabla_X Y, Z) - g(Y, \nabla_X Z) = 0$$

for all vector fields X, Y, Z.

2.1. Exercise. Write down explicitly the condition that a linear $U(n)$-connection on an almost Hermitian manifold satisfies.

However, in general, G-connections which, as linear connections, are *torsion free* may not exist. We will first investigate the question whether, given a finite group H of gauge transformations of the tangent bundle, there exists a torsion free connection which is H-invariant. We have already seen [Ch. 5, 2.12] that H acts by affine transformations on the space C of all linear connections. The linear part of this action on $(\mathcal{T}^* \otimes \mathcal{T}^* \otimes \mathcal{T})(M)$ is given by the trivial action on the first factor and the natural action on the other two.

In addition there is also an involution τ acting on C. For any linear connection ∇, we define $\tau(\nabla)$ by the formula

$$(\tau(\nabla))_X Y = \nabla_Y X + [X, Y]$$

for any two vector fields X, Y. In order to convince ourselves that $\tau(\nabla)$ is also a linear connection, the additivity in X and Y being obvious, we need only check that $(\tau(\nabla))_X(fY) = (Xf)Y + f(\tau(\nabla))_X Y$, and $(\tau(\nabla))_{fX} Y = f(\tau(\nabla))_X Y$ for any $f \in \mathcal{A}(M)$. But these follow directly from the definition. Moreover this action too is affine, the linear part being the transposition of the first two factors in $(\mathcal{T}^* \otimes \mathcal{T}^* \otimes \mathcal{T})(M)$. We will collect these facts in the following.

2.2. Proposition. *There is an affine involution acting on the affine space of all linear connections, whose linear part is the transposition of the first two factors of $(\mathcal{T}^* \otimes \mathcal{T}^* \otimes \mathcal{T})(M)$. The fixed point space of this involution is the space of torsion free connections.*

Proof. The only thing that remains to be checked is the last statement. But obviously, saying that a linear connection is fixed under τ is only a restatement that the torsion tensor is zero.

The question whether a torsion free linear connection invariant under H exists is therefore the same as finding a connection fixed under the affine action of H as well as τ. Now the action of H and τ can be put together. Let K be the free product $H * \mathbb{Z}/2$ of H and the group $\{1, \tau\}$ of order 2. Then the action of H and of τ together means that K acts by affine transformations of C. The linear part of this action of K on $T^* \otimes T^* \otimes T$ is characterised by a) H acting as the tensor product of the trivial action on the first factor and the natural (gauge) action on the other two, and b) τ acting as the transposition of the first two factors. If some element of K has linear part zero, then it acts as translation. If this translation is nontrivial, then there can be no K-fixed connection. So we will try and analyse the linear action.

We first show that the linear part of the action of K factors through to a smaller group. Indeed, consider the semidirect product P of $\mathbb{Z}/2$ by $H \times H \times H$ where the nontrivial element of $\mathbb{Z}/2$ acts by transposition of the first two factors. Thus P consists of elements of the form $((h_1, h_2, h_3), 0)$ and $((h_1, h_2, h_3), 1)$. It contains $H \times H \times H \times \{0\}$ as a normal subgroup with quotient $\mathbb{Z}/2$.

The homomorphism of H into P given by $h \mapsto ((1, h, h), 1)$ and the homomorphism $\mathbb{Z}/2 \to P$ given by $\tau \mapsto ((1, 1, 1), \tau)$ together define (by the universal property of free products) a homomorphism of K into P. Now P acts on $(\mathcal{T}^* \otimes \mathcal{T}^* \otimes \mathcal{T})(M)$ by $H \times H \times H$ acting by the tensor product of the natural actions, and τ acting by transposition of the first two factors. Clearly the action of K factors through to this action of P. In fact one can even compute the image of K in P.

2.3. Proposition. *Consider the semidirect product P of $\mathbb{Z}_2 = \{1, \tau\}$ by $H \times H \times H$ for the action of τ on $H \times H \times H$ given by the transposition of the first two factors. The subgroup Q of $H \times H \times H$ consisting of elements of the form (a, b, c) with $c = ab \bmod([H, H])$ is invariant under the transposition. Let R be the corresponding semidirect product of $\mathbb{Z}/2$ by Q. The homomorphism of $K = H * \mathbb{Z}_2$ into P which maps h to $(1, h, h)$ for all $h \in H$ and the generator τ of $\mathbb{Z}_2$ to the transposition, has R as its image.*

Proof. If $c = ab \mod([H,H])$ and $c' = a'b' \mod([H,H])$, then $cc' = aba'b' = aa'bb' \mod([H,H])$ and thus one checks that Q is a subgroup. The transposition of the first two factors leaves it invariant, for (a,b,c) belongs to Q if and only if (b,a,c) does, since $ab = ba$ in $H/[H,H]$.

Consider the subgroup $\tilde{K}$ of K consisting of elements in $H * \mathbb{Z}_2$ where τ occurs an even number of times.

In order to show that the image of K is R, it is enough to verify that Q is the image of $\tilde{K}$. Clearly, $\tilde{K}$ contains H, and is in fact generated by H and elements of the type $\tau x \tau$ with $x \in H$. The image of H is clearly contained in Q and the image of $\tau x \tau$ is $(x, 1, x)$ which again belongs to Q. Hence the image of $\tilde{K}$ is contained in Q.

Any element of Q is a product of elements of the form (a, b, ab) and $(1, 1, aba^{-1}b^{-1})$ by definition. But $(a, b, ab)(a^{-1}, b^{-1}, a^{-1}b^{-1}) = (1, 1, aba^{-1}b^{-1})$ and $(a, b, ab) = (a, 1, a)(1, b, b)$, so that Q is actually the image of $\tilde{K}$, proving the proposition.

Let N be the kernel of the surjective homomorphism of K into R. By definition, in the affine action of K on the space of linear connections, all elements of N have linear parts 0, that is to say, we have a homomorphism of N into the group of translations, namely the space of sections of $T^* \otimes T^* \otimes T$. Since this group is abelian, this homomorphism factors through to the derived group $N/[N,N]$. In fact it actually gives an $\mathbb{R}$-linear map tor of the vector space $N/[N,N] \otimes \mathbb{R}$ into the space of tensor fields of the above type. We have already made the following simple observation.

2.4. Proposition. *A necessary condition for the existence of an H-invariant torsion free linear connection is that the linear map* tor $: N/[N,N] \otimes \mathbb{R} \to (T^* \otimes T^* \otimes T)(M)$ *be 0.*

2.5. Remark. The action of the gauge group on C is not effective. The subgroup consisting of nonzero scalars acts trivially on C. So if the group H contains nontrivial scalars, it is better to pass to the quotient H' of H, by the intersection of H with $\mathbb{R}^\times$. Correspondingly K needs to be replaced by $H' * \mathbb{Z}/2$ and R by the quotient of R by the subgroup consisting of (a, b, ab) with a, b scalars in H.

The question arises whether the necessary condition in Proposition 2.4 is also sufficient. In fact, if tor is 0, then the *affine* action of the group K on C also factors to R. If H is finite, R is also finite. Now a finite group of affine transformations always has a fixed point. For example, we may take any connection ∇ and take the barycentre of the polygon $\langle g(\nabla) \rangle, g \in R$. Thus we have

2.6. Theorem. *If H is a finite subgroup of the gauge group, a necessary and sufficient condition for the existence of an H-invariant torsion free connection is that the linear map* tor *vanishes.*

2.7. Remark. Notice that if H is finite, the subgroup N of $H * \mathbb{Z}_2$ is of finite index since R is finite. Hence it is finitely generated, implying that $N/[N, N] \otimes \mathbb{R}$ is finite-dimensional. Thus the vanishing of tor is equivalent to the vanishing of finitely many tensor fields.

2.8. Exercise. Show that if the subgroup H of the gauge group contains any element of the form $m(f)$, namely multiplication by a nonzero (non-constant) function, then there does not exist any torsion free connection invariant under it.

3. Complex Manifolds

Let M be an almost complex manifold. We start by asking for conditions for the existence of a torsion free linear connection which leaves the almost complex structure invariant. This is the particular case of the discussion in the previous section, in which the subgroup H of the gauge group is $\mathbb{Z}/4$ and the action is given by the almost complex structure J. Then $H' = \mathbb{Z}/2$ and the nontrivial element acts via J. In this case, we will compute tor. Firstly we have to compute $N = \ker : K' = \mathbb{Z}/2 * \mathbb{Z}/2 \to Q'$. Let j, τ denote the generators of the two cyclic groups of order 2. Then $\tau j \tau$ maps to $(j, 1, j)$, while j maps to $(1, j, j)$. From this it is obvious that $(\tau j)^4$ is in the kernel. Indeed, this element generates the kernel, since the quotient of $\mathbb{Z}/2 * \mathbb{Z}/2$ by the normal subgroup generated by $(\tau j)^4$ is of order 8, which is also the order of the group R'.

If ∇ is any connection, then $\tau j(\nabla)$ is the connection $\nabla^{(1)}$ defined by $\nabla^{(1)}_X(Y) = [X, Y] + (j\nabla)_Y X = [X, Y] - J(\nabla_Y(JX))$. Hence $(\tau j)^2(\nabla) = \nabla^{(2)}$ is given by $\nabla^{(2)}_X Y = [X, Y] - J(\nabla^{(1)}_Y(JX)) = [X, Y] - J([Y, JX] - J\nabla_{JX}(JY)) = [X, Y] - J[Y, JX] - \nabla_{JX}(JY)$. From this we deduce that $((\tau j)^4(\nabla))_X Y = [X, Y] - J[Y, JX] - \nabla^{(2)}_{JX}(JY) = [X, Y] - J[Y, JX] - [JX, JY] - J[JY, X] + \nabla_X Y$. This is just the translation of ∇ by the tensor field $[X, Y] - J[Y, JX] - [JX, JY] - J[JY, X]$. In other words, the element $(\tau j)^4$ of N is mapped by tor to this tensor field.

3.1. Definition. If J is an almost complex structure, then the tensor field taking vector fields X, Y to

$$[X, Y] - J[Y, JX] - [JX, JY] - J[JY, X]$$

is called the *torsion tensor field* of the almost complex structure.

From Theorem 2.5 and the above computation, we conclude

3.2. Theorem. *The vanishing of the torsion tensor field of an almost complex structure J is a necessary and sufficient condition for the existence of a linear torsion free connection invariant under J.*

If M is an almost complex manifold, then the complexified tangent bundle breaks up into a direct sum of two bundles $T^{1,0}$ and $T^{0,1}$ according to comments following Definition 1.14. In the following we will extend the bracket operation to complex vector fields $\mathbb{C}$-bilinearly. If the torsion is zero and if X, Y are vector fields of type $(1,0)$, then $[X,Y]$ is also of type $(1,0)$. For $JX = iX$ and $JY = iY$ and the torsion tensor field takes the value $[X,Y] - iJ[Y,X] + [X,Y] - iJ[Y,X]$ on (X,Y). Its vanishing implies therefore that $J[X,Y] = i[X,Y]$. Conversely, if the bracket of any two vector fields of type $(1,0)$ is again of type $(1,0)$, then take two real vector fields X and Y and note that the vector fields $X - iJX$ and $Y - iJY$ are of type $(1,0)$. By assumption, the vector field $[X - iJX, Y - iJY]$ is also of type $(1,0)$ and so we have $J[X - iJX, Y - iJY] = i[X - iJX, Y - iJY]$. Equating the real parts, we get $J[X,Y] - J[JX,JY] - [X,JY] - [JX,Y]$. This is the same as the value of the torsion tensor on $(X, -JY)$. Thus we have shown the following.

3.3. Proposition. *The vanishing of the torsion of an almost complex structure J is equivalent to the requirement that the bracket of any two vector fields of type $(1,0)$ is also of type $(1,0)$. Similarly the bracket of any two vector fields of type $(0,1)$ is again of type $(0,1)$.*

In the case of a complex manifold, J is given in local coordinates by $\frac{\partial}{\partial x_j} \mapsto \frac{\partial}{\partial y_j}$ and $\frac{\partial}{\partial y_j} \mapsto -\frac{\partial}{\partial x_j}$. Hence $J(\frac{\partial}{\partial x_j} - i\frac{\partial}{\partial y_j}) = \frac{\partial}{\partial y_j} + i\frac{\partial}{\partial x_j} = i(\frac{\partial}{\partial x_j} - i\frac{\partial}{\partial y_j})$. We denote the vector fields $\frac{1}{2}(\frac{\partial}{\partial x_j} - i\frac{\partial}{\partial y_j})$ by $\frac{\partial}{\partial z_j}$. They form a basis over $\mathcal{A}$ for vector fields of type $(1,0)$. Similarly the complex vector fields $\frac{1}{2}(\frac{\partial}{\partial x_j} + i\frac{\partial}{\partial y_j})$, which we will denote by $\frac{\partial}{\partial \bar{z}_j}$, form a basis for the space of vector fields of type $(0,1)$. Hence it is clear that their brackets are zero, and consequently, the bracket of any two vector fields of type $(1,0)$ is again of type $(1,0)$. We conclude therefore that a necessary condition for an almost complex structure to come from a complex structure is that the torsion should vanish.

3.4. Remark. It is true that the above condition is also sufficient for a given almost complex structure to come from a complex structure, thanks to a theorem of Newlander and Nirenberg [**13**].

3.5. The Dolbeault complex.

The (complexified) de Rham complex $\Lambda(T^*)$ of a complex manifold has a natural decomposition. We have seen that $\Lambda^i(T^*)$ is the direct sum of components $\Lambda^{(p,q)}(T^*)$ consisting of forms of type (p,q). For simplicity of notation, we will simply write $\Lambda^{p,q}$ for these spaces.

A form of type (p,q) has the expression $\sum f_{\alpha,\beta} dz_\alpha \wedge d\overline{z_\beta}$. Here, α is a multiindex of length p and β one of length q. If $\alpha = (i_1, \ldots, i_p)$, then as usual we denote by dz_α the type $(p,0)$ form $dz_{i_1} \wedge \cdots \wedge dz_{i_p}$. From this, we may conclude that actually the bundle $\Lambda^{p,q}$ is nothing other than $\Lambda^p(T^{(1,0)})^* \otimes \Lambda^q(T^{(0,1)})^*$. Now we wish to see the relationship of the exterior derivative to the type decomposition. From our definition, namely

$$
\begin{aligned}
d\omega(X_1, \ldots, X_n) \;=\; & \sum (-1)^{i+1} \omega(X_1, \ldots, \hat{X}_i, \ldots, X_{r+1}) \\
+ \; & \sum (-1)^{i+j} \omega([X_i, X_j], X_1, \ldots, \hat{X}_i, \ldots, \hat{X}_j, \ldots, X_{r+1}),
\end{aligned}
$$

we conclude that if ω is of type (p,q), the evaluation of $d\omega$ on $r+1$ vector fields of which more than $p+1$ (resp. more than $q+1$) are of type $(1,0)$ (resp. $(0,1)$) gives zero. This means that when ω is of type (p,q), we get only two terms in the type decomposition of $d\omega$, namely one of type $(p+1,q)$ and one of type $(p, q+1)$. We will denote these two components respectively by $d'\omega$ and $d''\omega$. On iteration of d which gives 0, we get the following identities:

$$
d'^2 = 0, \quad d''^2 = 0, \quad \text{and} \quad d'd'' + d''d' = 0.
$$

Consider the complex $\mathrm{Dol}(M)^\circ$, given by

3.6.
$$
0 \to \mathcal{A} \to \Lambda^{0,1} \to \cdots \to \Lambda^{0,n} \to 0
$$

and the inclusion $0 \to \mathcal{O} \to \mathrm{Dol}(M)^\circ$. Here the differentials are supposed to be d'', the sheaf $\mathcal{O}$ is the sheaf of holomorphic functions and the map is the inclusion of holomorphic functions in complex-valued differentiable functions.

3.7. Definition. The complex defined in 3.6 is called the *Dolbeault complex* of the complex manifold M.

Then we have the following analogue of the statement which we proved for the de Rham complex in [Ch. 2, Proposition 6.14].

3.8. Proposition. *The Dolbeault complex is a soft resolution of the sheaf* $\mathcal{O}$.

Proof. All the sheaves $\Lambda^{0,p}$ are $\mathcal{A}$-modules, and consequently soft [Ch. 4, Example 1.10, 1)]. Let f be a local section of the sheaf $\mathcal{A}$ such that $d''f = 0$, i.e. $\frac{\partial}{\partial \bar{z}_j} f = 0$ for all j in local coordinates. Then it is holomorphic, proving that the sequence at the left end is exact. We have then to show in a polydisc U in $\mathbb{C}^n$ that if $\omega \in \Lambda^{0,p}$ satisfies $d''\omega = 0$, then it is of the form $d''\alpha$ where $\alpha \in \Lambda^{0,p-1}(V)$ and V is a neighbourhood of 0 contained in U. Following the same inductive argument as in [Ch. 2, 6.14] we can complete the proof if we show the following (in one variable).

3.9. Lemma. *If f is any differentiable function of one variable z in the unit disc, depending differentiably on real parameters t and holomorphically on complex parameters s, there exists a function g on some disc D around 0, depending differentiably on z and t and holomorphically on s such that $\frac{\partial}{\partial \bar{z}} g = f$ on D.*

Proof. This is proved by using what is called the Cauchy kernel. We will ignore the parameters in what follows since from our construction of g, its differentiable dependence on t and holomorphic dependence on s will be obvious. By multiplication by a differentiable function which is 1 in a neighbourhood of 0 and zero outside a bigger disc with closure contained in the unit disc, we may assume that f has support inside the unit disc U. Then with $D(\epsilon)$ denoting the disc of radius ϵ around a, we have $f(a) = \lim_{\epsilon \to 0} \frac{1}{2\pi i} \int_{\partial D(\epsilon)} \frac{f(z)}{z-a} dz$. This integral can also be written as $f(a) = -\frac{1}{2\pi i} \int_{U \setminus D(\epsilon)} \frac{d}{d\bar{z}}(\frac{f(z)}{z-a}) dz d\bar{z}$. This is because the latter integral is the difference of the integrals of $\frac{f(z)}{z-a} dz$ over ∂U and over $\partial D(\epsilon)$ and the former is zero since f vanishes on ∂U. But $\frac{1}{z-a}$ is holomorphic in the domain of integration and so we may write the integrand also as $\frac{df(z)}{d\bar{z}} \frac{1}{z-a} dz d\bar{z}$. Making the substitution $z - a = w$, we get

$$f(a) = \int \frac{df(w+a)}{d\bar{w}}(1/w) dw d\bar{w}.$$

Now if a is allowed to vary, then $\frac{df(w+a)}{d\bar{w}}$ can also be written as $\frac{df(w+a)}{d\bar{a}}$. So we deduce that the function $g(a)$ defined to be $\int \frac{f(w+a)}{w} dw d\bar{w}$ has the property that $\frac{dg(a)}{d\bar{a}} = f(a)$.

Note that d'' satisfies $d''(f\omega) = d''f \wedge \omega + f d''\omega$ and in particular, all the differentials in the complex are $\mathcal{O}$-linear.

As in the case of differentiable vector bundles and locally free $\mathcal{A}$-sheaves, we have an identity (proved in exactly the same way) between locally free $\mathcal{O}$-sheaves and *holomorphic vector bundles*. If E is a holomorphic vector bundle over a complex manifold, then we denote the associated $\mathcal{O}$-module by $\mathcal{E}_h$, retaining the notation $\mathcal{E}$ for the corresponding $\mathcal{A}$-module. Thus we have $\mathcal{E} \simeq \mathcal{E}_h \otimes_\mathcal{O} \mathcal{A}$. We can tensor the Dolbeault resolution over $\mathcal{O}$ by $\mathcal{E}_h$ and obtain a resolution of the following kind:

$$0 \to \mathcal{E}_h \to \mathcal{E} \to \Lambda^{0,1} \otimes \mathcal{E} \to \cdots \to \Lambda^{0,n} \otimes \mathcal{E} \to 0.$$

This will be called the *Dolbeault resolution* for $\mathcal{E}_h$. Since it is a soft resolution as well, we may apply global sections to this complex of sheaves and compute the cohomology of the resulting complex of vector spaces in order to compute $H^i(M, \mathcal{E}_h)$ [Ch. 4, 3.1, 3.3]. Thus we have

3.10. Theorem. *Let E be a holomorphic vector bundle over a complex manifold M of dimension n. Then the cohomology vector spaces of E are canonically isomorphic to the cohomology spaces of the Dolbeault complex, in which the components are $(0, q)$-forms with values in E and the differentials are d''. In particular, $H^i(M, \mathcal{E}_h) = 0$ for $i > n$.*

3.11. Definition. The bundle $T^{1,0}$ has a natural holomorphic structure, and it is called the *holomorphic tangent bundle* and the corresponding $\mathcal{O}$-sheaf is denoted by $\mathcal{T}_h$.

We can tensor the Dolbeault complex by the holomorphic tangent bundles $\Lambda^p(T^{1,0})$ and get the complex

$$0 \to \Lambda^p(\mathcal{T}^*)_h \to \Lambda^{p,0} \to \cdots \to \Lambda^{p,n} \to 0.$$

3.12. Remark. The direct sum of all these complexes over all the p's consists of the same components as the de Rham complex but with the differential d'', instead of d.

4. The Outer Gauge Group

There is a group which contains the gauge group as a subgroup of index 2. In order to define this, we will first associate to any (finite-dimensional) vector space V a group which contains $GL(V)$ as a subgroup of index 2. In fact, consider the (set) union of $GL(V)$ and the set I of all isomorphisms $V \to V^*$. On this set we will introduce a group structure. The group structure on the subset $GL(V)$ will be the same as the standard structure on $GL(V)$. On the other hand, $GL(V)$ acts on the right on I by composition. It acts on the left by composition with the action $A \to (A')^{-1}$ of $GL(V)$ on V^*. Finally we define composition of $B_1, B_2 \in I$ by setting $B_1.B_2 = (B_1')^{-1} \circ B_2$. It is easy to check that this makes the union a group, which we call the *outer linear group* and denote $OL(V)$. This group realizes the outer automorphism $A \mapsto (A')^{-1}$ (defined after choosing a basis) as the restriction of an inner automorphism.

Now the group $GL(V)$ acts on $V^* \otimes V$ by $g \mapsto (g'^{-1} \otimes g)$. As for the other coset, any element b in it takes $V^* \otimes V$ to $V \otimes V^*$ by the same map, namely $b \mapsto (b'^{-1} \otimes b)$. We can compose it with the transposition t of the two factors to get an automorphism of $V^* \otimes V$. We will denote this by $g \mapsto L(g)$. The element $L(bg)$ is then $t \circ ((bg)'^{-1} \otimes bg) = t \circ (b'^{-1}g'^{-1} \otimes bg) = t \circ (b'^{-1} \otimes b) \circ (g'^{-1} \otimes g) = L(b)L(g)$. Also $L(gb) = t \circ ((g'^{-1} \circ b)'^{-1} \otimes g'^{-1} \circ b)$ by our definition of the group structure on $OL(V)$. Hence $L(gb) = t \circ (g \circ b'^{-1} \otimes g'^{-1} \otimes b) = t \circ (g \otimes g'^{-1}) \circ (b'^{-1} \otimes b) = (g'^{-1} \otimes g) \circ t \circ (b'^{-1} \otimes b) = L(g)L(b)$. Similarly one also checks that $L(b)L(b') = L(bb')$ whenever both b, b' do not belong to $GL(V)$. In other words we have a representation L of $OL(V)$ on $V^* \otimes V$. We will modify it a little. Define $\rho(x) = L(x)$ if $x \in GL(V)$ and $\rho(x) = -L(x)$ if not. This is the representation of interest to us.

4.1. Exercise. Imbed the group $GL(V)$ as a subgroup of $GL(V \oplus V^*)$ by mapping any $A \in GL(V)$ to the automorphism $(A \oplus (A^t)^{-1})$. Show that its normaliser in $GL(V \oplus V^*)$ is generated by $OL(V)$ and the scalars.

Now it is clear how to define the corresponding *outer gauge group $OL(E)$* of a (real) vector bundle E. It is simply the assignment to each point m of M, of an element of $OL(E_m)$ (depending differentiably on the point m). If M is connected, it is the union of the gauge group with the set of differentiable isomorphisms of E with E^*. From the above linear algebraic consideration we also see that there is a natural action ρ of $OL(E)$ on $E^* \otimes E$.

4.2. Proposition. *The outer gauge group $OL(E)$ of a vector bundle E acts by affine transformations on the space C of connections on E. The linear part of the action on $T^* \otimes E^* \otimes E$ is the tensor product of the trivial action on T^* and the action ρ on $E^* \otimes E$ described above.*

Proof. The action of $GL(E)$ on C has already been discussed earlier. If φ is an isomorphism of E with E^*, then it carries a connection ∇ on E to one on E^*, which in turn induces a connection on its dual, namely E. This is the action envisaged. Explicitly, let b be an element of $OL(E) \setminus GL(E)$ and B a nondegenerate bilinear form on E representing b. In other words, b is the isomorphism $E \to E^*$ which maps v to the linear form $b(v) : w \mapsto B(v, w)$. Then the action of b on C is determined by the following. For any connection ∇, a vector field X and sections s, t of E, we have

$$B(s, (b(\nabla))_X t) = XB(s, t) - B(\nabla_X s, t).$$

From this it follows that the map $\nabla \to b(\nabla)$ is an affine transformation whose linear part, which is an endomorphism $\alpha \mapsto b(\alpha)$ of $T^* \otimes E^* \otimes E$, satisfies the characterizing condition

$$B(s, (b(\alpha))(X)t) + B(\alpha(X)s, t) = 0.$$

Finally one can check that this is a group action of the outer gauge group with the linear part as claimed.

4.3. Remark. When we consider a linear connection, that is to say when we take $E = T$, then α may also be interpreted as a bilinear map $T \otimes T \to T$, and the above equality is then the following:

$$B(\alpha(X, Y), Z) = -B(\alpha(X, Z), Y).$$

Our generalisation consists in taking a subgroup $\tilde{H}$ of $OL(T)$ instead of $GL(T)$. We will denote by $\tilde{H}'$ its image modulo scalars and by H its intersection with the gauge group. We would like to investigate the question of existence of a torsion free connection invariant under the $\tilde{H}$-action, just as we did above.

Again this depends on the affine action of the group $\tilde{K} = \tilde{H} * \mathbb{Z}_2$ on C, where the second factor represents the action by τ. Exactly as before we have the homomorphism ρ of this group into the group of linear transformations of $(\mathcal{T}^* \otimes \mathcal{T}^* \otimes \mathcal{T})(M)$, determined by the action ρ of $OL(T)$ and the above action of τ.

If $\tilde{N}$ is the kernel of ρ, then it acts on C by translations. If we denote by tor the $\mathbb{R}$-linear map $\tilde{N}/[\tilde{N}, \tilde{N}] \otimes \mathbb{R} \to (\mathcal{T}^* \otimes \mathcal{T}^* \otimes \mathcal{T})(M)$, then we have the following conclusion, exactly as above.

4.4. Proposition. *The vanishing of the $\mathbb{R}$-linear map* tor $: \tilde{N} \to \mathcal{T}^* \otimes \mathcal{T}^* \otimes \mathcal{T}(M)$ *is a necessary condition for the existence of a torsion free connection which is $\tilde{H}$-invariant. If $\tilde{H}$ is finite, it is also sufficient.*

Proof. The only thing that needs checking is that the linear action of $\tilde{H} * \tau$ factors through a finite group. It is clear that if b is an element of $\tilde{H}$ which is not in H, then we may use it to identify T^* with T. Then all the elements in $\tilde{H}$ are in H modulo b, and we see that the image action factors through $H \times H \times H$, the permutation group in 3 letters, and possibly ± 1.

4.5. Torsion free connections and bilinear forms.

We will now apply the above considerations to determine a necessary and sufficient condition for the existence of a torsion free connection leaving a given nondegenerate bilinear form $B : T \times T \to \mathcal{A}$ invariant. Treat B as an element b of $OL(T)$.

We will assume that b is either symmetric or skew-symmetric. If we use the bilinear form to identify T with T^*, then the action of $\rho(b)$ on $T^* \otimes T^* \otimes T^*$ is by $\mp(23)$, where (23) denotes the transposition of the second and the third factors. Thus we get the image under the linear action of $\tilde{H} * \mathbb{Z}/2 = \mathbb{Z}/2 * \mathbb{Z}/2$

to be $\{\pm 1\} \times S_3$ or S_3, according as b is symmetric or skew-symmetric. This is because τ acts as (12) and b as (23) in the latter case, and so the group generated is S_3. When b is symmetric, we note that the group contains $-(123)$, and raising it to the third power, that it contains $-\,\mathrm{Id}$. The image being of order 12 (resp. 6) when b is symmetric (resp. alternating), it follows that the kernel is generated by $(b\tau)^6$ (resp. $(b\tau)^3$).

So we will compute how $(b\tau)^3$ acts on connections.

4.6. Computation. Assume that $B(X,Y) = \pm B(Y,X)$. Then we have

$$
\begin{aligned}
B(X,((b\tau)\nabla)_Y Z) &= YB(X,Z) - B((\tau\nabla)_Y X, Z) \\
&= YB(X,Z) + B([X,Y],Z) - B(\nabla_X Y, Z) \\
&= YB(X,Z) + B([X,Y],Z) \mp B(Z,(\nabla_X Y)).
\end{aligned}
$$

Repeating this three times, we get the equality

$$
\begin{aligned}
B(X,((b\tau)^3\nabla)_Y, Z) = {}&YB(X,Z) \mp XB(Z,Y) + ZB(Y,X) + B([X,Y],Z) \\
&\mp B([Z,X],Y) + B([Y,Z],X) \mp B(X,\nabla_Y Z).
\end{aligned}
$$

Thus if B is symmetric, we repeat this once more to conclude that the action by $(b\tau)^6$ on C is *always* trivial. Hence the group fixes a connection. Moreover, if two connections are fixed, their difference is also fixed, but since the image of ρ contains $-\,\mathrm{Id}$ in this case, the two connections are the same.

Thus we have proved the following important fact.

4.7. Theorem *If g is a pseudo-Riemannian structure, then there exists a unique torsion free linear connection which leaves g invariant.*

4.8. Definition. The unique connection as in Theorem 4.7 above is called the *Levi-Civita* or the *Riemannian connection*.

Hence any connection which is fixed under $\tilde{H}$ and τ is also fixed under $(b\tau)^3$ and so such a connection is given by the formula

$$
\begin{aligned}
2B(\nabla_X Y, Z) &= XB(Y,Z) + YB(Z,X) - ZB(X,Y) \\
&\quad - B(X,[Y,Z]) + B(Y,[Z,X]) + B(Z,[X,Y]).
\end{aligned}
$$

In local coordinates, if we write as usual, $\nabla_{\frac{\partial}{\partial x_i}} \frac{\partial}{\partial x_j} = \sum_k \Gamma_{ij}^k \frac{\partial}{\partial x_k}$, then the above formula gives

4.9.
$$
\sum \Gamma_{ij}^l g_{kl} = \tfrac{1}{2}\left(\frac{\partial g_{jk}}{\partial x_i} + \frac{\partial g_{ki}}{\partial x_j} - \frac{\partial g_{ij}}{\partial x_k} \right).
$$

Riemannian manifolds have been studied intensely for over a century. In view of their importance, we will return to their study in Section 5, and prove some of their properties.

4.10. Exercises.

1) Assume that a tangent bundle is the direct sum of subbundles T_i. We say that a linear connection leaves this decomposition invariant if $\nabla_X Y$ is a section of T_i whenever Y is. Apply the above criterion (using the group generated by the involutions of T having T_i and $\bigoplus_{j \neq i} T_j$ as eigenspaces with eigenvalues 1 and -1 respectively) and show that there exists a torsion free connection leaving the decomposition invariant if and only if all the subbundles are involutive.

2) Assume that g is a Riemannian structure on M. Show that the involutions corresponding to the above decomposition commute with g in $OL(V)$ if and only if the decomposition is orthogonal for the Riemannian structure.

4.11. Symplectic structure.

Now suppose B is alternating. Then the formula gives

$$B(((b\tau)^3\nabla - \nabla)_X Y, Z) = \sum (XB(Y,Z) + B(X,[Y,Z])).$$

In this case, treating B as a 2-form, this computes the torsion tensor to be the exterior derivative dB of B. Hence we have

4.12. Theorem. *The necessary and sufficient condition for the existence of a torsion free linear connection leaving a symplectic form ω invariant is that it be closed.*

4.13. Remark. A theorem of Darboux states that any symplectic form ω which is closed can be written locally in a suitable coordinate system $(q_1, \ldots, q_n, p_1, \ldots, p_n)$ as $\sum dp_i \wedge dq_i$. Note that this is exactly how the symplectic form looks on the cotangent bundle of any bundle, in the natural coordinate system on it (Example 1.11, 2)).

In a symplectic manifold (M, ω) one can define a Lie bracket on the space of functions as follows. If f is a function, then using the isomorphism of T with T^* given by the symplectic form, we can identify the differential form df with a vector field X_f. The inner product using X_f will be denoted i_f. Then we define the *Poisson bracket* $\{f, g\}$ of $f, g \in \mathcal{A}(M)$, to be the function $i_f i_g \omega$. It is therefore given by the equation $\omega(X_f, X_g) = -df(X_g) = -X_g(f) = X_f(g)$. In the Darboux coordinates, $X_f = \sum a_i \frac{\partial}{\partial p_i} + \sum b_i \frac{\partial}{\partial q_i}$ is computed by the equation $\omega(X_f, \frac{\partial}{\partial p_i}) = b_i$ and $\omega(X_f, \frac{\partial}{\partial q_i}) = a_i$. Hence $(df)(\frac{\partial}{\partial p_i}) = b_i$ and $df(\frac{\partial}{\partial q_i}) = -a_i$. In other words, $b_i = \frac{\partial f}{\partial p_i}$ and $a_i = -\frac{\partial f}{\partial q_i}$, determining X_f to be $\sum(\frac{\partial f}{\partial p_i}\frac{\partial}{\partial q_i} - \frac{\partial f}{\partial q_i}\frac{\partial}{\partial p_i})$. This also computes the Poisson bracket in local coordinates. It is given by $\omega(X_f, X_g) = \sum \frac{\partial f}{\partial p_i}\frac{\partial g}{\partial q_i} - \frac{\partial f}{\partial q_i}\frac{\partial g}{\partial p_i}$.

From this we also conclude that $[X_f, X_g] = X_{\{f,g\}}$. Thus we have an exact sequence of Lie algebras:

$$0 \to \mathbb{R} \to \mathcal{A} \to \mathcal{T}(M).$$

The image in $\mathcal{T}(M)$ can also be checked to be the set of all vector fields X such that $L(X)(\omega) = 0$. Such vector fields are called *locally Hamiltonian*. These notions are fundamental in Hamiltonian mechanics.

4.14. Kähler structure.

Suppose we are given an almost complex structure as well as a Hermitian structure. The symmetry of g is equivalent to $g^2 = 1$ in $OL(V)$ and the Hermitian condition states that J and g commute. The element $-Jg$ represents the alternating 2-form $\omega : (X, Y) \mapsto g(X, JY)$. So for the existence of a torsion free connection that leaves J as well as g invariant, two necessary conditions are that the torsion tensor associated to the almost complex structure J is zero, and that $d\omega = 0$. One can show that these are also sufficient, again by determining the kernel $\tilde{N}$. But since we know that the Riemannian connection is unique, we have only to check that $d\omega = 0$ implies that ω is invariant under this connection. Thus we have the following result.

4.15. Proposition. *Let an almost complex structure J and a Hermitian structure g with respect to it, be given. A necessary and sufficient condition for the existence of a torsion free connection leaving J and g invariant, is that the torsion tensor of J should vanish and that $d\omega = 0$.*

In particular, such a manifold is a complex manifold as well as a symplectic manifold.

4.16. Definition. A complex manifold with an almost Hermitian structure such that the associated symplectic form is closed, is called a *Kähler manifold*. The corresponding real cohomology class in $H^2(M)$ is called its *Kähler class*.

4.17. Examples.

1) We have computed the Chern form of the Hopf bundle on the complex projective space in [Ch. 5, 6.2]. It is a purely imaginary form. It is easily verified that it is the Kähler form of a Hermitian form on the projective space. In other words the topological Chern class of the dual of the Hopf bundle is a Kähler class.

2) Any closed complex submanifold of a Kähler manifold is also Kähler. In particular, all closed submanifolds of the projective space are Kähler.

3) Any one-dimensional complex manifold is Kähler. In fact, we can take any Hermitian structure. The corresponding 2-form is closed since there are no nonzero 3-forms on the manifold.

4) Any complex torus A, namely the quotient of $\mathbb{C}^g$ by a lattice Γ, is Kähler. This is because firstly $\mathbb{C}^g$ is Kähler under the standard Hermitian metric. Secondly, the metric is invariant under translation by elements of Γ and induces a Hermitian structure on A. Hence the form on A associated to the Hermitian structure on it, has closed inverse image on $\mathbb{C}^g$. Therefore, it is itself closed.

5. Riemannian Geometry

We will use notions pertaining to a linear connection on a (pseudo-) Riemannian manifold, such as exponential mapping, geodesics, curvature, ..., implicitly assuming that the linear connection intended is the Levi-Civita connection, namely the unique torsion free connection with respect to which the Riemannian metric tensor is invariant.

5.1. Geodesics on a Riemannian manifold.

A Riemannian manifold (M, g), considered as a topological space, can be provided with a metric space structure. In fact, if p, q are in M and $\gamma : [a, b] \to M$ is a piecewise differentiable path connecting p and q, then we define the *length $l(\gamma)$* of the path γ by the formula

$$l(\gamma) = \int_a^b \sqrt{g(\gamma'(t), \gamma'(t))}\, dt$$

where $\gamma'(t)$ is the image of $\frac{d}{dt}$ under the differential of γ. One defines the *distance $d(p, q)$* to be the infimum of $l(\gamma)$ where γ runs through all paths connecting p and q. Note that we are not claiming that this infimum is attained.

5.2. Remark. If M is the configuration space, then a curve describes the dynamics of the system and the vector $\gamma'(t)$ may be thought of as the *velocity* of the system. From the point of view of physics, the number $E(\gamma)$ defined by the formula

$$E(\gamma) = \frac{1}{2} \int_a^b g(\gamma'(t), \gamma'(t))\, dt$$

is equally important and is known as the *energy* of the path. The evolution of the path ensures that the energy is minimal.

If γ is a geodesic, then the length of the velocity vector is the same at all points (since the velocity vectors at any two points are obtained by parallel translation along the curve on the one hand [Ch. 6, Remark 1.4, 2)], and the metric tensor field is invariant under parallel translation on the other). Hence the length of the curve from 0 to t is simply $(l.t)$ where l is the norm of the velocity vector. Thus, if we wish, we can re-parametrise the geodesic by requiring this length to be t, that is to say, *parametrise it by its length.*

5.3. Example. If M is the Euclidean space $\mathbb{R}^n$ provided with the Euclidean Riemannian structure, then this definition gives a distance function on $\mathbb{R}^n$. This is of course the usual Euclidean distance. To check this we may as well assume that p is the origin and that $\gamma(1) = q$. If $\gamma = (\gamma_i)$ is a path connecting p and q, then $l(\gamma)$ is $\int_0^1 \sqrt{\sum \gamma_i'(t)^2}\,dt$ in this case. Now we have $\frac{d}{dt}(\sqrt{\sum \gamma_i(t)^2}) = \frac{\sum \gamma_i(t)\gamma_i'(t)}{\sqrt{\sum \gamma_i(t)^2}}$ and the latter is at most $\sqrt{\sum \gamma_i'(t)^2}$, by Schwarz's inequality. Hence we have the inequality $\|q\| = \sqrt{\sum \gamma_i(1)^2} = |\int_0^1 \frac{d}{dt}(\sqrt{\sum \gamma_i(t)^2})\,dt| \leq \int_0^1 \sqrt{\gamma_i'(t)^2}\,dt = l(\gamma)$. In this case, of course, the infimum is attained when we take for γ the straight line segment connecting p and q. For the line segment σ is given by $\sigma_i(t) = tx_i$ where x_i are the coordinates of q. Now $\sigma_i'(t) = x_i$ and $l(\sigma) = \int_0^1 \sqrt{\sum x_i^2}\,dt = \|q\|$.

5.4. Proposition. *The distance function on a Riemannian manifold M as defined above, gives rise to a metric space structure on the set M. The topology of this metric structure is the same as that of the manifold.*

Proof. We will first check that the distance function that we defined above is indeed a metric. By definition, the distance function is symmetric in p, q. Also if $p, q, r \in M$, then the distances $d(p, q)$ and $d(q, r)$ can be approximated by $l(\gamma_1)$ and $l(\gamma_2)$ where γ_1 and γ_2 are two paths connecting p to q, and q to r respectively. The composite path from p to r has length $l(\gamma_1) + l(\gamma_2)$ which approximates $d(p, q) + d(q, r)$. Hence $d(p, r) \leq d(p, q) + d(q, r)$. If $p = q$, we take the constant path $\gamma : [0, 1] \to M$ taking all t to p, and note that in that case, $\gamma'(t) = 0$ for all t and so $l(\gamma) = 0$. Hence $d(p, p) = 0$. Conversely, assume that $p \neq q$. Choose a coordinate system (U, x) around p such that q does not belong to U. Let B_a be the closed ball around p obtained by transferring to U the closed ball in $\mathbb{R}^n$ around the origin with radius a. Now any path $\gamma : [0, 1] \to M$ connecting p to q, has to intersect the boundary S_a of B_a. For otherwise, $\{t : \gamma(t)$ is in the open ball$\}$ is open and closed in $[0, 1]$, contradicting the assumption that p belongs to this set and q does not.

Take the least s such that $\gamma(s) \in S_a$. Then the path γ is the composite of the part $\gamma|[0, s]$ and the part $\gamma|[s, 1]$. The length of γ is the sum of the lengths of these two parts and so the length of γ is at least the length of the first part. This first part has the property that the initial point is p, the end point is in S_a, and all other points are in the open ball. Its length is given by $\int_0^s \sqrt{g(\gamma'(t),\gamma'(t))}dt$. Now the following simple remark completes the proof, by comparing the distances in the Euclidean metric and the induced Riemannian metric on the open ball.

5.5. Lemma. *Let g_1, g_2 be two Riemannian metrics on a domain D in $\mathbb{R}^n$ and K a compact subset of D. Then there exist positive constants k, k' such that*

$$kg_2(v, v) \le g_1(v, v) \le k'g_2(v, v)$$

for all tangent vectors v at all points of K.

Proof. From the bilinearity of g_1 and g_2, we see that it is enough to prove such an inequality for v in the unit sphere, say for the metric g_2. The space of all tangent vectors v at all points of K with $g_2(v) = 1$ is a compact subset of the tangent bundle. The function $v \mapsto g_2(v)/g_1(v)$ is a strictly positive function on it and hence is bounded above and below by positive constants.

Completion of the proof of 5.4. From Lemma 5.5, applied to the given Riemannian metric g and the Euclidean metric on the compact set B_a, we conclude on integration that the length of γ over $[0, s]$ in the given metric is at least (resp. at most) k times (resp. k' times) its length with respect to the Euclidean metric. This implies that $d(p, q) \ge ka \ne 0$ and concludes our proof that the distance function defines a metric space structure on M. At the same time it also shows that that the metric space induced by the Riemannian structure is equivalent to the Euclidean metric, and so the induced topology is the same as the Euclidean one, on the coordinate open set.

Let m be a point of M and assume that the exponential map is a diffeomorphism of an open set W around 0 in $V = T_m(M)$ onto an open set U containing m. The tangent bundle on W has a natural trivialisation, identifying all tangent spaces with V. Hence the tangent space at any $p = \exp(q) \in U$ can be identified (by the differential of the exponential map at m) with V. But there are two Riemannian metrics on W. On the one hand, V has the Riemannian metric given by the symmetric form g on the tangent space at m and W inherits it. On the other hand the exponential diffeomorphism transports the Riemannian metric from U to W. These two are of course in general different, one of them being the flat metric. We will distinguish between the two by denoting them g_{fl} and g.

We wish to restrict ourselves to $W \setminus \{0\}$. On it we can introduce polar coordinates, and consequently we have a vector field $X = \frac{\partial}{\partial r}$, which has norm 1 at all points. Hence using the metric, we can decompose the tangent space at all points as the direct sum of the trivial bundle given by the above vector field and its orthogonal complement. The question arises which of the two metrics do we use for this decomposition. Actually, it does not matter! Firstly, any radial vector v of norm 1 at $x \in W$ is mapped by the exponential map into the tangent to the geodesic $t \mapsto \exp(tx)$ at $\exp(x)$. It is therefore the parallel translate, along the geodesic, of $v \in T_m(M)$. Hence its norm is 1 in the metric on $T_{\exp(v)}(M)$ as well. In other words the tangent vector v at x has norm 1 in both metrics. Our claim above is that its orthogonal complement for g_{fl} is also orthogonal for g. This is known as *Gauss' lemma*.

5.6. Lemma (Gauss). *The orthogonal complement of the radial vector for the flat metric at a point $x \in W \setminus \{0\}$ is also orthogonal for g transported from $T_{\exp(x)}(M)$ by the exponential map.*

Proof. Note that since the radial vector field X of norm 1 is invariant under parallel transport for g, it follows that $\nabla_X(X) = 0$. If v is any tangent vector at x orthogonal to X_x, then one might as well confine oneself to the two-dimensional subspace spanned by x and v. Let then Y be the angular vector field $\frac{\partial}{\partial \theta}$. Then Y_x is a multiple of v. Now $Y\langle X, X \rangle = 0$. Since the metric is invariant under ∇, this is $2\langle \nabla_Y X, X \rangle$. Since the torsion is zero, this implies that $\langle \nabla_X Y, X \rangle = 0$. Again since $\nabla_X X = 0$, we get $X\langle Y, X \rangle = 0$. In other words, $\langle X_v, Y_v \rangle$ is invariant along any radius. But since the vectors X_x and Y_x are orthogonal for the flat metric, they are zero for g at $T_m(M)$. This proves the assertion that Y_x is orthogonal to X_x for g.

5.7. Geodesics as minimising distances.

From Gauss' lemma, we will deduce that if $c(t)$ is any curve in W joining 0 to $x \in W$, then the length of $\exp(c(t))$, namely $\int_0^1 g(c'(t), c'(t))^{1/2} dt$, is at least $b = \|x\|$. We can and will assume that for all $t \neq 0$, $c(t)$ is not the zero vector. We write $c(t) \in W$ as $r(t)v(t)$ where $r(t) = \|c(t)\|$, and $g(v(t), v(t)) = 1$. Then $c'(t) = r(t)v'(t) + r'(t)v(t)$. Since we have assumed that $v(t)$ is of norm 1, it follows that $v(t)$ and $v'(t)$ are orthogonal for g_{fl}. Hence they are also orthogonal for g in $T_{\exp(c(t))}$. In particular, $g(c'(t), c'(t)) \geq |r'(t)|$. Hence the length of $c(t)$ from 0 to 1 is at least $\int_0^1 r'(t) dt = b$. But the length of the geodesic joining m to $\exp(x)$ is b. Thus the geodesic has the shortest length among curves joining m and any point in U.

5.8. Definition. A Riemannian manifold is said to be *complete* if it is complete as a metric space, the metric being that induced from the Riemannian structure.

5.9. Remark. It can be shown that in a complete Riemannian manifold, any two points can be connected by a geodesic.

6. Riemannian Curvature Tensor

The curvature of the Levi-Civita connection on a Riemannian manifold is called the *Riemannian curvature tensor*. It is a 2-form $R(X, Y)$ with values in $\operatorname{End} T$.

6.1. Local expression.

We computed the Levi-Civita connection in local coordinates in 4.9. From this we also get the following expression for the Riemannian curvature tensor. We will write $R_{ijkl} = \langle R(\frac{\partial}{\partial x_i}, \frac{\partial}{\partial x_j})\frac{\partial}{\partial x_k}, \frac{\partial}{\partial x_l}\rangle$. Then we have

$$
\begin{aligned}
R_{ijkl} &= \left\langle R(\tfrac{\partial}{\partial x_i}, \tfrac{\partial}{\partial x_j})(\tfrac{\partial}{\partial x_k}), \tfrac{\partial}{\partial x_l} \right\rangle \\[4pt]
&= \left\langle \nabla_{\frac{\partial}{\partial x_i}} \nabla_{\frac{\partial}{\partial x_j}} \tfrac{\partial}{\partial x_k} - \nabla_{\frac{\partial}{\partial x_j}} \nabla_{\frac{\partial}{\partial x_i}} \tfrac{\partial}{\partial x_k}, \tfrac{\partial}{\partial x_l} \right\rangle \\[4pt]
&= \left\langle \nabla_{\frac{\partial}{\partial x_i}} \sum_m (\Gamma_{jk}^m \tfrac{\partial}{\partial x_m}), \tfrac{\partial}{\partial x_l} \right\rangle - \left\langle \nabla_{\frac{\partial}{\partial x_j}} \sum_m (\Gamma_{ik}^m \tfrac{\partial}{\partial x_m}), \tfrac{\partial}{\partial x_l} \right\rangle \\[4pt]
&= \sum_{m,r} \Gamma_{jk}^m \Gamma_{im}^r g_{rl} + \sum_m \tfrac{\partial \Gamma_{jk}^m}{\partial x_i} g_{ml} - \sum_{m,r} \Gamma_{ik}^m \Gamma_{jm}^r g_{rl} - \sum_m \tfrac{\partial \Gamma_{ik}^m}{\partial x_j} g_{ml}.
\end{aligned}
$$

Here Γ_{ij}^k are determined by the equation

$$
\sum g_{kl} \Gamma_{ij}^l = \tfrac{1}{2}\left(\tfrac{\partial}{\partial x_i} g_{jk} + \tfrac{\partial}{\partial x_j} g_{ki} - \tfrac{\partial}{\partial x_k} g_{ij}\right).
$$

We, however, warn the reader that the traditional notation R_{ijkl} stands for $\langle R(\frac{\partial}{\partial x_k}, \frac{\partial}{\partial x_l})\frac{\partial}{\partial x_j}, \frac{\partial}{\partial x_i}\rangle$, which differs from the above by a factor of -1. See [**10**, p. 21].

We may try to get some idea of the Riemannian curvature tensor by finding the consequence of its vanishing. Recall that we defined a connection to be flat, if the curvature form is zero. So we call a (pseudo-) Riemannian manifold *flat* if the Riemannian curvature is zero. Notice first that locally this implies that the tangent bundle can be trivialised together with the linear connection. In other words, around any point $m \in M$, there exist a coordinate neighbourhood and a basis (X_i) for vector fields (over the algebra of functions) such that $\nabla_Y X_i = 0$ for all vector fields Y. Since the connection is torsion free, it follows that

$$
\nabla_{X_i} X_j - \nabla_{X_j} X_i - [X_i, X_j] = 0.
$$

Since the first two terms are zero, we have

$$
[X_i, X_j] = 0 \text{ for all } i, j.
$$

Hence we can introduce a coordinate system (U, x) such that $X_i = \frac{\partial}{\partial x_i}$ for all i. Finally, the invariance of the metric tensor implies that

$$
(\nabla_{\frac{\partial}{\partial x_i}} g)(\tfrac{\partial}{\partial x_j}, \tfrac{\partial}{\partial x_k}) = \tfrac{\partial}{\partial x_i} g(\tfrac{\partial}{\partial x_j}, \tfrac{\partial}{\partial x_k}) = 0.
$$

In other words, $g(\frac{\partial}{\partial x_j}, \frac{\partial}{\partial x_k})$ is a constant function. Now we can change the coordinate system by a real linear transformation so that the metric form is simply $\sum \pm (dx_i)^2$. In particular, we deduce that if a Riemannian metric is flat, then there is a coordinate diffeomorphism of U with an open set in $\mathbb{R}^n$ that takes the given metric to the Euclidean metric. This explains the origin of the name 'flat' for curvature free connections.

6.2. The space of curvature tensors.

The Riemannian curvature tensor has many symmetries. Firstly, since the linear connection is actually one on the reduced orthogonal bundle, it takes values in the adjoint bundle associated to this principal $O(n)$-bundle. In other words, it can be considered as a 2-form with values in $\mathrm{Skew}(T)$, the bundle of skew-symmetric endomorphisms of T (with respect to the metric form g). In any case, one can directly verify that the invariance under ∇ of g, namely

$$Xg(Y, Z) = g(\nabla_X Y, Z) + g(Y, \nabla_X Z),$$

implies that the curvature form R satisfies

$$g(R(X, Y)Z, T) + g(Z, R(X, Y)T) = 0.$$

It satisfies other identities as well. Consider the multilinear form in four variables on the tangent space, given by $(X, Y, Z, T) \mapsto R(X, Y, Z, T) = g(R(X, Y)Z, T)$. It satisfies the following identities:

i) $R(X, Y, Z, T) = -R(Y, X, Z, T)$.

　　This states that R is an alternating form in X, Y.

ii) $R(X, Y, Z, T) = -R(X, Y, T, Z)$.

　　This reflects the remark above that $R(X, Y)$ is a skew-symmetric endomorphism with respect to g.

iii) $R(X, Y, Z, T) + R(Y, Z, X, T) + R(Z, X, Y, T) = 0$.

　　This is the first Bianchi identity, valid for any torsion free linear connection [Ch. 6, 2.8].

As a consequence, one can also derive the following identity:

iv) $R(X, Y, Z, T) = R(Z, T, X, Y)$.

To see this, write out the Bianchi identity for (X, Y, Z, T), (Y, Z, T, X), (Z, T, X, Y) and (T, X, Y, Z), add the first two and subtract the other two, to get

$$R(X, Y, Z, T) - R(Z, T, X, Y) = 0.$$

Since all these identities are tensorial identities, namely identities on the tangent space at every point of M, it makes sense to consider the following purely linear algebraic setup. Let V be a (finite-dimensional) vector space with a nondegenerate quadratic form g on it. Consider the space $RC(V)$ of multilinear forms on V in four variables, satisfying the identities i), ii) and iii) above (where X, Y, Z, T are now vectors in V). Any $R \in RC$ can also be considered as an alternating 2-form in the variables $X, Y \in V$ with values in the space of skew symmetric endomorphisms of V. Let us call this space the *Riemannian curvature space*. In what follows we will denote, for convenience of notation, the inner product given by g on V by $(X, Y) \mapsto \langle X, Y \rangle$.

If M is a Riemannian manifold and $m \in M$, then we are interested in taking $V = T_m(M)$ and g to be the metric on the tangent space. In fact, there is a vector bundle on M (which we will denote by $RC(M)$) whose fibre at $m \in M$ can be identified with $RC(T_m)$ for all m. The Riemannian curvature tensor R is then a tensor field whose value at m belongs to $RC(T_m)$, that is to say, a section of $RC(M)$. It contains a great deal of geometric information. While this tensor is easier to handle from the point of view of algebraic manipulations, there is an equivalent notion which is more geometric even at the outset.

6.3. Sectional curvature

Let $R \in RC$. Consider the function $Q : (X, Y) \mapsto R(X, Y, X, Y)$. Then R can be recovered by polarisation from Q. In order to prove this, since R is a multilinear form in 4 variables, we may as well assume that V is 4-dimensional. Now R defines (and is determined by) a quadratic form on $\Lambda^2(V)$. Note that Q is essentially the restriction of R to the space of decomposable tensors, i.e. tensors of the form $X \wedge Y$. If R and R' are two elements of RC which coincide with Q on decomposable tensors, then $R - R'$ vanishes on decomposable tensors. But then any quadratic form which vanishes on decomposable tensors in the second exterior power of a 4-dimensional space, is unique up to a scalar multiple. This form can in fact be described as the wedge product $S^2(\Lambda^2(W)) \to \Lambda^4(W)$. So we have only to convince ourselves that this form does not belong to RC. As a 4-form it maps (X, Y, Z, T) to $X \wedge Y \wedge Z \wedge T$. If T is fixed and we take the cyclic sum in (X, Y, Z), we get a nonzero multiple of the same form, proving that it does not satisfy iii) in the definition of RC.

6.4. Exercise. Follow the above logic and actually compute R in terms of Q.

On the other hand, the square of the area of the surface determined by X and Y is defined to be $\|X\|^2 \|Y\|^2 - \langle X, Y \rangle^2$. Note that the given quadratic form on V induces a quadratic form on $\Lambda^2(V)$ and the above is just $\|X \wedge Y\|^2$. If g is positive definite, so is the induced quadratic form, and if X and Y are linearly independent, this number is nonzero. The *sectional curvature* of a two-dimensional subspace W spanned by the vectors X, Y, is defined to be the real number

6.5. $$\operatorname{Sec}(W) = -\frac{Q(X,Y)}{\|X \wedge Y\|^2} = -\frac{\langle R(X,Y)X,Y \rangle}{\|X \wedge Y\|^2}.$$

It depends only on W and not on X, Y. In the general case of a nondegenerate (not necessarily positive definite) quadratic form, it is well defined only when the quadratic form restricts to the two-dimensional subspace as a *nondegenerate form* (which is always the case, if g is positive definite). We have thus proved the following fact.

6.6. Proposition. *The Riemannian curvature is determined by the sectional curvature and conversely. In particular, the sectional curvature is zero if and only if the Riemannian curvature is zero.*

If M is a Riemannian manifold of dimension 2, that is to say, a surface, and $m \in M$ is any point, and we take $W = T_m(M)$, then the number given by the sectional curvature of W is called the *Gaussian curvature c_m* at the point m.

6.7. Curvature of a hypersurface in $\mathbb{R}^m$.

Suppose a differential manifold M is imbedded in $\mathbb{R}^m$, and we take the induced metric on it. The induced Riemannian tensor is called the *first fundamental form*. We therefore get a natural linear connection on M, namely the Levi-Civita connection of the induced Riemannian structure.

Consider the exact sequence

$$0 \to T_M \to \mathcal{A}^m \to \operatorname{Nor}(M, \mathbb{R}^m) \to 0.$$

There is a natural splitting of this sequence given by the Riemannian metric. According to [Ch. 6, 2.13] this gives rise to a linear connection ∇' on M. Indeed it simply amounts to the following. Given two vector fields X and Y on M, extend both to $\mathbb{R}^n$ and consider the Euclidean connection $\nabla_X Y$ and then project its restriction to M orthogonally to the tangent bundle of M. Let X, Y be vector fields tangential to M at points of M. Then $\nabla'_X Y - \nabla'_Y X$ differs from $\nabla_X Y - \nabla_Y X$ by a normal field on M. Hence we see that the torsion tensor of ∇' is zero. On the other hand, if X, Y, Z are vector fields on $\mathbb{R}^m$ which are tangential to M at points of M, then $g(\nabla'_X Y, Z)$ is the same as $g(\nabla_X Y, Z)$ and so we conclude that the Riemannian tensor is left invariant. Hence the induced linear connection is the same as the Levi-Civita connection of the induced metric.

Now we also have another form, namely the *second fundamental form S* as defined in [Ch. 5, 4.11]. This associates to vector fields X, Y on M the normal field given by projecting $\nabla_X Y$. In other words, we have $\nabla_X Y = \nabla'_X Y + S(X, Y)$.

We have remarked that a hypersurface in $\mathbb{R}^n$ is oriented and that its normal bundle is trivial. Let ν be a section of the normal bundle, giving the unit normal at all points. Then we can use the connection ∇ to define an endomorphism of the tangent bundle as follows. Notice first that $\langle \nabla_X \nu, \nu \rangle + \langle \nu, \nabla_X \nu \rangle = X \langle \nu, \nu \rangle = 0$. This implies that $\langle \nabla_X \nu, \nu \rangle = 0$, and hence that $\nabla_X \nu$ is tangential to M. The map $X \mapsto \nabla_X \nu$ gives an endomorphism w of the tangent bundle, called the *Weingarten map*. All our computations can be made in terms of w. Firstly, let X, Y be tangential to M. Then $\langle \nabla_X Y, \nu \rangle = X \langle Y, \nu \rangle - \langle Y, \nabla_X \nu \rangle$. But Y is orthogonal to ν so that we have $\langle \nabla_X Y, \nu \rangle = -\langle Y, w(X) \rangle$. This means that the normal component of $\nabla_X Y$ is $-\langle Y, w(X) \rangle \nu$. The tangential component is $\nabla'_X Y$ by definition. Thus we have

6.8.
$$\nabla'_X Y = \nabla_X Y + \langle w(X), Y \rangle \nu.$$

In particular, the second fundamental form is given by $(X, Y) \mapsto \langle w(X), Y \rangle \nu$. Thus the data of the second fundamental form and the Weingarten endomorphism are the same.

Interchanging X and Y in equation 6.8 and subtracting we get

$$\langle w(X), Y \rangle = \langle X, w(Y) \rangle,$$

that is to say, w is a symmetric endomorphism of T_M.

Let X, Y, Z be tangential to M. Then we will compute $R(X, Y)Z$ where R is the curvature tensor of ∇'. We have

$$\begin{aligned}
\nabla'_X \nabla'_Y Z &= \nabla_X(\nabla'_Y Z) + \langle w(X), \nabla'_Y Z \rangle \nu \\
&= \nabla_X(\nabla_Y Z + \langle w(Y), Z \rangle \nu) + \langle w(X), \nabla_Y Z \rangle \nu.
\end{aligned}$$

Interchanging X and Y and subtracting, and noting that $R(X, Y)Z$ is tangential to M, we get the equality

6.9.
$$R(X, Y)Z = \langle w(Y), Z \rangle w(X) - \langle w(X), Z \rangle w(Y).$$

An equivalent form of the above is

$$Q(X, Y) = \langle w(Y), X \rangle \langle w(X), Y \rangle - \langle w(X), X \rangle \langle w(Y), Y \rangle.$$

Since our definition of the Riemannian curvature is intrinsic, this expression, which involves the Weingarten map or what is the same the second fundamental form, is in fact independent of the imbedding. Originally this formula was the definition of curvature of an imbedded hypersurface. So the formula has, as a corollary, the fact that the curvature so defined is independent of the imbedding. In other words, if the same hypersurface is isometrically imbedded in some other manner, the curvatures defined in terms of the second fundamental form for the two surfaces are the same. Gauss, who discovered this fact, termed it 'the most excellent theorem', and therefore this theorem is known as the 'theorema egregium'.

We have derived the consequence of the vanishing of the sectional curvature. A weakening of this drastic assumption is to say that the sectional curvature is a constant. This means that the sectional curvature at any point is the same for all two-dimensional subspaces of the tangent space at that point. A priori, there are two notions possible here. One is that this constant depends on the point $m \in M$, and the other is that it is in addition the same at all points. The first condition is always satisfied for surfaces, while the second is not generally true. But the two notions coincide for higher-dimensional manifolds.

6.10. Theorem (Schur). *If the sectional curvature is a constant at every point of a connected Riemannian manifold of dimension at least 3, this constant is the same at all points.*

Proof. We would like to convert the assumption into one on the Riemannian curvature tensor. Define a tensor s by the formula $s(X, Y, Z, T) = g(Y, Z)g(X, T) - g(X, Z)g(Y, T)$. Then our assumption says that

$$R(X, Y, X, Y) = C(g(X, Y)^2 - g(X, X)g(Y, Y)) = Cs(X, Y, X, Y).$$

Thus we conclude that $R = Cs$, for some function C. Now we will use the *second* Bianchi identity, namely $(\nabla_X R)(Y, Z) = 0$. In our case, this simplifies to

$$\sum XC.s(Y, Z) + C \sum (\nabla_X s)(Y, Z) = 0,$$

where the summation is cyclic and X, Y, Z are any three vector fields, and s is considered as a bilinear form with values in $\mathrm{End}(T)$. From the definition of s and the invariance of g, it follows that $\nabla s = 0$. Hence the cyclic sum $\sum (XC)s(Y, Z)$ vanishes. Moreover, it is easily seen that if X, Y, Z are linearly independent at a point p, then $s(Y, Z)_p, s(Z, X)_p$ and $s(X, Y)_p$ are all linearly independent. Hence the above equation implies that $X_p C = 0$. This proves that C is a constant, and consequently our assertion.

6.11. Example. Consider the unit sphere $S^n = \{(x) \in \mathbb{R}^{n+1} : \sum x_i^2 = 1\}$ with the Riemannian metric induced from the Euclidean metric. We will compute the curvature tensor at the point $P = (0, \ldots, 0, 1)$. A coordinate neighbourhood is given by $\{(x) \in S^n : x_{n+1} \neq 0\}$. The local coordinates are simply $x_i, i \leq n$. In this coordinate system the metric tensor g is given by $\sum_{i=1}^{n}(dx_i)^2 + (d(1 - \sum_{i=1}^{n} x_i^2)^{1/2})^2$. This simplifies to

$\sum_{i=1}^{n}(1 + \frac{x_i^2}{1-\sum x_i^2})(dx_i)^2 + \sum \frac{x_i x_j}{1-\sum x_i^2}dx_i dx_j$. In other words, the metric tensor is given by $\sum g_{ij}dx_i dx_j$ with

$$g_{ij} = \frac{\delta_{ij} + x_i x_j}{1 - \sum x_i^2}.$$

Notice that in this expression, all the functions g_{ij} have the value δ_{ij} at P and their first partial derivatives at P are all zero. It follows that $\Gamma_{ij}^k(P)$ are zero as well. Now using the determination 6.1 in local coordinates, of the Riemannian curvature tensor, we conclude that

$$
\begin{aligned}
(R_{ijkl})(P) &= \frac{\partial \Gamma_{jk}^l}{\partial x_i}(P) - \frac{\partial \Gamma_{ik}^l}{\partial x_j}(P) \\
&= \frac{1}{2}\left(\frac{\partial^2 g_{jl}}{\partial x_i \partial x_k} + \frac{\partial^2 g_{ik}}{\partial x_j \partial x_l} - \frac{\partial^2 g_{il}}{\partial x_j \partial x_k} - \frac{\partial^2 g_{jk}}{\partial x_i \partial x_l} \right)(P).
\end{aligned}
$$

Incidentally, our calculation above is quite general, and is valid in the following generality.

6.12. Lemma. *If $g = \sum g_{ij}dx_i dx_j$ is a Riemannian metric on a domain in $\mathbb{R}^n$ with $g_{ij} - \delta_{ij}$ and $\frac{\partial g_{ij}}{\partial x_k}$ vanishing at a point P, then the Riemannian curvature tensor is given at the point P by the expression*

$$\frac{1}{2}\left(\frac{\partial^2 g_{jl}}{\partial x_i \partial x_k} + \frac{\partial^2 g_{ik}}{\partial x_j \partial x_l} - \frac{\partial^2 g_{il}}{\partial x_j \partial x_k} - \frac{\partial^2 g_{jk}}{\partial x_i \partial x_l} \right)(P).$$

Going back to our example, we conclude that the Riemannian curvature tensor on S^n is given at the point $(0, \ldots, 0, 1)$ by

$$R_{ijkl}(P) = \delta_{jk}\delta_{il} - \delta_{ik}\delta_{jl}.$$

Invariantly expressed, this means that

$$R(X, Y, Z, T) = \langle Y, Z \rangle \langle X, T \rangle - \langle X, Z \rangle \langle Y, T \rangle$$

for any four tangent vectors X, Y, Z, T at P, since it is true for the basic vectors $\frac{\partial}{\partial x_i}$ at the point. Hence the sectional curvature corresponding to the plane X, Y in T_P is given by $\frac{-R(X,Y,X,Y)}{\|X\|^2\|Y\|^2 - \langle X,Y \rangle^2}$ which is 1, independent of the plane. By Schur's theorem it is also independent of the point. In other words, the Riemannian structure on the unit sphere S^n is of constant sectional curvature 1. Incidentally, this explains why in the definition of sectional curvature, we introduced a negative sign. We preferred S^n to have positive curvature. We may also see the constancy of sectional curvature by noting that the orthogonal group acts transitively on S^n and that the isotropy group at P is the orthogonal group $O(n-1)$ which acts transitively on all the two-planes on the tangent space at that point. Since the

Riemannian metric on S^n is invariant under $O(n+1)$, it follows that the sectional curvature is also invariant under the action and hence independent of the plane and also of the point.

6.13. Example. Consider the closed submanifold M' of $\mathbb{R}^{n+1}$ given by

$$\left\{ (y_1, \ldots, y_{n+1}) : \sum_{i=1}^{n} y_i^2 - y_{n+1}^2 = -1 \right\}.$$

This is a hyperboloid of two sheets, and is not connected, since the continuous function y_{n+1} on it has image $\mathbb{R} \setminus (-1, 1)$. Consider the component M given by $y_{n+1} \geq 1$. The map which sends $(y_1, \ldots, y_{n+1})$ to $(y_1, \ldots, y_n)$ in $\mathbb{R}^n$ is in fact a homeomorphism of M with $\mathbb{R}^n$, the map $\mathbb{R}^n \to M$ given by $(y_1, \ldots, y_n) \mapsto (y_1, \ldots, y_n, (1 + \sum_{i=1}^{n} y_i^2)^{1/2})$ being its inverse (see Remark 1.6, 2)). Taking this isomorphism as a (global) coordinate system, we may express the *Riemannian metric* induced on M from the pseudo-Riemannian metric $\sum_{i=1}^{n} dy_i^2 - dy_{n+1}^2$ on $\mathbb{R}^{n+1}$ in terms of these coordinates. It is given by the form

$$\sum_{i=1}^{n} (dx_i)^2 - \left(d \left(1 + \sum x_i^2 \right)^{1/2} \right)^2.$$

This Riemannian manifold is called the *hyperbolic space*. As in the first example, this leads to the determination of the Riemannian metric on M as $\sum g_{ij} dx_i dx_j$, with

$$g_{ij} = \delta_{ij} - \frac{\sum x_i x_j}{(1 + \sum x_i^2)^{1/2}}.$$

Thanks to Lemma 6.12, which is applicable to this example also, we have at the point $P = (0, \ldots, 0, 1)$ in M,

$$R_{ijkl} = -(\delta_{jk}\delta_{il} - \delta_{ik}\delta_{jl}).$$

Hence in this case the sectional curvature turns out to be -1. Of course, as a consequence of Schur's theorem this implies that the sectional curvature is -1 at all points. The orthogonal group of the quadratic form $y_1^2 + \cdots + y_n^2 - y_{n+1}^2$, namely $O(n, 1)$, acts on M' transitively. In fact, even $SO(n, 1)$ acts transitively on it. Although this group is not connected, the connected component of 1, which is denoted $SO^o(n, 1)$, acts transitively on M, implying again the constancy of the sectional curvature.

6.14. Definition. A connected Riemannian manifold is said to be a *space form* if it has constant sectional curvature. It is said to be of *spherical, flat* or *hyperbolic* type according as the constant curvature is positive, zero or negative.

7. Ricci, Scalar and Weyl Curvature Tensors

We will now decompose the curvature space $RC(V)$ defined in 6.2 for a vector space provided with a nondegenerate quadratic form, into a direct sum of three subspaces all of which are invariant under the natural action on $RC(V)$ of the orthogonal group $O(g)$ of g.

Firstly, we have the following homomorphism of $S^2(V^*)$ into RC. Start with a symmetric bilinear form b on V. Then we can define an element $\rho(b)$ in RC by setting

$$\rho(b)(X, Y, Z, W) = b(X, Z)\langle Y, W\rangle - b(Y, Z)\langle X, W\rangle$$
$$+ b(Y, W)\langle X, Z\rangle - b(X, W)\langle Y, Z\rangle$$

for all elements $X, Y, Z, W \in V$. The first two (resp. the last two) terms taken together are clearly alternating in the variables X and Y. Again the first and last terms (resp. the second and third terms) taken together, are alternating in the variables Z and W. We will verify 6.2, iii). On taking the sum cyclically permuting X, Y, Z, the first two terms cancel out, thanks to the symmetry of b and the inner product. The same is true of the last two terms. Thus $\rho(b)$ belongs to RC.

We also have a linear map in the reverse direction. To see this we will interpret the elements of RC as maps $V \times V \to \mathrm{End}(V)$. For any $F \in RC$, consider the bilinear form on V which associates to (X, Y), the trace of $Z \mapsto F(X, Z)Y$. From 6.2, iv) we conclude that this is a symmetric form, thereby yielding a linear map $RC \to S^2(V^*)$. The symmetric form corresponding to F will be denoted $\mathrm{Ric}(F)$. The composite of Ric and ρ is *not* the identity on $S^2(V^*)$, but is nevertheless an automorphism (when $n = \dim(V)$ is at least 3). It is easy to compute this composite.

Let (e_i) be a basis of V and (e_i') the dual basis with respect to the inner product. We then have

$$\begin{aligned}
\mathrm{Ric}(\rho(b))(X, Y) &= \text{ trace of } Z \mapsto \rho(b)(X, Z)Y \\
&= \sum \langle \rho(b)(X, e_i)Y, e_i'\rangle \\
&= \sum b(X, Y)\langle e_i, e_i'\rangle - \sum b(e_i, Y)\langle X, e_i'\rangle \\
&\quad + \sum b(e_i, e_i')\langle X, Y\rangle - \sum b(X, e_i')\langle e_i, Y\rangle \\
&= nb(X, Y) - \langle X, \sum b(e_i, Y)e_i'\rangle + \mathrm{tr}(b)\langle X, Y\rangle \\
&\quad - \langle \sum b(X, e_i')e_i, Y\rangle \\
&= (n - 2)b(X, Y) + (\mathrm{tr}\, b)\langle X, Y\rangle.
\end{aligned}$$

In other words, $\mathrm{Ric}(\rho(b)) = (n - 2)b + \mathrm{tr}(b)g$. If $n \geq 3$, the map $b \mapsto \mathrm{Ric}(\rho(b))$ is easily seen to be an isomorphism.

We will call the kernel of Ric the *Weyl subspace* of RC and denote it by $\mathrm{Weyl}(V)$. From our remarks above, we have the following conclusion.

7.1. Proposition. *Assume that the rank of the vector space V is at least 3. Then the curvature space $RC(V)$ is a direct sum of the kernel $\mathrm{Weyl}(V)$ of $\mathrm{Ric} : RC \to S^2(V^*)$ and the image of $\rho : S^2(V^*) \to RC$.*

We can further decompose $\rho(S^2(V^*))$. Let $S_0^2(V^*)$ consist of trace free symmetric endomorphisms of V. Consider the following canonical element s in RC. To $X, Y \in V$ associate the endomorphism $s(X, Y)$ of V which takes Z to $\langle X, Z\rangle Y - \langle Y, Z\rangle X$. Clearly it is the same as $\frac{1}{2}\rho(g)$. We will call s the *scalar element*.

7.2. Proposition. *The image of ρ is the direct sum of the one-dimensional space spanned by the scalar element s, and the image of $\rho : S_0^2(V^*) \to RC$.*

Proof. The space $S^2(V^*)$ is the direct sum of the one-dimensional space spanned by g and $S_0^2(V^*)$. From our computation above, it follows that $\rho g = 2s$.

We can also get an explicit splitting of the map $\mathrm{Ric} : RC(V) \to S^2(V^*)$. Let us denote the automorphism of $S^2(V^*)$ which takes b to $(n-2)b + \mathrm{tr}(b)g$, by α. Then we just checked that $\mathrm{Ric} \circ \rho = \alpha$. So, in order to get a splitting of Ric, all we have to do is to replace ρ by $\rho \circ (\alpha)^{-1}$. Now one checks directly that α^{-1} is given by $b \mapsto \frac{b}{n-2} - \frac{\mathrm{tr}(b)}{(n-1)(n-2)}g$.

We will summarise our conclusions in the following proposition.

7.3. Proposition. *Let V be a vector space of dimension $n \geq 3$ and $g = \langle\ ,\ \rangle$, a nondegenerate symmetric bilinear form on it. There is a natural surjection $\mathrm{Ric} : RC(V) \to S^2(V^*)$, defined by $(\mathrm{Ric}(R))(X, Y) = $ trace of $Z \mapsto R(X, Z)Y$. A splitting of Ric is given by $b \mapsto \frac{\rho(b)}{n-2} - \frac{1}{(n-1)(n-2)}s$, where $\rho(b)$ is the element of $RC(V)$ determined by $\langle \rho(b)(X, Y)Z, W\rangle = b(X, Z)\langle Y, W\rangle - b(Y, Z)\langle X, W\rangle + b(Y, W)\langle X, Z\rangle - b(X, W)\langle Y, Z\rangle$, and $s = (1/2)\rho(g)$.*

Our interest in this linear algebraic computation is its application to the Riemannian curvature R of a pseudo-Riemannian manifold.

7.4. Definition. Let M be a Riemannian manifold and R the Riemannian curvature tensor field. The section of $S^2(T^*)$ given by

$$(X, Y) \mapsto -(\text{trace of the endomorphism } Z \mapsto R(X, Z)Y)$$

is called the *Ricci curvature tensor field*. The corresponding endomorphism of T is called the *Ricci endomorphism*. This can also be defined as the image, under the composition map $\mathrm{End}(T^*) \otimes \mathrm{End}(T^*) \to \mathrm{End}(T^*)$, of R

regarded as a section of $\mathrm{End}(T^*) \otimes \mathrm{End}(T^*)$. The component of R in the space $\mathrm{Weyl}(T_m)$ at each point $m \in M$ gives another tensor field called *Weyl's conformal curvature tensor field* or *Weyl curvature*.

7.5. Conformal structure.

Let V be a vector space and g, g' two metrics on it. We say that they are *conformally equivalent* if there exists a positive scalar a such that $g = ag'$.

7.6. Definition. A *conformal structure* on a differential manifold is an assignment of a (conformal) equivalence class of metrics on the tangent space at every point.

The set of all linear transformations T of V which satisfy $g(Tv) = \lambda g(v)$ for some $\lambda \neq 0$ form a group called the *conformal group*. The group $GL(V)$ acts on the set of all conformal equivalence classes of metrics on V transitively. The isotropy at a given conformal structure is the conformal group. Hence one concludes that the data of a conformal structure on M is just a reduction of the structure group of the tangent bundle to the conformal group.

7.7. Definition. Two Riemannian metrics g, g' on a differential manifold are said to be *conformally equivalent* if there exists an everywhere positive function φ such that $g = \varphi g'$.

7.8. Exercise. Show that any conformal structure comes from a Riemannian metric.

A conformal structure on a manifold is nothing but the data of a Riemannian metric, conformally equivalent metrics being considered the same.

7.9. Remark. If g, g' are conformally equivalent metrics, a routine, if long, computation can be made to express the Riemannian curvature of g' in terms of that of g. From this one can conclude that the Weyl curvature is the *same* for g and g'. This is the reason for calling it Weyl's conformal curvature tensor field. In particular, the vanishing of this tensor is equivalent to saying that there exist coordinate neighbourhoods in which g is conformally equivalent to the Euclidean metric $\sum dx_i^2$.

Since the Weyl curvature is a conformal invariant, one may wonder if there is a direct definition of this tensor starting with a conformal structure on M. In fact, one can show that there exists a unique Cartan connection based on [Ch. 6, Example 2.16, 2)] such that the curvature is in the Weyl space, and that explains the invariance of the Weyl curvature under conformal equivalence.

7.10. Riemannian density.

If V is a vector space over $\mathbb{R}$ of dimension n and g is a nondegenerate quadratic form on it, then there is a natural nondegenerate quadratic form on all the spaces $\Lambda^p(V)$ as well. For, g gives an isomorphism between V and V^* and hence a natural isomorphism of $\Lambda^p(V)$ with $\Lambda^p(V^*)$. Explicitly, this associates to the element $(v_1, \ldots, v_p)$, the p-form given by $(w_1, \ldots, w_p) \mapsto \det(g(v_i, w_j))$. In particular there is also a canonical quadratic form on the one-dimensional space $\Lambda^n(V)$, that is to say an element of $\bigotimes^2(\Lambda^n(V^*))$. It is called the *discriminant* of g.

We are interested in the case when V is the tangent space at a point m of a pseudo-Riemannian manifold M. Thus the pseudo-Riemannian structure gives a section of $\bigotimes^2(\Lambda^n(T^*)) = K^2$. If g has the expression $\sum g_{ij}dx_i dx_j$ in a local coordinate system (U, x), then the above section of $\mathcal{K}^2$ is obviously given by $\det(g_{ij})(dx_1 \wedge \cdots \wedge dx_n)^2$. If we take the sheaf $\mathcal{S} = \mathcal{K} \otimes OR$ of densities of M, and if dx is the Lebesgue measure in the coordinate system, then this element is given by $\det(g_{ij})(dx)^2$.

Let us now assume that g is actually a Riemannian metric, that is to say positive definite. Then one can actually find a section of $\mathcal{S}$ itself, whose square is the above. It has the local expression $\sqrt{\det(g_{ij})}dx$. Since it is uniquely characterised as the positive measure whose square is the above section of $\mathcal{S}^2$, it is determined globally as a section of $\mathcal{S}$.

7.11. Definition. The unique positive density whose square is the discriminant section of $\mathcal{S}^2$ is called the *Riemannian density*.

The Riemannian density, being nonzero everywhere, trivialises the sheaf $\mathcal{S}$. In the case of a Riemannian manifold, we will always trivialise $\mathcal{S}$ in this fashion. In the case of an oriented Riemannian manifold, $\mathcal{K}$ is the same as $\mathcal{S}$ and is therefore trivialised. It is obvious that its square as a section of K^2 is invariant under the Levi-Civita connection, and therefore the Riemannian density itself is invariant under the connection.

7.12. The star operator.

Let V be an n-dimensional vector space. There is then a canonical bilinear map $\Lambda^p(V) \times \Lambda^{n-p}(V) \to \Lambda^n(V)$, given by the wedge product. This map gives rise to an isomorphism of $\Lambda^p(V)$ with $(\Lambda^{n-p}(V))^* \otimes \Lambda^n(V)$. In the case of differential manifolds, this leads to an isomorphism of bundles

$$\Lambda^p(T^*) \to (\Lambda^{n-p}(T^*))^* \otimes \mathcal{K}.$$

Tensoring with the orientation system OR, we get an isomorphism

$$(\Lambda^p(T^*)) \otimes OR \to (\Lambda^{n-p}(T^*))^* \otimes \mathcal{S}.$$

If M is Riemannian, then the bundle $\mathcal{S}$ is canonically trivial on the one hand, and the bundles $\Lambda^p(T^*)$ are self-dual on the other. Hence we obtain an isomorphism of $\Lambda^p(T^*)$ with $\Lambda^{n-p}(T^*) \otimes OR$.

7.13. Definition. The linear map $* : \Lambda^p(T^*) \to \Lambda^{n-p}(T^*) \otimes OR$ defined as above, using a Riemannian metric g, is called the *star operator*.

By definition, the star operator is determined by the equality

$$\int g(*\alpha, \beta)\,dm = \int \alpha \wedge \beta$$

where α (resp. β) is a (resp. twisted) differential form of degree p (resp. $n - p$), and dm is the Riemannian density.

7.14. Computation. $*^2 = (-1)^{p(n-p)} \operatorname{Id}$ *on* $\Lambda^p(V)$.

Proof. For this computation we may use a local coordinate neighbourhood and use the orientation given locally. Let V be the tangent space of any point in this neighbourhood and $(e_1, \ldots, e_n)$ an orthonormal basis. Then the star operator on $\Lambda^p(V)$ can be computed to be $*_p(e_{i_1} \wedge \cdots \wedge e_{i_p}) = \epsilon(\sigma)e_{j_1} \wedge \cdots \wedge e_{j_{n-p}}$ where $j_1, \ldots, j_{n-p}$ are the complementary indices in increasing order and σ is the permutation of $\{1, \ldots, n\}$ which is defined by $\sigma(r) = i_r$ if $r \le p$ and $\sigma(r) = j_{r-p}$ for $r \ge p + 1$. On the other hand, $*_{n-p}(e_{j_1} \wedge \cdots \wedge e_{j_{n-p}}) = \epsilon(\tau)e_{i_1} \wedge \cdots \wedge e_{i_p}$ where τ is the permutation given by $\tau(r) = j_r$ for all $r \le n - p$ and $\tau(r) = i_{r-(n-p)}$ for all $r \ge n - p + 1$. But then it is clear that $\sigma \circ \alpha = \tau$ where α is the permutation given by $\alpha(r) = p + r$ for $r \le n - p$ and $\alpha(r) = r - (n - p)$ for all $r \ge n - p + 1$. Hence we conclude that $*_{n-p} \circ *_p(e_{i_1} \wedge \cdots \wedge e_{i_p}) = \epsilon(\alpha)e_{i_1} \wedge \cdots \wedge e_{i_p}$. This proves our assertion.

In [Ch. 3, 3.8, Example 1)], we saw that the adjoint of the exterior derivation $d_i : \Lambda^i(T^*) \to \Lambda^{i+1}(T^*)$ is $(-1)^{i+1}d_{n-i-1} : \Lambda^{n-i-1}(T^*) \otimes \mathcal{S} \to \Lambda^{n-i}(T^*) \otimes \mathcal{S}$. This uses the canonical isomorphism between $(\Lambda^j(T^*))^*$ and $\Lambda^{n-j}(T^*) \otimes \mathcal{K}$.

On the other hand, in the case of Riemannian manifolds, we have just remarked that the Riemannian form g gives rise to a nondegenerate quadratic form on $\Lambda^j(T^*)$. So in this case, there is a natural adjoint $\Lambda^{i+1}(T^*) \to \Lambda^i(T^*)$. The canonical duality transforms the adjoint to $(-1)^{i+1}d$. Since $*_j$ is the composite of the canonical duality and the self-duality given by the metric on Λ^j, it follows that the adjoint operator is the transform by $*$ of $(-1)^{i+1}d$, namely $(-1)^{i+1} * d *^{-1}$. Substituting $(-1)^{i(n-i)}*$ for the inverse of $*$, we finally get the adjoint operator to be $(-1)^{ni+1} * d *$. This differential operator from $\Lambda^{i+1}(T^*)$ to $\Lambda^i(T^*)$ is denoted ∂_{i+1}. In other words, ∂ on Λ^j is defined to be $(-1)^{nj+n+1} *_{n-j+1} \circ d_{n-j} \circ *_j$.

We will now compute it when $M = \mathbb{R}^n$. On functions it is zero, by definition. We will compute it on 1-forms. Let $\alpha = \sum f_i dx_i$ be a differential 1-form. Then $*\alpha = \sum (-1)^{i-1} f_i dx_1 \wedge \cdots \wedge \widehat{dx_i} \wedge \cdots \wedge dx_n$. Therefore $d * \alpha = (\sum \frac{\partial f}{\partial x_i}) dx_1 \wedge \cdots \wedge dx_n$. Finally $\partial \alpha = -\sum \frac{\partial f_i}{\partial x_i}$.

7.15. Proposition. *The lift of the symbol $T^* \otimes \Lambda^j(T^*) \to \Lambda^{j-1}(T^*)$ given by $(X, \alpha) \mapsto -i_X \alpha$ is ∂.*

Proof. Since $*$ and the Riemannian density are invariant under the Levi-Civita connection, it is clear that ∂ is also a lift of its symbol. The adjoint symbol of d assigns to $v \in T^*$ the negative of the transpose of the symbol of d. In computing the transpose we need to use the metric on the exterior powers. If σ is the symbol, then we have $\langle \sigma(\alpha), \beta \rangle = \langle \alpha, v \wedge \beta \rangle$. Taking decomposable tensors for α and β, we easily conclude that σ is $-i_v$.

7.16. Definition. The operator $\Delta = (d\partial + \partial d)$ from the sheaf of differential forms (resp. twisted differential forms) of degree i into itself is called the *Laplacian* of the Riemannian manifold. Any (possibly twisted) differential form α such that $\Delta \alpha = 0$ is said to be a *harmonic form*.

From the above computation of ∂ on 1-forms on $\mathbb{R}^n$ we see that $\Delta(f) = \partial df = \partial(\sum \frac{\partial f}{\partial x_i} dx_i) = -\sum \frac{\partial^2 f}{\partial x^2}$. This is the negative of the usual Laplace operator on the Euclidean space. For this reason Δ is sometimes defined to be $-(d\partial + \partial d)$.

8. Clifford Structures and the Dirac Operator

8.1. Clifford algebra. Let V be a real vector space of dimension n, and q a quadratic form on it. We will denote the associated symmetric bilinear form by b. The quotient of the tensor algebra $T(V)$ of V, by the two-sided ideal generated by $v \otimes v - q(v).1$, $v \in V$, is called the *Clifford algebra $C(q)$* of the quadratic form. The inclusion of V in $T(V)$ gives a canonical linear map $V \to C(q)$. The algebra $C(q)$ is characterised by the universal property that any linear map f of V into any algebra A satisfying $f(v)^2 = q(v).1$ for all $v \in V$ has a unique extension $\tilde{f}$ as an algebra homomorphism of $C(q)$ into A. Every v gives rise to two linear endomorphisms of the exterior algebra, namely a) the wedge product $\lambda_v : \Lambda^{r-1} V \to \Lambda^r V$, and b) the inner product $\iota_v : \Lambda^r V \to \Lambda^{r-1} V$ given by $(v_1, \ldots, v_r) \mapsto \sum (-1)^{i+1} b(v, v_i) v_1 \wedge \cdots \wedge \widehat{v_i} \wedge \cdots \wedge v_r$. Then one easily verifies that $\lambda_v^2 = 0$, $(\iota_v)^2 = 0$, and $\iota_v \circ \lambda_v + \lambda_v \circ \iota_v = q(v).\mathrm{Id}$. As a consequence the map $f : v \mapsto \lambda_v + \iota_v$ of V into $\mathrm{End}(\Lambda(V))$ satisfies $(f(v))^2 = q(v).1$. By the universal property this gives an algebra homomorphism $\tilde{f} : C(q) \to \mathrm{End}(\Lambda(V))$. Note that $f(v)$

acts on the scalars in $\Lambda(V)$ as $a \mapsto av$. In particular, f is injective on V. This proves also that the canonical map $V \to C(q)$ is injective.

In fact we have more. Consider the map $x \mapsto \tilde{f}(x)(1)$ of $C(q)$ into $\Lambda(V)$. The algebra $C(q)$ has a natural filtration and a $\mathbb{Z}/2$-gradation as well (coming from those of $T(V)$). The above map respects the filtration and the $\mathbb{Z}/2$-gradation on $\Lambda(V)$ induced by its gradation. Therefore there is a companion homomorphism of the associated graded algebras. Now $Gr(C(q))$ comes with a linear map $V \to F_1(C(q))/F_0(C(q))$ and all elements in the image have square 0. By the universal property of the exterior algebra we have therefore an algebra homomorphism of $\Lambda(V)$ into $Gr(C(q))$. The map $v \to (\tilde{f}(v))(1)$ of $C(q)$ given in the first paragraph of this section also respects the filtration and induces a linear map $GrC(q) \to \Lambda(V)$. It is a trivial computation to see that this provides an inverse.

8.2. Theorem. *The Clifford algebra of a quadratic form q on V is a $\mathbb{Z}/2$-graded filtered algebra whose associated graded algebra is canonically isomorphic to the exterior algebra $\Lambda(V)$. There is also a canonical linear isomorphism of $\Lambda(V)$ with $C(q)$ preserving the $\mathbb{Z}/2$-gradation and filtration, whose companion map at the associated graded level gives the inverse of the above map.*

8.3. Structure of $C(q)$.

We will hereafter assume that q is a nondegenerate form and recall some of the properties of $C(q)$. They are quite easy to prove, and one may consult [**12**] for details. The algebra $C(q)$ itself is a central simple algebra when $n = 2m$ is even. Hence it is isomorphic to the $(2^m, 2^m)$ matrix algebra over $\mathbb{R}$ or a $(2^{m-1}, 2^{m-1})$ matrix algebra over the quaternion algebra. The even part $C^+(q)$ has a nontrivial centre and is either isomorphic to a matrix algebra over $\mathbb{C}$ or breaks up into the product of two simple algebras, each isomorphic to a matrix algebra as above over the reals or the quaternions.

The *Clifford group* $\Gamma(q)$ is the group of invertible elements x in the Clifford algebra which satisfy $xVx^{-1} \subset V$. The group $\Gamma^+(q)$ of elements in $C^+(q)$ which satisfy the same condition is called the *even Clifford group*.

If n is even, there are up to isomorphism two complex irreducible $C^+(q)$-modules denoted $Spin^+$ and $Spin^-$. In particular, the group Γ^+ has representation in these spaces. These are irreducible representations, called the *half-spin representations*. The group Γ^+ has also a representation on V which is called the *vector representation*. This is simply the action $\rho(x)v = xvx^{-1}$. Since $q(xvx^{-1}) = (xvx^{-1})^2 = xv^2x^{-1} = x.q(v).x^{-1} = q(v)$, it follows that the induced automorphisms $\rho(x)$ preserve the quadratic form. In other words, we have a natural homomorphism $\Gamma \to O(q)$ and $\Gamma^+ \to SO(q)$. The kernel of this representation is $\mathbb{C}^\times$ or $\mathbb{R}^\times \times \mathbb{R}^\times$. The representations of

$\Gamma(q)$ on $Spin^+$ and $Spin^-$ *do not* go down to representations of the orthogonal or the special orthogonal groups.

Now if we start with a pseudo-Riemannian manifold of dimension $n = 2m$, we get in the above manner a bundle called the *Clifford bundle*. But there are in general no bundles corresponding to the two half-spin representations.

8.4. Definition. A pseudo-Riemannian manifold M of signature (p, q) is said to have a *Clifford structure* if the structure group of its tangent bundle can be lifted from $O(p, q)$ to the Clifford group $\Gamma(p, q)$. If M is oriented, the structure can then be lifted to the group $\Gamma^+(p, q)$.

8.5. Remark. Here we are interested only in the Riemannian case. Even there we actually have two choices. We may use the Riemannian metric g or its negative $-g$ and then take the Clifford group. It *does* make a difference. It is usually taken to be the latter.

Assume given a Clifford structure. Associated to the lifted principal bundle Σ, are the bundles $Spin^+(M)$, associated to the spin representations $Spin^+$ and $Spin^-$ of $\Gamma^+(n)$. The $C(q)$-module structure on $Spin^+ \oplus Spin^-$ gives linear maps $V \otimes Spin^+ \to Spin^-$ and $V \otimes Spin^- \to Spin^+$. These are clearly $\Gamma^+(q)$-homomorphisms. Hence they induce vector bundle homomorphisms $T \otimes Spin^+(M) \to Spin^-(M)$, and $T \otimes Spin^-(M) \to Spin^+(M)$. Since T is naturally isomorphic to T^*, these are potential symbols of differential operators of order 1 from one half-spin bundle into another.

In order to lift these into differential operators, what we need, according to the prescription in Chapter 5, is a connection in the $\Gamma^+(q)$-bundle. Accordingly we assume given a connection on this principal bundle which on extension of the structure group to $O(n)$ gives the Levi-Civita connection.

8.6. Definition. A *Clifford structure* on an even-dimensional oriented Riemannian manifold is a principal Γ^+-bundle, together with a connection which induces under the orthogonal representation of Γ, the tangent bundle with the Riemannian connection. The differential operators of order 1 taking one half-spin bundle into another on a Clifford manifold which lift the above symbols are called *Dirac operators*.

There is a canonical antiautomorphism β of $C(q)$ which is the identity on V. It is simply the extension of the natural inclusion of V in $C(q)$ considered as a linear map into $C(q)^{opp}$ as an algebra homomorphism $C(q) \to C(q)^{opp}$. Then one can show that $a\beta(a)$ is a scalar for every $a \in \Gamma(q)$. Moreover the map $a \mapsto a\beta(a)$ is a homomorphism called the *norm* of a. The kernel of the norm homomorphism of $\Gamma^+(q)$ into $\mathbb{R}^\times$ is called the *Spin-group* of q.

The restriction of the vector representation of $\Gamma(q)$ to $Spin(q)$ gives a homomorphism $Spin(q) \to SO(q)$. Any $v \in V \setminus \{0\}$ is obviously invertible in $C(q)$, $\frac{v}{q(v)}$ being its inverse. Moreover $vwv^{-1} = vwv/q(v) = b(v,w)v - wv^2/q(v) = b(v,w)v - w$ for all $w \in V$. Hence by definition v belongs to $\Gamma(q)$. Also, in the vector representation v is represented by the reflection in the hyperplane orthogonal to v up to sign. From this it follows that $Spin(q) \to SO(q)$ is surjective. Moreover the kernel consists of nonzero scalars of norm 1, namely ± 1. Let v, w be two vectors of norm 1. The element vw in $C(q)$ acts as the composite of two reflections, which is actually a rotation in the two-dimensional space spanned by v and w. Assuming that v, w are orthogonal, consider the one-parameter group γ in $Spin(q)$ taking t to $\cos(t) + \sin(t)vw$. Its image in $SO(q)$ is the one-parameter group η taking t to the rotation through an angle of $2t$ in the plane of v and w. The inverse image of η contains γ, and since the map $\pi : Spin(q) \to SO(q)$ is two-sheeted, and $\gamma \to \eta$ is already two-sheeted, we deduce that γ is the total inverse image of η and that $Spin(q)$ is connected.

The tangent vector at 1 to γ is the element vw in the Clifford algebra. It maps to $2E_{v,w}$ where $E_{v,w}$ is the skew-symmetric endomorphism $v \mapsto w$, $w \mapsto -w$ and 0 on vectors orthogonal to both v and w. Now the Spin representation is obtained as the restriction of an even Clifford module structure. Hence the element E_{ij} in the Lie algebra of $SO(q)$ acts in the Spin representation through the element $(1/2)vw$ in $C(q)$. Elements of the form $E_{v,w}$ generate the Lie algebra and hence we have identified its action on the Spin representation.

8.7. Definition. An oriented Riemannian manifold is said to be a *Spin-manifold* if a lifting of the structure group of the tangent bundle from the special orthogonal group to the *Spin*-group is given.

8.8. Remark. There is not much difference between the groups Γ^+ and *Spin*, since the latter is the kernel of a homomorphism into $\mathbb{R}^+$. Moreover Γ^+ is not compact. However, one can concoct another group, which is more interesting. This is called $Spin_c(q)$. Take the product of $Spin(q)$ and $U(1) = S^1$ and take the quotient of this group by the imbedding of $\mathbb{Z}/2$ in the product, by mapping the nontrivial element to $(c, -1)$ where c is the nontrivial element in the kernel $Spin(q) \to SO(q)$. We still have a vector representation of this group given by the vector representation of $Spin(q)$ and the trivial representation of S^1. A reduction (or lift) of the structure group of an oriented Riemannian manifold to this group is called a $Spin_c$ structure. The group $Spin_c$ acts on the two-spin representations, where $Spin(q)$ acts via the Clifford algebra, while S^1 acts by multiplication by the corresponding complex scalars. The corresponding bundles will still be denoted $Spin_+$ and $Spin_-$.

Chapter 8

Local Analysis of Elliptic Operators

In this chapter, we give a quick account of the L^2 properties of an elliptic operator. Basic to this is the theory of Fourier transforms, Schwartz spaces, Sobolev spaces and the like. The main aim is to prove the theorems of Sobolev and Rellich, and the interior regularity of elliptic operators.

1. Regularisation

The notion of a kernel function $K(x, y)$ in two (sets of) variables is used for transforming functions f of y into those of x by the prescription $f \mapsto \int K(x, y) f(y) dy$. If the kernel is nice, then since we are sort of averaging the values of f with weights coming from K, the resulting function is well behaved. Convolution with functions is one such operation. If φ is a nice function, we take the kernel function $K(x, y) = \varphi(x - y)$. We will now explain how this procedure leads to what is called *regularisation*.

Let φ be any (infinitely) differentiable function in $\mathbb{R}^n$ with compact support. Then for any locally summable function f, the function

1.1.
$$x \mapsto \int \varphi(x - y) f(y) dy$$

is well defined. It is said to be the *convolution product* $\varphi * f$ of φ and f. Then we claim that $\varphi * f$ is *differentiable*. In fact, for any $v \in \mathbb{R}^n$, consider

$$\frac{(\varphi * f)(x + tv) - (\varphi * f)(x)}{t} - (\partial_v \varphi * f)(x)$$

$$= \int \left(\frac{\varphi(x + tv - y) - \varphi(x - y)}{t} - \partial_v \varphi(x - y) \right) f(y) dy.$$

Since φ has compact support, the integration (for every fixed x, v and bounded t) needs to be performed only over a compact set K. We may assume that f does not vanish on K. The differentiability of φ implies that for any $\epsilon > 0$, we have the inequality

$$\left| \frac{\varphi(x + tv - y) - \varphi(x - y)}{t} - \partial_v \varphi(x - y) \right| < \frac{\epsilon}{\int_K |f|},$$

for small enough t and all y in the compact set K. This shows that $\partial_v(\varphi * f)$ exists and is in fact equal to $(\partial_v \varphi) * f$. By iteration, we conclude the following.

1.2. Proposition. *Let φ be any (infinitely) differentiable function with compact support. Then for any locally summable function f, the function $\varphi * f$ is also (infinitely) differentiable. For any differential operator D of the form $\sum a_\alpha \frac{\partial^\alpha}{\partial x^\alpha}$ with a_α constants, we have $D(\varphi * f) = D\varphi * f$.*

1.3. Remarks.

1) Notice that if both φ and f have compact supports, say K_1 and K_2, the convolution product $\varphi * f$ also has compact support contained in the sum $K_1 + K_2$ of the two supports, namely,

$$\{x : \text{ there exist } a \in K_1 \text{ and } b \in K_2 \text{ such that } x = a + b\}.$$

2) The convolution product is defined whenever the integral in question makes sense, and is often useful in this greater generality. For example, if φ and f are square summable, this product makes sense.

Suppose we can choose a sequence of differentiable functions φ_k with compact support such that $\varphi_k * f$ converges to f. Then we will have proved that f can be approximated by differentiable functions. How do we find such a sequence? As a matter of fact, it turns out to be very easy. Take *any* differentiable function φ with values in $[0, 1]$, which is 1 in a neighbourhood of 0, has compact support and is such that $\int \varphi(x) dx = 1$. We wish to construct a sequence of such functions with supports in smaller and smaller neighbourhoods of 0 but with the same properties. For example, we may cut down the support to half its size, not affecting the integral, by defining $\varphi_2(x) = 2^n \varphi(2x)$. Now it is clear how to define φ_k. Take $\varphi_k = k^n \varphi(kx)$. The idea of choosing such a sequence is the following. The integrand in

$\varphi_k * f$ has the factor $\varphi_k(x-y)$ and so for large k, the integration needs to be done only over y very close to x. If f were continuous, the other factor $f(y)$ will be close to $f(x)$. Hence the integral tends to $\int \varphi_k(x-y)f(x)dy = f(x)$.

It remains to verify that our idea works. Assume that f is continuous. We have

$$(\varphi_k * f)(x) = \int k^n \varphi(k(x-y))f(y)dy.$$

Substitute $y' = k(x-y)$ to get $(\varphi_k * f)(x) = \int \varphi(y')f(x - \frac{y'}{k})dy'$. For all y' in the support of φ, the integrand tends to $\varphi(y')f(x)$ uniformly as k tends to infinity. Hence the integral tends to $f(x)$. From the same consideration it is clear that the convergence is uniform on compact sets. Also if a function has support K, and K' is any compact neighbourhood of K, then the approximating functions $\varphi_k * f$ have support contained in K' for large k by (Remark 1.3, 1). We have thus shown

1.4. Proposition. *The space of differentiable functions with compact support in an open domain of $\mathbb{R}^n$ is dense in the space of continuous functions with compact support provided with the topology of uniform convergence.*

1.5. Corollary. *Any linear form μ on the space of differentiable functions with compact support which is continuous, in the sense that if $\{f_k\}$ is a sequence with support contained in a compact set K that tends to zero uniformly, then $\{\mu(f_k)\}$ tends to zero, can be extended uniquely to a linear form on the space of continuous functions with compact support with the same continuity property.*

This reconciles the definition of measure we gave in Chapter 3 with the usual one.

Let us now start with a square summable function f (i.e. $f \in L^2$) and see where the regularisation procedure leads us.

1.6. Remark. If $f \in L^2$ and φ is one of the functions φ_k as above, then

$$
\begin{aligned}
|(\varphi * f)(x)| &\leq \int |\varphi(x-y)|^{1/2} \cdot |\varphi(x-y)|^{1/2}|f(y)|dy \\
&\leq \|\varphi(x-y)^{1/2}\| \cdot \|\varphi(x-y)^{1/2}f(y)\| \\
&= \left(\int \varphi(x-y)dy \right)^{1/2} \cdot \|\varphi(x-y)^{1/2}f(y)\| \\
&= \|\varphi(x-y)^{1/2}f(y)\|.
\end{aligned}
$$

Hence $\int |(\varphi * f)(x)|^2 dx \leq \int \varphi(x-y)|f(y)|^2 dxdy$. Setting $(x',y') = (x-y,y)$ the latter integral becomes $\int \varphi(x')|f(y')|^2 dx'dy'$, which by Fubini's theorem equals $\|f\|^2$. This shows that $\varphi * f$ is actually in L^2, and that $\|\varphi * f\| \le \|f\|$.

Either by definition of L^2 or as a standard theorem, we know that continuous functions with compact support form a dense subspace of L^2. Approximating any $f \in L^2$ by a sequence of continuous functions with compact support, we get on regularisation, the following

1.7. Proposition. *In any domain in $\mathbb{R}^n$, differentiable functions with compact support form a dense subspace of L^2.*

2. A Characterisation of Densities

2.1. Theorem. *If μ is a differentiable measure, that is to say, a (Borel) measure in a domain in $\mathbb{R}^n$ such that the Lie derivatives of all orders exist as Borel measures, then it is of the form $f dx$ where f is Lebesgue summable and dx is the Lebesgue measure.*

Proof. In view of the local nature of the problem, it is enough to prove the theorem for μ with compact support. We show that μ is absolutely continuous with respect to the Lebesgue measure. In other words, if S is a set of Lebesgue measure 0, then $\mu(S) = 0$ as well. We will use the following.

2.2. Lemma. *If S is a set of Lebesgue measure 0, then there exists a sequence (x_k) of points in $\mathbb{R}^n$, tending to zero as k tends to infinity, such that $T_k(S) \cap S = \emptyset$ for all k. Here T_k denotes translation by x_k.*

Proof. This is simply a reformulation of the following.

2.3. Characterisation of sets of positive Lebesgue measure. *A subset of $\mathbb{R}^n$ is of positive Lebesgue measure if and only if the set*

$$\{x \in \mathbb{R}^n : \text{there exist } a, b \in S \text{ such that } x = a - b\}$$

contains an open neighbourhood of $\{0\}$.

If μ is not absolutely continuous with respect to the Lebesgue measure, then by the Lebesgue-Nikodym decomposition theorem, there exists a set S of Lebesgue measure zero such that $\mu(\mathbb{R}^n \setminus S) = 0$. We may assume for our purposes that S is bounded by replacing it by its intersection with the support of μ. Let (x_k) be a sequence of points as in Lemma 2.2. We may assume, by passing to a subsequence, that there exist $t_k > 0$ such that $\{\frac{x_k}{t_k}\}$ has a limit, say v, in S^{n-1}. Note that $\{t_k\}$ is a sequence tending to zero. Since $T_k(S) \subset \mathbb{R}^n \setminus S$, we have $\mu(T_k S) = 0$ for all k. Hence $|(T_k(\mu) - \mu)(S)|$ is independent of k and the sequence $\frac{(T_k(\mu) - \mu)(S)}{t_k}$ cannot have a limit. On the other hand, in view of the supposed differentiability of μ, we see that $\frac{(T_k(\mu)-\mu)(S)}{t_k}$ does have a finite limit, namely $\lim \frac{(T_{t_k v}(\mu)-\mu)(S)}{t_k} = \partial_v(\mu)(S)$. This proves by contradiction that μ is absolutely continuous with respect to the Lebesgue measure.

2.4. Remark. By Lebesgue's theorem, there exists a Lebesgue summable function f such that $\mu = f dx$. The assumption that μ admits iterated Lie derivatives can be translated into the existence of a measurable function g such that $\int f \partial^\alpha \varphi / \partial x^\alpha dx = \int g \varphi dx$ for all differentiable functions φ with compact support. Under such a condition one can show that f is itself differentiable. This is known as Sobolev's theorem, and a version of it will be proved in 6.6.

3. Schwartz Space of Functions and Densities

Let V be a real vector space of dimension n. In this section we wish to define an algebra $SF(V)$ of functions on V as well as a module SD over it. Any choice of a Lebesgue measure will provide an isomorphism of the latter with $SF(V)$ as an $SF(V)$-module.

There is of course no canonical choice of Lebesgue measure. However notice that on $V \times V^*$, there is a canonical nondegenerate alternating bilinear form, namely $((v, f), (w, g)) \mapsto \langle w, f \rangle - \langle v, g \rangle$. This bilinear form may also be considered as a translation invariant exterior 2-form on the differential manifold $V \times V^*$. With respect to a linear coordinate system $(x_1, \ldots, x_n)$ in V, and the dual coordinate system $(\xi_1, \ldots, \xi_n)$ in V^*, this form is given by $\sum_{i=1}^n dx_i \wedge d\xi_i$. Its nth exterior power gives a top exterior form which is translation invariant. In the above coordinate system this form has the expression $n! dx_1 \wedge d\xi_1 \wedge \cdots \wedge dx_n \wedge d\xi_n$. Thus we have a canonical translation invariant $2n$-form which has the local expression $dx_1 \wedge d\xi_1 \wedge \cdots \wedge dx_n \wedge d\xi_n$. Hence *there is also a canonically defined positive, translation invariant measure σ on $V \times V^*$.* Whenever we fix a Lebesgue measure dx on V there is a Lebesgue measure dy on V^* such that $\sigma = dx \otimes dy$.

3.1. Definition of SF and SD.

Any $v \in V$ gives rise to a vector field ∂_v on V treated as a differential manifold. These are the translation invariant vector fields on V. Any two of these vector fields commute and hence by repeated operation of these we get an isomorphism $Q \mapsto \partial_Q$ of the symmetric algebra $S(V)$ with the algebra of translation invariant differential operators on V. These may also be called *differential operators with constant coefficients.* On the other hand, we can regard elements of $S(V^*)$ as polynomial functions on V. Thus we get a map of $S(V^*) \otimes S(V)$ into the space of differential operators in V, induced by

$(P, Q) \mapsto P\partial_Q$. It is easily seen that this is an injective linear map. Elements in the image are differential operators with polynomial coefficients and will be called simply *polynomial differential operators*. The set of polynomial differential operators form a subalgebra $P(V)$ of the algebra of differential operators.

3.2. Definition. A differentiable function f such that for every $D \in P(V)$, Df is bounded, is said to be a *Schwartz function*. The space of Schwartz functions will be denoted $SF(V)$ or simply SF, called the *Schwartz space of functions*.

3.3. Exercise. Show that Schwartz functions tend to zero at infinity.

3.4. Example. Let q be a positive definite quadratic form on V. Then the function $e^{-q(v)}$ is a Schwartz function. In fact, if we apply any polynomial differential operator to this function, the result is a finite linear combination of functions of the type $p(v)e^{-q(v)}$, where $p(v)$ is a polynomial function. So we have only to check that $p(v)e^{-q(v)}$ is bounded, which is obvious.

The set $SF(V)$ is obviously a vector space closed under multiplication and is thus a commutative algebra. Note however that it *does not have* an identity element, since the constant function 1 is *not* a Schwartz function. (Why not?) Clearly differentiable functions with compact support form an ideal in SF.

3.5. Remark. Let q be as in the above example. We will show that $(\frac{1}{1+q(v)})^{(n+1)/2}$ is integrable. Let Q be the quadric $\{v \in V : q(v) = 1\}$. We identify $\mathbb{R}^+ \times Q$ with $V \setminus \{0\}$ by restricting the map $(t, v) \mapsto tv$ of $\mathbb{R}^+ \times V \to V$. The differential of the above map takes a tangent vector $(\lambda, x) \in T_{(t,v)}$ to the vector $\lambda v + tx \in T_{tv}$. Its nth exterior power takes $\frac{d}{dt} \wedge a + b$ with $a \in \Lambda^{n-1}$ and $b \in \Lambda^n$, to $t^{n-1}v \wedge a + t^n b$. From this we conclude that the pull-back of a translation invariant n-form ω on V to $\mathbb{R}^+ \times V$ at the point (t, v) is $t^n p_2^*(\omega) + t^{n-1}dt \wedge p_2^*(i_v\omega)$. Its restriction to $\mathbb{R}^+ \times Q$ is therefore $t^{n-1}i_v(\omega)|T_v(Q)$. This means that the image under the product map of the product measure of dt on $\mathbb{R}^+$ and $t^{n-1}\mu$ on Q is the restriction of the Lebesgue measure on V to $V \setminus \{0\}$, where μ is the measure given by the $(n-1)$-form on Q which associates to v the $(n-1)$-differential $i_v(\omega)$. This is clearly the Riemannian measure on Q. This measure is referred to as 'the Lebesgue measure in spherical polar coordinates'.

The pull-back of the function $(\frac{1}{1+q(v)})^{(n+1)/2}$ to $\mathbb{R}^+ \times Q$ is the function $(t, v) \mapsto (\frac{1}{1+t^2})^{(n+1)/2}$. This function is summable with respect to the above measure if and only if $\frac{t^{n-1}}{(1+t^2)^{(n+1)/2}}$ is integrable on $\mathbb{R}^+$ with respect to the measure dt. It is clearly integrable in $\{t \leq 1\}$, since it extends to 0 as a continuous function. On the other hand, substituting $s = 1/t$ one sees that the transformed differential is $-1/(1 + s^2)ds$, which is also integrable in $\{s \leq 1\}$.

Write any $f \in SF$ as $f.(1 + q(x))^N(\frac{1}{1+q(x)})^N$, and note that the function $f.(1 + q(x))^N$ belongs to SF for large N and is therefore bounded. Taking $N = \frac{n+1}{2}$, we conclude that f is integrable. Applying the same to any power of f, we deduce that f is in L^p for all p.

3.6. Definition. If μ is a density on V such that $D\mu$ is bounded for every $D \in P(V)$, then it is called a *Schwartz density*. The set of all Schwartz densities is denoted $SD(V)$ and is called the *Schwartz space of densities*.

The set SD, besides being a vector space, is also an SF-module. In fact, if $f \in SF$ and $\mu \in SD$, the density $f\mu$ is again easily seen to be in SD.

3.7. Remark. We have seen that any density μ is of the form $f\,dx$, where dx is a Lebesgue measure. Since the Lie derivatives $L(v)$ of dx are all zero, it follows that $L(v)(f\,dx) = \partial_v(f)dx$. Hence μ is a Schwartz density if and only if it is of the form $f\,dx$ where f is a Schwartz function. In other words, the identification of Schwartz densities and Schwartz functions is compatible with the action by polynomial differential operators. This shows that although there is no canonical identification, the two SF-modules SF and SD are isomorphic.

3.8. Convolution of densities.

We will now define the *convolution product* of two densities. Let μ, ν be two densities. Then we define their convolution $\mu * \nu$ to be the image of the product measure $\mu \otimes \nu$ on $V \times V$ under the addition map $V \times V \to V$. More concretely, it is given by the prescription $(\mu * \nu)(f) = \int f(x + y)d\mu(x)d\nu(y)$, for every $f \in C_c^\infty$. This makes sense, if we assume that one of μ, ν has compact support, or that both of them are bounded (i.e. of finite measure). In particular, if both μ and ν are in SD, then $\mu * \nu$ is well defined and is actually in SD. In fact, if we show that on applying any polynomial differential operator, the resulting measure is a linear combination of measures of the same form, our assertion will follow. For any $v \in V$, the linear form $f \mapsto -(\mu * \nu)(\partial_v f)$ is simply $f \mapsto ((\partial_v\mu) * \nu)(f)$ from the definition. On the other hand, if l is a linear function on V, then we have

$$(\mu * \nu)(lf) = \int l(x + y)f(x + y)d\mu(x)d\nu(y)$$

$$= \int f(x + y)l(x)d\mu(x)d\nu(y)$$

$$+ \int f(x + y)d\mu(x)l(y)d\mu(y).$$

In other words, $(\mu * \nu)(lf) = (l\mu * \nu)(f) + (\mu * l\nu)(f)$ for all f. This implies that $l(\mu * \nu) = (l\mu) * (\nu) + \mu * (l\nu)$. We deduce that $\mu * \nu \in SD$. Since it is obvious that the map $(\mu, \nu) \mapsto \mu * \nu$ is bilinear, we have thus defined a linear map $SD \otimes SD \to SD$.

This multiplication is commutative and associative and makes the space of densities a commutative algebra. If $\mu \in SD(V)$ and $f \in SF(V)$, we can define a convolution product $\mu * f \in SF(V)$ to be the function $x \mapsto \int f(x - y)d\mu(y)$. This makes $SF(V)$ a module over $SD(V)$.

If we identify Schwartz densities with Schwartz functions (which we can do on fixing a Lebesgue measure dx on V), then the space of Schwartz functions also acquires the structure of a commutative algebra under convolution. However, it is already an algebra under the usual product of functions. In order to distinguish the two structures, even when we fix a Lebesgue measure on V and identify densities with functions, we will continue to denote the algebra under convolution by $SD(V)$ and under ordinary product by $SF(V)$. We will now compute the formula for this product for the convolution of two functions after identification of SF with SD. Let $f, g, h \in SF$. We identify f and g with the measures $f dx$ and $g dx$ respectively. Then the convolved measure $(f dx) * (g dx)$ gives on h the value

$$\int f(y)g(z)h(y + z)dydz.$$

Substituting $(y', z') = (y + z, z)$, we see that the integral is the same as

$$\int f(y' - z')g(z')h(y')dy'dz' = \int (f * g)(y')h(y')dy'$$

where $(f * g)(x) = \int f(x - y)g(y)dy$. In other words, we have

3.9. Proposition. *If f, g are in SF, then the convolution of the measures $f dx$ and $g dx$ is the measure $(f * g)dx$, where $f * g$ is the function $x \mapsto \int f(x - y)g(y)dy$.*

3.10. Remark. This formula is in conformity with the convolution product of two functions which we defined in 1.1 in connection with regularisation. Note that unlike the convolution of measures or that of a measure and a function, which does not require any Lebesgue measure on V, the convolution of functions does depend on such a choice.

3.11. Topology on SF and SD.

We will now topologise SF. There is a unique topology on it in which a sequence $\{f_k\}$ tends to f if and only if for every polynomial differential operator P, the sequence $\{Pf_k\}$ tends to Pf uniformly. If Q is any polynomial differential operator, it induces a map $Q : SF \to SF$, and the above topology is designed to make these maps continuous. In fact we have only to check that for every polynomial differential operator $P : SF \to SF$ the composite $P \circ Q$ is continuous into SF provided with the uniform topology. This follows from the definition since $P \circ Q$ is also a polynomial differential operator. Moreover, it is clear that multiplication of functions is continuous in this topology. We have only to show that if (f_k) (resp. (g_k)) tends to f (resp. g) in SF and P is a polynomial differential operator, then $(P(f_k g_k))$ tends to $P(fg)$ uniformly. But $P(f_k g_k)$ is a finite linear combination of terms of the form $P_i(f_k)Q_i(g_k)$ where P_i, Q_i are polynomial differential operators. Now $(P_i(f_k)Q_i(g_k))$ tends uniformly to $P_i(f)Q_i(g)$. This proves our assertion.

3.12. Approximate identity.

We remarked that the algebra SF does not have an identity element. With the introduction of this topology, we can make good this deficiency somewhat, by the notion of an *approximate identity*, namely a sequence $\{f_k\}$ such that $\{f_k.g\} \to g$ in the above topology, for every $g \in SF$. In order to construct such a sequence, we start with a function $f \in SF$ with values in $[0, 1]$ which restricts to the constant function 1 on the unit ball and has support inside a ball of radius 2. Then we define $f_k(x) = f(x/2^k)$. It is clear that for any g, and any compact set K, we have $(f_k.g)|K = g|K$ for large enough k. For any $\epsilon > 0$, $|g|$ is smaller than ϵ outside some compact set K, and $sup_K|f_k.g - g|$ can be made arbitrarily small by taking large enough k. Thus the sequence $\{f_k\}$ constitutes an approximate identity.

It is clear that SD can also be provided with a topology in a similar way. The algebra structure on SD (under the convolution product) also has an approximate identity. Take any density μ in SD such that the total measure is 1. Define μ_k by $\mu_k(f) = \mu(f_k)$, where $f_k(x) = f(x/2^k)$ for any $f \in C_c^\infty$. Then $(\mu_k * \nu)(f) = \int f(x + y)d\mu_k(x)d\nu(y) = \int f(2^{-k}x + y)d\mu(x)d\nu(y)$. As k tends to infinity, the integrand tends to $f(y)$ uniformly (having support contained in a fixed compact set) and therefore the integral tends to $\int f(y)d\mu(x)d\nu(y) = \int f(y)d\nu(y) = \nu(f)$. Hence $\mu_k * \nu \to \nu$ as $k \to \infty$.

If we fix a Lebesgue measure dx and identify SD with the space of functions, and take μ to be $\varphi(x)dx$, then the measure μ_k is given by $\mu_k(f) = \int f(2^{-k}x)\varphi(x)dx = f(y)\varphi(2^k y)2^k dy$ on substituting $y = 2^{-k}x$. Otherwise stated, the approximate identity μ_k is simply $f^{(k)}dx$, where $f^{(k)} = 2^k f(2^k x)$.

In addition to all these structures, there is also the action of the group V on SF by translation, that is to say, if $f \in SF$ and $v \in V$, then we define $T_v(f)$ to be the function taking $w \in V$ to $f(w-v)$. Clearly the latter function also belongs to SF. In a similar way we can also make the group $GL(V)$ act on SF.

4. Fourier Transforms

Let V be a finite-dimensional real vector space and V^* its dual. The Fourier transform is defined as explained at the beginning of this chapter, using the kernel $K(x,y) = e^{-i\langle x,y\rangle}$ as a function on $V \times V^*$. This function is called the *Fourier kernel*.

Fourier transform of Schwartz densities.

4.1. Definition. The *Fourier transform* of a Schwartz density on V, namely an element μ of $SD(V)$, is the function $\hat{\mu}$ on V^* given by the formula

$$\hat{\mu}(\xi) = \int e^{-i\langle x,\xi\rangle} d\mu(x).$$

Since any bounded measurable function is integrable with respect to μ, our definition makes sense. It is also clear that the function $\hat{\mu}$ is bounded (by the bound of the measure μ). We will now prove that $\hat{\mu}$ is actually a Schwartz function on V^*. Since $P(V^*)$ is generated by operators of the form $\partial_l, l \in V^*$ and multiplication by linear functions, namely elements $v \in V$, it will be enough to show that $\partial_l \hat{\mu}$ and $v\hat{\mu}$ are again of the same form.

4.2. Lemma. *If $\mu \in SD(V)$ and $v \in V$, then the Fourier transform of $\partial_v \mu$ is the function $iv.\hat{\mu}$.*

Proof. By definition, $\widehat{\partial_v \mu}$ is given by

$$\xi \mapsto \int e^{-i\langle x,\xi\rangle} d(\partial_v \mu)(x) = -\int \partial_v(e^{-i\langle x,\xi\rangle})d\mu(x)$$

$$= \int e^{-i\langle x,\xi\rangle} iv(\xi)d\mu(x),$$

where v is considered as a linear function on V^*. The right side is clearly $(iv\hat{\mu})(\xi)$.

4.3. Exercise. Note that the equality $\varphi(x)d(\partial_v\mu)(x) = -\int \partial_v\varphi(x)d\mu(x)$ is the definition of ∂_v, but one had to take φ in C_c^∞ for this. We have used it for the function $e^{-i\langle x,\xi\rangle}$. Justify it.

4.4. Lemma. *If $\mu \in SD(V)$ and $\eta \in V^*$, then the Fourier transform of $\eta\mu$ is $i\widehat{\partial_\eta\mu}$ (where η is regarded as a function on V).*

Proof. We have $\widehat{\eta\mu}(\xi) = \int e^{-i\langle x,\xi\rangle}\eta(x)d\mu(x) = -i\int \partial_\eta(e^{-i\langle x,\xi\rangle})d\mu(x) = i\int e^{-i\langle x,\xi\rangle}d(\partial_\eta\mu)(x) = i\widehat{\partial_\eta\mu}(\xi)$, as claimed.

These two simple lemmas show that the image of $SD(V)$ under Fourier transform is contained in $SF(V^*)$ for all $\mu \in SD(V)$.

We will see next how Fourier transform behaves with respect to the various structures on the Schwartz spaces that we defined above. To start with, it is obvious that it is $\mathbb{C}$-linear.

4.5. Lemma. *If μ and ν are two elements of $SD(V)$, then the Fourier transform of the convolution $\mu * \nu$ is the product of the functions $\hat{\mu}$ and $\hat{\nu}$.*

Proof. We have $\widehat{\mu * \nu}(\xi) = \int e^{-i\langle x+y,\xi\rangle}d\mu(x)d\nu(y)$. This simplifies, thanks to the theorem of Fubini, to the product of $\int e^{-i\langle x,\xi\rangle}d\mu(x) = \hat{\mu}(\xi)$ and $\int e^{-i\langle y,\xi\rangle}d\nu(y) = \hat{\nu}(\xi)$.

In other words, $\mu \mapsto \hat{\mu}$ is a homomorphism of the algebra $SD(V)$ into the algebra $SF(V^*)$.

4.6. Remark. If we fix a Lebesgue measure dx on V and identify SD with the space of Schwartz functions, then we get $(\widehat{f(x)dx})(\xi) = \int e^{-i\langle x,\xi\rangle}f(x)dx$. Hence we may *define* the Fourier transform of a Schwartz function f on V to be the *Schwartz function $\hat{f}$* on V^* given by

$$\hat{f}(\xi) = \int e^{-i\langle x,\xi\rangle}f(x)dx.$$

Fourier transform of Schwartz functions.

4.7. Definition. The *Fourier transform* from $SF(V)$ into $SD(V^*)$ associates to any $f \in SF(V)$, the linear form $\hat{f}$ on $C_c^\infty(V^*)$ defined by $g \mapsto \int e^{-i\langle x,\xi\rangle}f(x)g(\xi)d\sigma$. The measure used on the product is the canonical Lebesgue measure on $V \times V^*$.

It is clear that since f and g are integrable, the above integral exists. Moreover it it easy to check that the linear form is continuous for the usual topology on C_c^∞. Let us choose Lebesgue measures dx, $d\xi$ on V, V^* respectively in such a way that the product measure is canonical. Then we will show (Remark 4.9) that for $f \in SF(V)$, the measure $\hat{f}$ as defined in 4.7 is the same as $\hat{f}(\xi)d\xi$ in the notation of 4.6. This shows that the measure $\hat{f}$ actually belongs to $SD(V^*)$. Finally it is also true that

$$\widehat{\mu * f} = \hat{\mu}\hat{f}.$$

4.8. Exercise. Prove the above identity. Thus the $SD(V)$-module structure of $SF(V)$ is taken by the Fourier transform to the $SF(V^*)$-module structure of $SD(V^*)$.

4.9. Remark. If we identify a function f in $SF(V)$ with the density $f(x)dx$, then $\widehat{f(x)dx}$ is a function in $SF(V^*)$. Choosing the Lebesgue measure $d\xi$ on V^* such that $dx \otimes d\xi$ is canonical, we can identify it with a density in V^*. Let us now compute this density. Firstly, $\widehat{f(x)dx}(\xi) = \int e^{-i\langle x,\xi\rangle} f(x)dx$. Hence the corresponding density is the linear functional which takes $\varphi(\xi)$ to $\int e^{-i\langle x,\xi\rangle}\varphi(\xi)f(x)dxd\xi$. This is simply $\hat{f}(\varphi)$ by definition. Thus the Fourier transforms $SD(V) \to SF(V^*)$ and $SF(V) \to SD(V^*)$ are the same if we use dx to identify $SD(V)$ with $SF(V)$ and simultaneously $SD(V^*)$ with $SF(V^*)$ using $d\xi$.

As far as the topological structure is concerned, we have the following

4.10. Lemma. *The Fourier transform is a continuous map from $SD(V)$ (resp. $SF(V)$) into $SF(V^*)$ (resp. $SD(V^*)$).*

Proof. In view of Lemmas 4.2 and 4.4, we need only prove that the map is continuous when the target is provided with the topology of uniform convergence. Now if $\mu \in SD(V)$, then $|\hat{\mu}(\xi)| \leq \int |e^{-i\langle x,\xi\rangle}|d|\mu| = |\mu|(\mathbb{R}^n)$. Hence if $\{\mu_k\} \to \mu$, we see that $\sup|\hat{\mu}_k - \hat{\mu}|$ can be made arbitrarily small for large k. The proof that $SF(V) \to SD(V^*)$ is continuous is similar.

4.11. Remark. Note that since the Fourier transform is an algebra homomorphism, the above lemma implies that any approximate identity for the usual multiplication of functions is taken to an approximate identity of the convolution algebra and conversely. Indeed, for the former, we took the approximate identity $f^k(x) = f(x/2^k)$, for a suitable f. It has as Fourier transform, the density

$$g \mapsto \int e^{-i\langle x,v\rangle} f(x/2^k)g(v)d\sigma(x,v) = \int e^{-i\langle y,w\rangle} f(y)g(w/2^k)d\sigma(y,w)$$

on setting $(y,w) = (x/2^k, 2^k v)$. Hence it is $\mu_k(g)$ with the notation of 3.12, where μ is the density $\hat{f}$.

4.12. Lemma. *For $f \in SF(V)$ and $v \in V$, the Fourier transform of the translate $T_v(f)$ is given by $e^{-iv}\hat{f}$ where v is considered a linear function on V^*.*

Proof. In fact, $\widehat{T_v f}(\varphi) = \int e^{-i\langle x,\xi\rangle} f(x-v)\varphi(\xi)d\sigma(x,\xi)$. On substituting $(y,\eta) = (x-v,\xi)$ this becomes $\int e^{-i\langle y,\eta\rangle} f(y)e^{-i\langle v,\eta\rangle}\varphi(\eta)d\sigma(y,\eta) = (e^{-iv}\hat{f})(\varphi)$.

4.13. Theorem (Fourier Inversion Formula). *The composite of the two Fourier transforms $SF(V) \to SD(V^*)$ and $SD(V^*) \to SF(V)$ is the map $f(x) \mapsto f(-x)$.*

Proof. Let $f \in SF(V)$. We will first show that for any $g \in SF(V^*)$, we have $\widehat{g\hat{f}}(x) = \tilde{f} * \hat{g}$, where $\tilde{f}$ is the function $x \mapsto f(-x)$. In any case, if the assertion of the theorem were true, this would be an obvious consequence. Conversely, if this is proved, then we can apply it to an approximate identity (g_k) in place of g, and since $(\hat{g}_k)$ is an approximate identity in SD (4.8), the theorem would follow on letting k tend to infinity.

By definition we have

$$
\begin{aligned}
\widehat{g\hat{f}}(x) &= \int e^{-i\langle x,\xi\rangle} d(g\hat{f})(\xi) \\
&= \int e^{-i\langle x,\xi\rangle} g(\xi)d\hat{f}(\xi) \\
&= \int e^{-i\langle x,\xi\rangle} g(\xi)e^{-i\langle y,\xi\rangle} f(y)d\sigma(y,\xi).
\end{aligned}
$$

On the other hand, we have

$$
\begin{aligned}
\tilde{f} * \hat{g}(x) &= \int \tilde{f}(x-y)d\hat{g}(y) \\
&= \int f(-x+y) \int e^{-i\langle y,\xi\rangle} g(\xi)d\sigma(y,\xi) \\
&= \int f(y')g(\xi')e^{-i\langle x+y',\xi'\rangle} d\sigma(y',\xi')
\end{aligned}
$$

on substituting $(y',\xi') = (-x+y,\xi)$.

4.14. Corollary. *Fourier transform is a linear homeomorphism (as well as an algebra isomorphism) of $SD(V)$ with $SF(V^*)$ and $SF(V)$ with $SD(V^*)$.*

Proof. The map $F' : \mu \mapsto \hat{\mu}(-x)$ composed with the Fourier transform $SD(V) \to SF(V^*)$ is Id on $SD(V)$ by the above theorem. In view of Remark 4.6, the composite of the Fourier transform $SF(V^*) \to SD(V)$ with F' is also the identity on $SF(V^*)$.

The expression for the Fourier transform of a function, namely $\hat{f}(\xi) = \int e^{-i\langle x,\xi\rangle} f(x)dx$, makes sense even if f is just summable. This satisfies the identity $\int f\hat{g}dx = \int \hat{f}gd\xi$ for all $g \in SF$. In fact, since g is also summable, the function $f(x)g(\xi)e^{-i\langle x,\xi\rangle}$ is summable with respect to the measure $\sigma = dxd\xi$. By Fubini's theorem, we see that the above expressions are the same.

To make things more symmetric, we may look at the pairing between $SF(V)$ and $SF(V^*)$, given by $\mathcal{F}(f,g) = \hat{f}(g) = \int e^{-i\langle x,\xi\rangle} f(x)g(\xi)d\sigma(x,\xi)$. It is clear that this is a symmetric pairing. In other words, we have

4.15. Proposition. *If $f \in SF(V)$ and $g \in SF(V^*)$, then $\int fd\hat{g} = \int gd\hat{f}$.*

Let us fix a Lebesgue measure dx and take $g(x) = \hat{h}(-x)$ in the above proposition. Then we get $\int f(x)h(x)dx = \int \hat{h}(-x)\hat{f}(x)dx$. Now the Fourier transform of $\overline{h}$ is given by $\int e^{-i\langle x,\xi\rangle}\overline{h(x)}dx = \overline{\int e^{i\langle x,\xi\rangle}h(x)dx} = \overline{\hat{h}(-\xi)}$. In terms of the L^2 scalar product, the above formula therefore reduces to $\langle f,h\rangle = \langle \hat{f},\hat{h}\rangle$ on replacing h by $\overline{h}$. We have thus proved

4.16. Theorem (Plancherel). *The Fourier transform respects the L^2 scalar product on Schwartz functions.*

The Fourier transform can consequently be extended to a unitary isomorphism of $L^2(V,dx)$ into $L^2(V^*,d\xi)$ where $dx \otimes d\xi$ is the measure σ. Note however that if f is square summable, then the expression $\int e^{-i\langle x,\xi\rangle} f(x)dx$ does not make sense, since the integrand is not summable in general. For that we need f itself to be summable. If f is summable as well as square summable, then we thus have two notions of Fourier transform, namely the expression given above, and the extension of the map $SF \to SD$ that we have given, using Plancherel theorem. But we know that in this case, $\hat{f}$ satisfies $\langle \hat{f},\varphi\rangle = \langle f,\overline{\hat{\overline{\varphi}}}\rangle$, which is also satisfied by the L^2-extension of Fourier transform. Hence the two extensions coincide.

5. Distributions

Let U be a domain in V. Consider the space C_c^∞ of differentiable functions on U with compact support. We say that a sequence $\{f_k\}$ in this space tends to f if all f_k have support contained in a fixed compact set $K \subset U$, and for every differential operator D with constant coefficients, the sequence $\{Df_k\}$ tends to Df uniformly. A similar definition can be given also for densities with compact support, at least by identifying densities with functions and checking that the topology is independent of the Lebesgue measure chosen. We will call this space D_c^∞.

5.1. Remark. If we fix a compact subset K, then the space C_K^∞ of differentiable functions with support in K may be topologised by defining the neighbourhood filter at 0 to be the one generated by the sets of the form $U_{D,a} = \{f : \sup |Df| < a\}$ where D is a differential operator with constant coefficients and a a positive real number. This makes C_K^∞ a topological vector space.

5.2. Definition. A *distribution* (resp. an *n-current*) in U is a continuous linear form T on D_c^∞ (resp. C_c^∞), that is to say, if $\{\mu_k = f_k dx\} \to \mu = f dx$ (resp. $\{f_k\} \to f$) in the above sense, then $\{T\mu_k\} \to T\mu$ (resp. $\{Tf_k\} \to Tf$).

In view of the remark above, a distribution is a linear form T on D_c^∞ such that for every compact subset K, there is an inequality of the form

5.3. $$\sup |Tf| \leq C_D \sup |Df|$$

for every differential operator D with constant coefficients and for all $f \in D_K^\infty$.

If we fix a Lebesgue measure, we may identify n-currents and distributions.

5.4. Examples.

1) It is clear that if μ is any measure, it can be considered a current as well. In particular, for any $v \in V$, the measure $f \mapsto f(v)$ is a current, known as the *Dirac current* at v.

2) If f is a measurable function defined on U, we say that it is *locally square summable* if for any compact subset K of U, it is square summable on K. If f is locally square summable, then it defines a distribution T_f. Indeed, for every $\mu \in D_c^\infty$ we define $T_f(\mu) = \int f(x) d\mu$. One can easily check that the map $f \mapsto T_f$ is an injective linear map. Hence we may identify such functions with the distributions they define.

3) Let f be as above. For any $v \in V$ and $\mu \in D_c^\infty$, define $T(\mu) = -\int f(x)(\partial_v \mu)(x)$. It is easy to check that this is also a distribution. Moreover, if $\partial_v f$ exists as a function which is locally square summable, then T is the same as the distribution defined by $\partial_v f$. Therefore it is apt to define the *distributional derivative with respect to v* of a locally square summable function f (whether $\partial_v f$ exists or not, as a function) to be the distribution defined above. A good notation for this is therefore $\partial_v(T_f)$ or even $\partial_v(f)$.

In fact, we will generally define the derivative of a distribution as follows.

5.5. Definition. If T is a distribution (resp. current), then we define its *derivative ∂_v with respect to $v \in V$* to be the distribution (resp. current) given by $(\partial_v T)(\mu) = -T(\partial_v \mu), \mu \in D_c^\infty$ (resp. $(\partial_v T)(f) = -T(\partial_v f), f \in C_c^\infty$).

When we say a distribution *is* a function, we mean that it is the distribution associated to a function.

5.6. Definition. Any continuous linear functional on $SD(V)$ (resp. $SF(V)$) is called a *tempered twisted distribution* or *tempered current*. The space of tempered distributions will be denoted by $\Sigma D(V)$.

From the fact that $D_c^\infty \to SD$ is continuous, it follows that a tempered distribution gives rise to a distribution. Moreover, since the image of D_c^∞ in SD is dense, we see that tempered distributions form a subspace of the space of distributions.

The significance of tempered distributions is in the following. We have seen that the Fourier transform defines an isomorphism of $SD(V)$ with $SF(V^*)$ and of $SF(V)$ with $SD(V^*)$. But Schwartz functions are too restrictive, and we would like to extend the notion of Fourier transform to a large class of functions. Explicitly, we have the following definition.

5.7. Definition. The *Fourier transform $\hat{T}$* of a tempered distribution (resp. tempered current) T on V is the tempered current (resp. tempered distribution) on V^* defined by the prescription $\hat{T}(g) = T(\hat{g})$ for all $g \in SF(V^*)$ (resp. $g \in SD(V^*)$).

Since the Fourier transform is a homeomorphism, it is obvious that $\hat{T}$ is tempered when T is. If $f \in SF(V)$ then $\hat{T}_f$ is defined therefore by $g \mapsto T_f(\hat{g})$ for $g \in SF(V^*)$. But $T_f(\hat{g}) = \int f d\hat{g} = \int g d\hat{f}$ by Proposition 4.12. Thus we get $\hat{T}_f(g) = T_{\hat{f}}(g)$, or what is the same, $\hat{T}_f = T_{\hat{f}}$. This shows that our definition of the Fourier transform of tempered distributions is compatible with the definition of the Fourier transform of Schwartz functions.

Since the Fourier transform of the Schwartz spaces is a continuous linear isomorphism, it also defines by transposition, a linear isomorphism of the space $\Sigma D(V)$ with $\Sigma F(V^*)$ and of $\Sigma F(V)$ with $\Sigma D(V^*)$. The point of this is that large classes of functions can be considered as subspaces of ΣF and so Fourier transform theory becomes applicable to them.

Let f be a square summable function on V. The distribution it gives rise to is a tempered distribution. In fact, to every $\mu \in SD$, associate $\int f d\mu$. Since μ is of the form $g dx$ with $g \in SF$, it is the same as $\int f g dx$. We have seen in 3.5 that g is also square summable. Hence the integral makes sense. We have thus given a linear functional on $SD(V)$.

It remains to show that it is continuous. In other words, if (g_k) is a sequence in $SF(V)$ such that (Dg_k) tends to 0 uniformly for all polynomial differential operators D, then we have to show that $(\int f g_k dx)$ tends to 0. Notice first that by assumption, for every N, the sequence $\{(1+q(x))^N g_k\}$ is uniformly bounded. Choosing N so large that $\frac{1}{(1+q(x))^N}$ is square summable, we conclude that (g_k) is uniformly majorised by a square summable function. Hence (g_k) tends to 0 in L^2 by the dominated convergence theorem. Since f is also in L^2, we conclude that $\int f g_k dx$ tends to zero.

Thus we have given an injective linear map of L^2 into ΣF. In particular the Fourier transform of a square summable function makes sense. If (f_k) is a sequence of functions in C_c^∞ tending to $f \in L^2$, then we have already remarked that the Plancherel theorem ensures that $(\hat{f}_k)$ tends to a function φ in L^2. Hence for any $g \in L^2$, the sequence $\{\langle \hat{f}_k, g \rangle\}$ tends to $\langle \varphi, g \rangle$. In particular, the sequence of distributions $T_{\hat{f}_k}$ tends to T_φ. It is obvious that $T_{\hat{f}_k} = \hat{T}_{f_k}$. Thus we conclude that $T_{\hat{f}} = T_\varphi$. In other words, the continuous extension to L^2 of the Fourier transform defined on Schwartz functions, coincides with the restriction of the Fourier transform on tempered distributions to L^2.

5.8. Remark. If P is a polynomial differential operator and T a tempered distribution, then PT is also a tempered distribution. Formulas 4.2, 4.4 are valid for the extended Fourier transform as well.

6. Theorem of Sobolev

We will hereafter fix a positive invariant measure on $\mathbb{R}^n$.

Let r be any fixed nonnegative integer. Then one defines the *Sobolev space* of order r to be the space $\mathcal{H}_r$ of functions f on U which are square summable so that the distributions Df are square summable functions, for all differential operators D with constant coefficients of order $\leq r$. The meaning of the local Sobolev spaces $\mathcal{H}_r^{loc}$ is therefore clear.

Sobolev spaces come equipped with Hilbert space structures, with the scalar product

$$\langle f, g \rangle = \int \sum_{|\alpha| \leq r} D_\alpha f \overline{D_\alpha g}.$$

6.1. Remark. We saw in 1.7 that differentiable functions with compact support form a dense subspace of L^2. One can actually show that they also form a dense subspace of $\mathcal{H}_m$ in $\mathbb{R}^n$, for all m. In other words, any function in $\mathcal{H}_m$ can be approximated by differentiable functions with compact support. However this is *not true* for functions in an arbitrary domain U in $\mathbb{R}^n$. If a function on U belonging to $\mathcal{H}_m$ has compact support K, then it can be extended to the whole of $\mathbb{R}^n$ by setting it to be zero outside U. It can then be approximated by differentiable functions, with support still compact (which may be slightly bigger than K) and contained in U. This reduces the problem to the case of $\mathbb{R}^n$. The proof is, as in 1.1, by regularisation.

We will use the same notation and show that if f belongs to $\mathcal{H}_1$ for example, then for all $v \in \mathbb{R}^n$, the sequence $\{\partial_v(\varphi_k * f)\}$ tends to $\partial_v f$ in L^2. Note that by definition, $\partial_v f$ belongs to L^2 and hence so does $\varphi_k * \partial_v f$. As one might expect (in view of 1.2), it is indeed true that $\partial_v(\varphi * f) = \varphi * \partial_v f$ as distributions. For any $g \in C_c^\infty$, we have

$$
\begin{aligned}
(\partial_v(\varphi * f))(g) &= -\int (\varphi * f)(x).\partial_v g(x)dx \\
&= -\int f(x-y)\varphi(y).\partial_v g(x)dxdy \\
&= \int \partial_v f(x-y)\varphi(y)g(x)dxdy \\
&= \int (\varphi * \partial_v f)(x)g(x)dx \\
&= (\varphi * \partial_v f)(g).
\end{aligned}
$$

Now replacing φ by φ_k and letting k tend to infinity, we deduce that $\partial_v f$ is the limit in L^2 of $\varphi_k * \partial_v f = \partial_v(\varphi_k * f)$. In other words, $\varphi_k * f$ tends to f in $\mathcal{H}_1$. It is clear by induction that if $f \in \mathcal{H}_m$, then the convergence is actually in $\mathcal{H}_m$. Thus we have proved

6.2. Theorem. *Differentiable functions with compact support form a dense subspace of $\mathcal{H}_m$ in $\mathbb{R}^n$.*

The above argument proves the following stronger statement.

6.3. Proposition. *If u is in $\mathcal{H}_m$ and such that $Du \in L^2$ for some differential operator D with constant coefficients, then there is a sequence $\{f_k\}$ of differentiable functions with compact support, which tends to u in $\mathcal{H}_m$ so that $\{Df_k\}$ tends to Du in L^2 as well.*

Proof. We only have to verify that $D(\varphi * u) = \varphi * Du$ and repeat the argument above. But this follows from the assumption that D has constant coefficients.

6.4. Remark. Actually it will turn out that we can even take *any* differential operator D which has constant coefficients outside a compact set, in the above proposition. The proof is more involved since it is no longer true that $D(\varphi * u) = \varphi * Du$. We will now discuss this in greater detail.

Let $u \in \mathcal{H}_m$ such that $Du \in L^2$. We will approximate u by convolving with the functions φ_k. Clearly then $\varphi_k * u$ are all differentiable, and we have seen that the sequence $\{\varphi_k * u\}$ tends to u in $\mathcal{H}_m$. Since Du belongs to L^2 by assumption, the sequence $\{\varphi_k * Du\}$ converges to Du in L^2. But we wish to show that $D(\varphi_k * u)$ converges to Du in L^2. We are therefore naturally led to considering the difference $\delta_\varphi = D(\varphi * u) - \varphi * Du$.

If we can show that the map $\mathcal{H}_m \to L^2$ which takes u to $\delta_\varphi u$ is continuous and has uniformly bounded norm for all φ_k, then we will be through. For if u is infinitely differentiable, then $\{\varphi_k * u\}$ tends to u in $\mathcal{H}_r$ (where r is the order of D) as well, so that $\{D(\varphi_k * u)\}$ tends to Du in L^2. Hence the difference $D(\varphi_k * u) - \varphi_k * Du$ tends to 0 in L^2. Let u be any element of $\mathcal{H}_m$, and $\{u_k\}$ a sequence of differentiable functions with compact support, tending to u in $\mathcal{H}_m$. For any $\epsilon > 0$, there exists R such that $\|u_r - u\|_m < \frac{\epsilon}{2M}$, for all $r \geq R$, where M is a uniform bound for the norms of all the operators δ_{φ_k}. Hence $\|\delta_{\varphi_k}(u_R - u)\| < \epsilon/2$ for all k. But we have just noted that since u_R is differentiable, we have $\|\delta_{\varphi_k}(u_R)\| < \epsilon/2$ for large enough k. This implies that $\|\delta_{\varphi_k}(u)\| < \epsilon$, showing that the sequence $\{D(\varphi_k * u) - \varphi_k * Du\}$ tends to zero in L^2, as required.

Notice that by assumption, the coefficient functions in the expression for D are constant outside a compact set. By subtracting the corresponding operator with constant coefficients, we get an operator whose coefficients have *compact support*. This does not change the commutator that we are considering since any operator with constant coefficients commutes with convolution with φ_k. Thus we need to prove only the following.

6.5. Lemma. *Let D be a differential operator with compact support. The continuous linear map $\mathcal{H}_m \to L^2$ given by $u \mapsto D(\varphi_k * u) - \varphi_k * Du$ has bounded norm, independent of k.*

Proof. Since differentiable functions with compact support are dense in $\mathcal{H}_m$, we may compute norms by restricting the operators to C_c^∞. Any differential operator D is given by the expression $\sum a_\alpha(x)\frac{\partial}{\partial x_\alpha}$. To prove the lemma, we may assume that D is just $a_\alpha(x)\frac{\partial}{\partial x_\alpha}$. We will estimate the L^2-norm of $D(\varphi_k * u) - \varphi_k * Du$ by applying the Fourier transform and using the Plancherel theorem. Then we are led to the following computation:

$$\left|\widehat{D(\varphi_k * u)} - \widehat{(\varphi_k * Du)}\right| = \left|\hat{a}_\alpha(\xi) * (\xi^\alpha \hat{\varphi}_k(\xi)\hat{u}(\xi)) - \hat{\varphi}_k(\xi)\hat{a}_\alpha(\xi) * (\xi^\alpha \hat{u}(\xi))\right|$$

$$= \left|\int \hat{a}_\alpha(\xi - \eta)\eta^\alpha \hat{\varphi}_k(\eta)\hat{u}(\eta)d\eta\right.$$

$$\left. - \hat{\varphi}_k(\xi)\int \hat{a}_\alpha(\xi - \eta)\eta^\alpha \hat{u}(\eta)d\eta\right|$$

$$= \left|\int \hat{a}_\alpha(\xi - \eta)\eta^\alpha \hat{u}(\eta)(\hat{\varphi}_k(\eta) - \hat{\varphi}_k(\xi))d\eta\right|.$$

We need to get an L^2 estimate for the above function of ξ. Before we set out to do that, we make a few comments. Since $a_\alpha(x)$ has compact support, it and its Fourier transform are Schwartz functions and so remain bounded on multiplication by $(1 + q(\xi))^N$ for any positive integer N. We will replace η^α by $\sum \binom{\alpha}{\beta}(\eta - \xi)^{\alpha - \beta}\xi^\beta$. The powers of $\eta - \xi$ can then be clubbed with $\hat{a}_\alpha(\xi - \eta)$. We will denote by b_α the functions $\hat{a}_\alpha(\xi).\xi^N$. Therefore we wish to estimate functions of the type

$$\xi \mapsto \xi^\beta \int b_\alpha(\xi - \eta)\hat{u}(\eta)(\hat{\varphi}(\eta) - \hat{\varphi}(\xi))d\eta.$$

We are looking for an L^2-estimate in terms of the norm of u in $\mathcal{H}_m$. We will multiply ξ^β by a suitable power of $\frac{1}{(1+q(\xi))}$, so that the product is in $\mathcal{H}_m$. Once that is done, we need only to bound the integral

$$\int b(\xi - \eta)\hat{u}(\eta)(1 + q(\xi))^L(\hat{\varphi}(\xi) - \hat{\varphi}(\eta))d\eta$$

where b is a product of $\hat{a}_\alpha$ and a polynomial. The term causing difficulty in the estimate is the factor $(1 + q(\xi))^L$. If we had an inequality of the type $q(\xi - \eta) \geq \lambda q(\xi)$, then we would have $(1 + q(\xi))^L \leq (1 + q(\xi - \eta)/\lambda)^L$ and so we may replace b by a similar term b' and get the estimate. Accordingly, we split the integral as a sum $I_1 + I_2$ of the integrals over two subsets of $\mathbb{R}^n$, namely, one in which $q(\xi - \eta) \geq \lambda q(\xi)$, and the other in which the opposite inequality holds, and estimate each separately. In the former case, the term $|\hat{\varphi}(\eta) - \hat{\varphi}(\xi)|$ can be majorised by $C\|\eta - \xi\|$ where $C = \sup|\partial_v \hat{\varphi}|$, which is finite since $\hat{\varphi}$ is a Schwartz function. The integral of $b'(\xi - \eta)\hat{u}$ can be bounded in terms of the L^2-norm of $\hat{u}$ or, what is the same, that of u, since the term b' is a Schwartz function. This estimate is valid for all φ_k.

As for I_2, we note as above that $|\hat{\varphi}(\xi) - \hat{\varphi}(\eta)| = (\xi - \eta)(\partial_v \hat{\varphi}(\zeta))$, where ζ is some point in the segment $[\xi, \eta]$. Since $\partial_v \hat{\varphi}$ is a Schwartz function, we have the inequality, for any N,

$$|\hat{\varphi}(\xi) - \hat{\varphi}(\eta)| \leq C_2 \frac{\|\xi - \eta\|}{(1 + q(\zeta))^N}$$

(where C_2 of course depends on N). This is not good enough for us yet, since we do not have $(1 + q(\xi))^N$ in the denominator. However, we note that if λ is small, the ball $\{v : q(v) \leq \lambda q(\xi)\}$ with the origin as centre and the ball $\{v : q(\xi - v) \leq \lambda q(\xi)\}$ with ξ as centre do not intersect. Since we have assumed that η belongs to the second of these, the segment $[\xi, \eta]$ is contained in it and in particular ζ belongs to it. Thus we have $q(\zeta) \geq \lambda q(\xi)$ and consequently $\frac{1}{1+q(\zeta)} \leq \frac{1}{1+\lambda q(\xi)}$. This implies that $\frac{(1+q(\xi))^L}{(1+q(\zeta))^L} \leq (\frac{1}{\lambda})^L$ and we are done. Thus we have shown

6.6. Lemma (Friedrichs). *If $u \in \mathcal{H}_m$ is such that $Du \in L^2$ for some differential operator D whose coefficients are constant outside a compact set, then there exists a sequence $\{f_k\}$ of differentiable functions with compact support, which tends to u in $\mathcal{H}_m$ so that Df_k tends to Du in L^2.*

There are also other spaces which are of interest to us. Consider the space $\mathcal{G}_m$ consisting of all measurable functions f such that Pf is in L^2 for all polynomials P of degree $\leq m$. It is obvious that the Fourier transform takes $\mathcal{H}_m$ bijectively onto $\mathcal{G}_m$. It is a simple matter to see that the norm in $\mathcal{H}_m$ is equivalent to the L^2-norm of $f(1 + q(x))^{m/2}$. It is this latter norm that we will use below.

Similarly we can also provide $\mathcal{G}_m$ with the norm given by $\|f\|_m^2 = \sum \|x^\alpha f\|^2$. From the Plancherel theorem and Lemma 4.2, it follows that the Fourier transform is actually a metric equivalence of $\mathcal{H}_m$ with $\mathcal{G}_m$.

6.7. Theorem (Sobolev). *Suppose a function belongs to all the spaces $\mathcal{H}_m^{loc}$. Then it is differentiable.*

Proof. The assumption and conclusion are both local. Therefore it is enough to show that the function is differentiable in a neighbourhood of 0. We may multiply the function by a differentiable function with compact support which is 1 in a neighbourhood of 0 and obtain a function f which belongs to $\mathcal{H}_m$ for all m. It is enough to show that f is differentiable under this hypothesis.

We will first illustrate how to prove this by showing that f is continuous. Applying the Fourier transform to f, we see that it satisfies $P\hat{f} \in L^2$ for all polynomials P. In particular, $\hat{f}(1 + q(x))^N \in L^2$ for all positive integers N. Now $f(x) = \int e^{i\langle x,\xi\rangle} \hat{f}(-\xi)(1+q(\xi))^N \cdot (\frac{1}{1+q(\xi)})^N d\xi$. Take N large enough (for example $N > n/2$) so that $(\frac{1}{1+q(\xi)})^N$ is square summable. Then by the Schwarz inequality, we get $|f(x)| \leq C\|\hat{f}(1 + q(\xi))^N\|$, for some constant C. Approximating f in $\mathcal{H}_m$ by a sequence of differentiable functions with support in a compact neighbourhood K' of K, and applying the above inequality to the Cauchy differences, we see that the convergence is actually uniform. Hence f is continuous.

As for the general case, following the same procedure, we get the inequality $\|\partial_\alpha f\| \le C\|\hat{f}(1 + q(\xi))^{N+r}\|$ if $|\alpha| \le r$. Now if $m \ge N + r$, then $\|\hat{f}(1 + q(\xi))^{N+r}\| \le C\|f\|_m$. It follows that the sequence $f_k \in C_c^\infty$ which tends to f in $\mathcal{H}_m$ for large m satisfies $\sup|\partial_\alpha f_k - \partial_\alpha f_l| \le C\|f_k - f_l\|_m$. This shows that f_k tends to an r times continuously differentiable function.

6.8. Remark. In the above argument we could have taken N to be any integer greater than $n/2$. Thus we have actually shown the following more precise result.

6.9. Proposition. *If f and all its derivatives of order $\le m$ are locally in L^2, and $m > n/2$, then f is continuously differentiable r times where $r < m - n/2$.*

We also showed above that under the same hypothesis, we have the inequality (for any compact set K and for all $r < m - n/2$)

6.10.
$$\sup\left|\frac{\partial}{\partial x^\alpha}f\right|_K \le C_K\|f\|_m$$

for all α of order $\le r$.

From this we derive the following.

6.11. Corollary. *If T is any distribution, then its restriction to C_K^∞, where K is compact, extends to a continuous linear form on $\mathcal{H}_r$, for sufficiently large r.*

Proof. In fact, this follows from the above inequality and 5.3.

This can again be reformulated as follows. Denote the dual of $\mathcal{H}_r$ by $\mathcal{H}_{-r}$. Choose a Lebesgue measure and introduce the L^2-metric in $\mathcal{H} = \mathcal{H}_0$. We will identify the dual of $\mathcal{H}$ with itself. Then it is clear that we have continuous inclusions

$$\cdots \subset \mathcal{H}_{-(r-1)} \subset \mathcal{H}_{-r} \subset \cdots \subset \mathcal{H} \subset \cdots \subset \mathcal{H}_r \subset \mathcal{H}_{r+1} \subset \cdots .$$

This is called the *Sobolev chain*. We will fix a compact set K and consider the above chain taking spaces of functions with support in K. For some integer m, a distribution T satisfies (by 5.3) the inequality $|Tf| \le C\sup|Df|$ for differential operators of order $\le m$ for all f in $\mathcal{H}_m$ in the closure C_K^∞. This implies the inequality $|Tf| \le C'\|f\|_r$ by 6.10. Hence T belongs to one of the spaces in the Sobolev chain. We may paraphrase this by saying that the intersection of all the spaces in the local Sobolev chain is the space of differentiable functions, and the union is the space of distributions.

7. Interior Regularity of Elliptic Solutions

7.1. Definition. Let L be a matrix of differential operators of order r, in a domain in $\mathbb{R}^n$. We say that it is *elliptic* if its rth symbol, interpreted as a matrix of polynomial functions in $\mathbb{R}^n$, is invertible, for every $r \in \mathbb{R} \setminus \{0\}$.

One of the most important properties of elliptic operators is the following.

7.2. Theorem. *Let u be a distribution solution of an elliptic equation $Lu = f$, with f differentiable. Then u is differentiable.*

Proof. For notational simplicity, we consider L to be an operator on functions and not a matrix of operators, but the general case is identical. To start with, we know that locally u belongs to some Sobolev space $\mathcal{H}_m$ and that it is enough to show that it belongs to every Sobolev space, in view of the theorem of Sobolev.

We observe that the operator $1 + \Delta = 1 - \sum \frac{\partial^2}{\partial x_i^2}$ actually gives an isomorphism of $\mathcal{H}_k$ with $\mathcal{H}_{k-2}$. In fact, we apply Fourier transform on both sides, use the Plancherel theorem and reduce to proving that the map $\mathcal{G}_k = \widehat{\mathcal{H}}_k \to \mathcal{G}_{k-2} = \widehat{\mathcal{H}}_{k-2}$ given by multiplication by $(1 + \sum x_i^2)$ is an isomorphism. But this is obvious.

Going back to our question, let $u \in \mathcal{H}_m$ be a solution of $Lu = f$. We claim that we may assume that $m \geq 0$. If m is negative, consider $s \in \mathcal{H}_{-m}$ such that $(1 + \Delta)^m s = u$. Now the differential operator $L(1 + \Delta)^m$ is obviously elliptic, since its symbol is the composite of the (elliptic) symbol of L and $(\sum x_i^2)^m$. The element s is in $\mathcal{H}_{-m}$ and satisfies $L(1 + \Delta)^m s = f$. Now we will show by induction that s is in all $\mathcal{H}_k$. Thus we have reduced the problem to showing the following.

7.3. Proposition. *If $u \in \mathcal{H}_{m-1}$ for some $m \geq 1$, and $Lu \in \mathcal{H}_{m-r}$ for some elliptic operator L of order $\leq r$, then $u \in \mathcal{H}_m$.*

Proof. Suppose that u is, a priori, a differentiable function. Then we will prove the following inequality.

7.4. A priori inequality. *Let u be a differentiable function with support in a compact set K, and L an elliptic operator of order r in a compact neighbourhood K' of K. Then we have the inequality*

$$\|u\|_m \leq C(\|Lu\|'_{m-r} + \|u\|'_{m-1})$$

for some constant C. Here the norm on the left side is the norm in the Sobolev space $\mathcal{H}_m$ with respect to the set K, and the ones on the right side are the norms in Sobolev spaces with respect to K'.

7.5. Derivation of regularity from the a priori inequality.

Note that u can be approximated in $\mathcal{H}_{m-1}$ by a sequence $\{u_k\}$ of differentiable functions with support in K', ensuring at the same time (Lemma 6.6) that Lu_k tends to $Lu = f$ in $\mathcal{H}_{m-r}$. Now $\|Lu_k - Lu_l\|'_{m-r}$ is arbitrarily small for large k and l. On the other hand, $\|u_k - u_l\|'_{m-1}$ is also small for large enough k and l. Hence, from the above inequality we conclude that $\{u_k\}$ is a Cauchy sequence in $\mathcal{H}_m$ as well. It is clear that its limit in $\mathcal{H}_m$ is u, since $\mathcal{H}_m \to \mathcal{H}_{m-1}$ is a continuous inclusion.

7.6. Proof of the a priori inequality.

We write $L = L_1 + L_2$, where L_1 is the top homogeneous term of the operator L and L_2 is an operator of lower order. Then L_2 gives a continuous map from $\mathcal{H}_{m-1}$ to $\mathcal{H}_{m-r}$. Hence there exists a constant C_1 such that $\|L_2u\|_{m-r} \leq C_1\|u\|_{m-1}$. If we show that the inequality is valid for homogeneous operators (in particular for L_1), then there exists a constant C_2 such that $\|u\|_m \leq C_2(\|L_1u\|'_{m-r} + \|u\|'_{m-1})$. Hence we obtain

$$\begin{aligned}
\|u\|_m &\leq C_2(\|L_1u\|'_{m-r} + \|u\|'_{m-1}) \\
&\leq C_2(\|Lu\|'_{m-r} + \|L_2u\|'_{m-r} + \|u\|'_{m-1}) \\
&\leq C_2(\|Lu\|'_{m-r} + (C_1 + 1)\|u\|'_{m-1}) \\
&\leq C_2(C_1 + 1)(\|Lu\|'_{m-r} + \|u\|'_{m-1}).
\end{aligned}$$

Thus it is enough to prove 7.4 for homogeneous operators. We will prove it first for homogeneous operators with constant coefficients and then deduce the general case.

7.7. Proposition.
Let L be a homogeneous elliptic operator of order r, with constant coefficients. Then there is some $a > 0$ such that, for every differentiable function u whose support is contained in a closed ball of radius a, we have the inequality

$$\|u\|_m \leq C(\|Lu\|_{m-r} + \|u\|_{m-1}).$$

Proof. If L is the differential operator $\sum_{|\alpha|=r} a_\alpha \frac{\partial}{\partial x^\alpha}$ and P the polynomial $\sum a_\alpha \xi^\alpha$ where a_α are (r,r) matrices, then the ellipticity assumption implies that $P(\xi)$ is nonsingular for all real nonzero vectors ξ. If C is the infimum of the norms of $P(\xi)^{-1}, \|\xi\| = 1$, we have the obvious inequality $\|P(\xi)^2\| \geq C\|\xi\|^{2r}$ for all ξ. If $\|\xi\| \leq 1$, then we derive the inequality

$$(1 + \|\xi\|^2)^r \leq C'(1 + \|P(\xi)\|^{2r})$$

for a suitable C'. If $\|\xi\| \geq 1$, we have $(1 + \|\xi\|^2)^r \leq (2\|\xi\|^2)^r \leq 2C\|P(\xi)^2\|$. So we have the following inequality, valid for all ξ:

7.8. $$(1 + \|\xi\|^2)^r \leq C(1 + \|P(\xi)\|^2).$$

We now prove the proposition by multiplying 7.8 by $(1 + \|\xi\|^2)^{m-r}\|v(\xi)\|^2$ and applying the Fourier transform. We then get, thanks to the Plancherel theorem, the following inequality:

$$\|u\|_m^2 \leq C(\|u\|_{m-r}^2 + \|Lu\|_{m-r}^2)$$

where $u = \hat{v}$. But we have $\|u\|_{m-r} \leq \|u\|_{m-1}$ so that we may as well write this inequality as

$$\|u\|_m^2 \leq C(\|Lu\|_{m-r}^2 + \|u\|_{m-1}^2).$$

Clearly this implies the proposition.

Let L be any homogeneous elliptic operator and L_0 the operator with constant coefficients, which equals the operator L evaluated at a point, say 0. Both these operators are elliptic, and we wish to compare them. Notice first that the difference $L - L_0$ is an operator which vanishes at 0, and hence can be written in the form $\sum x_i D_i$, where D_i are homogeneous differential operators of order r. Now we make the following observation.

7.9. Lemma. *Let B_a be the closed ball of radius a and D a homogeneous operator of the form $\sum x_i D_i$ of order r. Denoting the closure of C_c^∞ in $\mathcal{H}_k$ by $\mathcal{H}_{k,a}$, the norm of the induced continuous map $\mathcal{H}_{k+r,a} \to \mathcal{H}_{k,a}$ tends to zero as a tends to zero.*

Proof. It is only a matter of checking the statement for an operator $x_i D$. It is clear that the norm of this operator is the product of the norm of D and the norm of the operator of multiplication by x_i in $\mathcal{H}_{k,a} \to \mathcal{H}_{k,a}$. The latter clearly tends to zero as a tends to zero.

Let L_0 be the system with constant coefficients, obtained by evaluating the coefficients of L at 0. Then L_0 is also elliptic and we have the inequality

$$\|u\|_m \leq C(\|L_0 u\|_{m-r} + \|u\|_{m-1}).$$

for all functions $u \in \mathcal{H}_{m,a}$. By Lemma 7.7 there exists $a > 0$ such that for all u with support in the ball of radius a, we have $\|(L - L_0)u\|_{m-r} \leq$

$(1/2C)\|u\|_m$. Then for all such u we get the inequality

$$\begin{aligned}
\|u\|_m &\le C(\|L_0 u\|_{m-r} + \|u\|_{m-1}) \\
&\le C(\|Lu\|_{m-r} + \|(L - L_0)u\|_{m-r} + \|u\|_{m-1}) \\
&\le C(\|Lu\|_{m-r} + (1/2C)\|u\|_m + \|u\|_{m-1}).
\end{aligned}$$

This implies the inequality in 7.4 with the constant $2C$ instead of C for all functions in $\mathcal{H}_{m,a}$.

Finally, given any compact subset K in U, we can use the above observation and cover U with open balls in which the assertion is valid. Since K is compact, finitely many of these balls cover K. Hence the a priori inequality persists for all $u \in C_c^\infty$ proving 7.4. As a consequence the regularity theorem is proved as well.

7.10. Rellich's theorem

We recall first a standard theorem in analysis and sketch its proof.

7.11. Theorem (Ascoli). *If $\{f_k\}$ is a sequence of differentiable functions which, together with their first order derivatives, are uniformly bounded, then there is a subsequence which converges in the topology of uniform convergence on compact sets.*

Proof. We remark first that there is a subsequence which converges pointwise (using Tychonoff's theorem to the effect that the product of a family of compact spaces is compact). If we show that this sequence is equicontinuous on compact subsets, it will follow that the convergence is actually uniform on compact sets. But equicontinuity follows from the assumption of uniform boundedness of first derivatives, thanks to the mean value theorem.

Rellich's theorem is similar in spirit.

7.12. Theorem (Rellich). *If $\{f_k\}$ is a sequence of functions in a domain U with support contained in a compact set $K \subset U$ such that they and their first derivatives belong to L^2 and are bounded in the L^2-norm, then there is a subsequence which converges in the L^2-norm.*

Proof. We remark that we may extend the functions to the whole of $\mathbb{R}^n$ by setting them equal to 0 in the complement of U. By regularising and replacing K by a bigger compact set if necessary, we may also assume without loss in generality that f_k are all differentiable. By the Plancherel theorem, the sequences $\{\hat{f}_k\}$ and $\{(1 + q(x))^{1/2}\hat{f}_k\}$ are bounded in L^2. From the equality $\hat{f}_k(\xi) = \int_K f_k(x)e^{-i\langle x,\xi\rangle}dx$, we deduce that

$$\sup |\hat{f}_k| \le \|f_k\| \operatorname{vol}(K)$$

from Schwarz's inequality. Similarly, we get

$$\sup |\partial_j \hat{f}_k| \leq \|f_k\| \|x_j\|_K$$

for any j. From this we conclude that $\hat{f}_k$ and $\partial_j \hat{f}_k$ are uniformly bounded. This implies, in view of the Ascoli theorem, that after replacing $\{f_k\}$ by a subsequence, we may assume that the sequence $\{\hat{f}_k\}$ is uniformly convergent on compact sets.

So far we used only that the sequence $\{f_k\}$ is bounded in L^2. Using the fact that $(1 + q(\xi))^{1/2} \hat{f}_k$ is also bounded in L^2 we wish to conclude that $\{f_k\}$ is a Cauchy sequence in the sup-topology and hence in L^2. This will prove our assertion.

Thanks to the Fourier inversion formula, we have

$$(f_k - f_l)(x) = \int (\hat{f}_k - \hat{f}_l)(-\xi) e^{i\langle x,\xi\rangle} d\xi.$$

Given any $\epsilon > 0$, we need to show that for large enough k and l, the absolute value of the above integral is less than ϵ. Let M be a positive real number and let $S_1 = q^{-1}[0, M]$ and $S_2 = q^{-1}[M, \infty]$. Then we write the expression above as

$$\left| \int (\hat{f}_k - \hat{f}_l)(-\xi) e^{i\langle x,\xi\rangle} d\xi \right|$$

$$\leq \int_{S_1} |(\hat{f}_k - \hat{f}_l)(-\xi)| d\xi$$

$$+ \int_{S_2} (|\hat{f}_k| + |\hat{f}_l|)(-\xi)(1 + q(\xi))^{1/2}(1 + q(\xi))^{-1/2} d\xi.$$

The second sum is smaller than $\|(|\hat{f}_k(\xi)| + |\hat{f}_l(\xi)|)(1 + q(\xi)^{1/2}\|.(1 + M)^{-1/2}$. Therefore, we may choose M so large that the second term is less than $\epsilon/2$. Having chosen such an M, we may choose p large enough so that $|\hat{f}_k - \hat{f}_l| \leq \epsilon/2 \operatorname{vol}(S_1)$ if k and l are greater than p. Then the first term is also less than $\epsilon/2$. This completes the proof.

Chapter 9

Vanishing Theorems and Applications

Many operators which occur naturally on differential manifolds (with additional structures perhaps) are elliptic. The local theory that we have developed in Chapter 8 lead to global results on solutions of elliptic equations on compact manifolds. We will see here how the theory of lifting symbols leads to a criterion for vanishing of solutions of elliptic equations. We will give specific examples and some applications.

1. Elliptic Operators on Differential Manifolds

1.1. Definition. A differential operator D from a vector bundle E to F on a differential manifold M is said to be *elliptic* if its symbol, interpreted to be a polynomial map of the vector bundle T^* into $\mathrm{Hom}(E, F)$, assigns to any nonzero vector in T_m^* an isomorphism of the fibre of E with that of F, for any $m \in M$.

Clearly this definition is compatible with the definition of an elliptic operator on open domains in $\mathbb{R}^n$ given in [Ch. 8, 7.1].

1.2. Remark. From the definition, it is clear that if there exists an elliptic operator from E to F, then the ranks of E and F are the same.

1.3. Examples.

1) Let $M = \mathbb{R}$. Then the operator $\frac{d}{dx}$ on real- or complex-valued functions is elliptic. Its symbol associates to any element a of T_m^* (which is canonically isomorphic to $\mathbb{R}$), multiplication by a. Hence it is an isomorphism for nonzero a.

2) Take a second order operator on $\mathbb{R}^2$ of the form $f\frac{\partial^2}{\partial x^2} + g\frac{\partial^2}{\partial x \partial y} + h\frac{\partial^2}{\partial y^2}$, where f, g, h are real-valued functions. Its symbol at any point m is given by $v = (a, b) \mapsto f(m)a^2 + g(m)ab + h(m)b^2$. It is therefore elliptic if and only if the above quadratic form is positive definite or negative definite at all points.

More generally any second order operator on $\mathbb{R}^n$ with real-valued functions as coefficients, is elliptic if and only if the quadratic form which its symbol at any point represents, is positive or negative definite. Such polynomials are *elliptic*, i.e. have ellipsoids as level surfaces, and this is the origin of the term 'elliptic operators'.

An operator of order k with real coefficients on $\mathbb{R}^n$ is elliptic if and only if its symbol at any point of $\mathbb{R}^n$ (which may be looked upon as a homogeneous polynomial of degree k in n variables) has no nontrivial real zeros. In particular, if $n \geq 2$ and the operator has real coefficients, it has to be of even order.

3) Let M be any Riemannian manifold. Then the Laplacian [Ch. 7, 7.16] $\Delta : \Lambda^i(T^*) \to \Lambda^i(T^*)$ is elliptic. In fact, its symbol is a quadratic map of T^* into $\mathrm{End}(\Lambda^i(T^*))$ given by $v \mapsto -g(v).\mathrm{Id}$, where g is the quadratic form on T^* given rise to by the metric. Hence it is an isomorphism for nonzero v. Notice that in the case of a pseudo-Riemannian manifold, this is no longer so.

4) If M is in addition a $Spin_c$-manifold [Ch. 7, 8.8], then we recall that there are three associated vector bundles in that case. These are the bundles associated to the vector representation and the half-spin representations of $Spin_c$. These are the tangent bundle and two other bundles $Spin^+(T)$ and $Spin^-(T)$. Moreover, the Clifford module structure on the Spin representation space gives a map $T \otimes Spin^+(T) \to Spin^-(T)$. Since there is a metric on T, we map replace T by T^* and treat the above as a first order symbol. This can be lifted into an operator from $Spin^+$ to $Spin^-$. This is the Dirac operator. Since any nonzero v is invertible in the Clifford algebra, this operator is elliptic.

From the definition and the formula in [Ch. 2, 7.23] for the symbol of a composite of two operators, it follows that the composite of an elliptic operator from E into F and another from F into G is again elliptic. Moreover the adjoint of an elliptic operator is elliptic, in view of the symbol calculation of the adjoint in [Ch. 3, 3.10].

We will now give global analogues of the results of Chapter 8 and derive consequences for general elliptic operators on manifolds. In order to do this, we need to set up the machinery of Sobolev spaces in the context of sections of vector bundles.

We first define *distribution sections* of a vector bundle E. Consider the space $\Gamma_c^\infty(E)$ of differentiable sections with compact support, of a vector bundle E. If $\Gamma_K(E)$ denotes the subspace of sections with support in a fixed compact set K, then we can define a topology on it by saying that a sequence $\{s_n\}$ tends to zero if the sequence $\{Ds_n\}$ of functions tends to zero uniformly, for every differential operator D from $\mathcal{E}$ to $\mathcal{A}$. Now we consider the adjoint bundle $\mathrm{adj}(E) = \overline{E}^* \otimes \mathcal{S}$ and define a *distribution section* of E to be a linear form on $\Gamma_c^\infty(\mathrm{adj}(E))$ whose restriction to any $\Gamma_K(\mathrm{adj}(E))$ is continuous in the above topology. A differential operator from E to F has an adjoint $\mathrm{adj}(F) \to \mathrm{adj}(E)$. It induces a continuous map of $\Gamma_K(\mathrm{adj}(F))$ into $\Gamma_K(\mathrm{adj}(E))$ and hence also a linear map of the space of distribution sections of E into that of F.

1.4. Remark. According to our definition in [Ch. 8, 5.2], a distribution on M is simply a distribution section of $\mathcal{A}$. Similarly an n-current is a distribution section of $\mathcal{S}$.

Next we would like to define the Sobolev space of sections. The first step is to define square summable sections. We may fix a global, everywhere positive density μ on a differential manifold M and a Hermitian metric along the fibres of E. Then for any measurable section s of E, the meaning of $\|s\|^2$ as a function on M is clear, and we require that this be integrable with respect to the fixed density. We can introduce the norm $\|s\|_M$ by setting $\|s\|_M^2 = \int \|s\|^2 d\mu$. We will denote this space by $L^2(E)$. If M is compact, the norms corresponding to different choices of the metric and of the density are clearly equivalent. Square summable sections of E (after making the above choices) can be considered distribution sections, since they can be paired with sections with compact support of the adjoint bundle $\mathrm{adj}(E)$.

1.5. Definition. Let E be a vector bundle over a compact differential manifold M. Elements u of $L^2(E)$ with the property that Du is a distribution in L^2 for every differential operator $\mathcal{E} \to \mathcal{A}$ of order $\leq m$, form a vector space $\mathcal{H}_m(E)$, called the *Sobolev space of sections of E*.

If M is compact, we can cover M by finitely many coordinate open sets and choose a partition of unity (φ_i) with respect to the covering. It is then easy to check that u belongs to $\mathcal{H}_m$ if and only if $\varphi_i u$ belongs to the corresponding Sobolev space in the domain in $\mathbb{R}^n$ given by the coordinate chart, for all i.

Thus we can imbed the Sobolev space $\mathcal{H}_m(E)$ in the product of the Sobolev spaces $\mathcal{H}_m$ in domains in the Euclidean space by the map $s \mapsto (\varphi_i.s)$. We can therefore provide $\mathcal{H}_m$ with the induced Hilbert space structure. It is again clear that this norm is, up to equivalence, independent of the coordinate charts chosen.

Since the assumption and the conclusion of the regularity theorem are both of a local nature, it implies the following.

1.6. Regularity Theorem. *Let $L : E \to F$ be an elliptic operator on vector bundles over a differential manifold. If s is a distribution section of E such that Ls is a differentiable section (in particular, if $Ls = 0$), then s is actually a differentiable section.*

We have the following global analogue of Rellich's theorem.

1.7. Theorem. *Let E be a vector bundle on a compact differentiable manifold. The inclusion of $\mathcal{H}_m(E)$ in $\mathcal{H}_{m-1}(E)$ takes bounded subsets into relatively compact subsets.*

We also have the following globalisation of the inequality in [Ch. 8, 7.4].

1.8. Theorem. *Let $L : E \to F$ be an elliptic operator of order r on a compact differential manifold. Then for every m, there exists a constant $C > 0$ such that*
$$\|s\|_m \le C(\|Ls\|_{m-r} + \|s\|_{m-1}).$$
The map $u \mapsto (Lu, u)$ of $\mathcal{H}_m(E)$ into $\mathcal{H}_{m-r}(F) \times \mathcal{H}_{m-1}(E)$ is a homeomorphism onto a closed subset. In particular, if s satisfies $Ls = 0$, then we have
$$\|s\|_m \le C\|s\|_{m-1}.$$

Proof. Since the map $u \mapsto (Lu, u)$ is injective, and is a continuous linear operator, the inequality implies that the map is actually a metric equivalence. In particular, the image is complete, and hence closed.

From these, we are in a position to deduce the following finiteness statement.

1.9. Finiteness Theorem. *If M is a compact differential manifold, and $L : E \to F$ an elliptic operator, then the space S of solutions, i.e. $\{s \in \mathcal{E}(M) : Ls = 0\}$, is finite-dimensional.*

Proof. Consider the continuous map of $\mathcal{H}_m(E)$ into $\mathcal{H}_{m-r}(F)$ induced by L (denoting by r the order of the operator). By Theorem 1.6, its kernel is the same as S. In view of Theorem 1.8, the kernel is mapped homeomorphically onto the subspace S of $\mathcal{H}_{m-1}(E)$ by the restriction to S of the inclusion of $\mathcal{H}_m$ in $\mathcal{H}_{m-1}$. By Rellich's theorem, the latter takes bounded sets to compact sets. In particular, the closed unit ball in S is compact. Hence the Hilbert space S cannot admit an infinite orthonormal set, proving that it is finite-dimensional.

We also deduce the following important fact about elliptic operators.

1.10. Theorem. *Let $L : E \to F$ be an elliptic operator on a compact manifold. Then the image of $\mathcal{H}_m(E)$ in $\mathcal{H}_{m-r}(F)$ is closed.*

Proof. Let $S^\perp$ be the orthogonal complement of S in $\mathcal{H}_m(E)$. Clearly, the restriction T of L to $S^\perp$ is injective and the images of $\mathcal{H}_m(E)$ and $S^\perp$ in $\mathcal{H}_{m-r}(F)$ are the same. In order to prove that the image is closed, it is enough to show that the inverse T^{-1} is a bounded operator from $T(S^\perp)$ to $S^\perp$. In other words, we must show that there exists $C > 0$ such that $\|v\|_m \leq C$ for all $v \in S^\perp$ with $\|Tv\|_{m-r} = 1$. If this were not true, there would exist a sequence $\{v_k\}$ in $S^\perp$ with $\|Tv_k\|_{m-r} = 1$ and $\|v_k\|_m \geq k$. Let $x_k = v_k/\|v_k\|_m$. Then we have $\|x_k\|_m = 1$ and $\|Tx_k\|_{m-r} = 1/\|v_k\|_m \leq 1/k$. Thus $\{x_k\}$ is a sequence in $S^\perp$ such that $\|x_k\|_m = 1$ and $\{Tx_k\}$ tends to zero in $\mathcal{H}_{m-r}$. The inclusion of $\mathcal{H}_m$ in $\mathcal{H}_{m-1}$ takes $\{x_k\}$ to a relatively compact subset. Therefore we can assume (on passing to a subsequence) that $\{x_k\}$ has a limit in $\mathcal{H}_{m-1}$. Since $\{Tx_k\}$ tends to 0 in $\mathcal{H}_{m-r}$ and $\{x_k\}$ has a limit in $\mathcal{H}_{m-1}$, it follows from Theorem 1.8 that $\{x_k\}$ has a limit in $\mathcal{H}_m$ as well. This limit clearly belongs to $S^\perp$ and is in fact zero since T is injective. But this is in contradiction to the assumption that $\|x_k\|_m = 1$.

2. Elliptic Complexes

2.1. Definition. Suppose we have a sequence of vector bundles E^i and differential operators $d^i : E^i \to E^{i+1}$. Assume that $d^i \circ d^{i-1} = 0$ for all i. In other words, E° is a complex of vector bundles where the differentials are given by differential operators. Then we say that it is an *elliptic complex* if for every nonzero cotangent vector v at any point $m \in M$, the symbol sequence $E^{i-1} \xrightarrow{\sigma^{i-1}(v)} E^i \xrightarrow{\sigma^i(v)} E^{i+1}$ is exact.

We will generally assume that the E^i's are zero for all but finitely many i's.

2.2. Remarks.

1) From the definition, it follows that if a single operator is considered to be a complex consisting of only two vector bundles, then the complex is elliptic if and only if the operator is elliptic.

2) If E° is a complex

$$\cdots \to E^{i-1} \xrightarrow{d^{i-1}} E^i \xrightarrow{d^i} E^{i+1} \to \cdots$$

with differentials given by differential operators, then we have a similar complex, namely the *adjoint complex* adj(E°), given by

$$\cdots \to \mathrm{adj}(E^{i+1}) \xrightarrow{\partial_i} \mathrm{adj}(E^i) \xrightarrow{\partial_{i-1}} \mathrm{adj}(E^{i-1}) \to \cdots$$

where ∂_i is the adjoint of d^i. For any nonzero element $v \in T_m^*$ the symbols of the adjoint complex give, up to sign of differentials, the sequence

$$\cdots \to \mathrm{adj}(E^{i+1}) \xrightarrow{(\sigma^i(v))^*} \mathrm{adj}(E^i) \xrightarrow{(\sigma^{i-1}(v))^*} \mathrm{adj}(E^{i-1}) \to \cdots .$$

Hence the adjoint complex adj(E°) is elliptic if and only if E° is.

2.3. Examples.

1) The de Rham complex of a differential manifold is elliptic. In fact, to see this we need to check the exactness of the symbol sequence for all $v \in T_m^* \setminus \{0\}$. This is a fibrewise check and amounts to the following. If V is a vector space, and $v \in V \setminus \{0\}$, then the sequence

$$\cdots \to \Lambda^{i-1}(V) \to \Lambda^i(V) \to \Lambda^{i+1}(V) \to \cdots$$

is exact, where the maps are given by wedging with v. In fact, we may assume that $v = e_1$ is the first member of a basis (e_i). Any element ω of $\Lambda^i(V)$ can then be written as $e_1 \wedge \alpha + \beta$ where the expressions in terms of the basis for α and β, do not involve e_1. If $e_1 \wedge \omega = 0$, then $e_1 \wedge \beta = 0$ which implies that β itself is zero. In other words, ω is of the form $e_1 \wedge \alpha$ with $\alpha \in \Lambda^{i-1}(V)$. This proves our assertion.

2) The Dolbeault complex

$$\cdots \to \Lambda^{p,q-1}(T^*) \to \Lambda^{p,q}(T^*) \to \Lambda^{p,q+1}(T^*) \to \cdots$$

of a complex manifold is elliptic. Recall that the Dolbeault differential on $\Lambda^{p,q}(T^*)$ is obtained by composing the exterior derivative with the projection to the $\Lambda^{p,q+1}(T^*)$. Hence its symbol for any $v \in T_m^*$ is given by wedging with its $(0,1)$ component. So the exactness of the symbol sequence for any nonzero v follows from the above observation in which we take $V = (T_m^{(0,1)})^*$ and tensor the wedging sequence by $\Lambda^p(T_m^{(1,0)})^*$.

Let

$$\cdots \to E^{i-1} \xrightarrow{d^{i-1}} E^i \xrightarrow{d^i} E^{i+1} \to \cdots$$

be an elliptic complex. We choose Hermitian metrics along the fibres of the bundles and a positive density on the manifold. For convenience, we will denote all the d^i's by d. Then the adjoint bundles adj(E^i) can be identified with E^i and we will denote all the adjoint operators $\partial_i : E_{i+1} \to E_i$ by ∂. It is obvious that this complex

$$\cdots \to E_{i+1} \xrightarrow{\partial_i} E_i \xrightarrow{\partial_{i-1}} E_{i-1} \to \cdots ,$$

which, by abuse of language, will also be called the *adjoint complex*, is elliptic.

We are interested in the cohomology spaces of an elliptic complex. To start with, *we will assume that all the operators in the complex are of the same order r.*

2.4. Definition. The operator $\Delta : E^i \to E^i$ defined by $\Delta = \partial_i d^i + d^{i-1} \partial_{i-1}$ is called the *Laplacian* of the elliptic complex. Sections s of E^i which satisfy $\Delta s = 0$ are called *harmonic sections.*

2.5. Theorem. *The Laplacians of an elliptic complex are elliptic.*

Proof. We take a nonzero cotangent vector v at a point $m \in M$ and consider the induced symbol maps (denoting $\sigma^i(v)$ by s^i), $(E^{i-1})_m \xrightarrow{s^{i-1}} (E^i)_m \xrightarrow{s^i} (E^{i+1})_m$. Using the computation of the symbol of the adjoint in [Ch. 3, 3.10] and that of a composite in [Ch. 2, 7.22], we compute the symbol of Δ to be the linear endomorphism $(-1)^r((s^i)^* \circ s^i + s^{i-1} \circ (s^{i-1})^*)$ of $(E^i)_m$. Here s^* denotes the usual linear adjoint defined by $\langle x, s^* y \rangle = \langle sx, y \rangle$ with $\langle \ , \ \rangle$ representing the Hermitian metric. Let τ be the map $E^i \to E^i$ given by $(-1)^r$ times the symbol of Δ evaluated on v. Then for any x in the fibre of E_i at m, we have $\langle \tau x, x \rangle = \langle ((s^i)^* \circ s^i + s^{i-1} \circ (s^{i-1})^*)x, x \rangle = \langle s^i x, s^i x \rangle + \langle (s^{i-1})^* x, (s^{i-1})^* x \rangle$. Hence if $\tau x = 0$, then in view of the positive definiteness of the scalar product, we conclude that $s^i x = 0$ and $(s^{i-1})^* x = 0$. From the exactness of the symbol sequence, there exists $y \in E_m^{i-1}$ such that $x = s^{i-1} y$. But then $\langle x, x \rangle = \langle x, s^{i-1} y \rangle = \langle (s^{i-1})^* x, y \rangle = 0$. It follows that $x = 0$. In other words, the linear map τ is injective and hence an isomorphism. This shows that Δ is elliptic.

2.6. Theorem. *Let M be a compact, connected manifold and E° an elliptic complex on it. A section $s \in \mathcal{E}^i(M)$ is harmonic if and only if it satisfies $ds = 0$ and $\partial s = 0$. The natural map of the space $\mathcal{V}$ of harmonic sections to the cohomology space $H^i(\mathcal{E}^\circ(M))$ of the induced complex*

$$\cdots \to \mathcal{E}^{i-1}(M) \xrightarrow{d^{i-1}} \mathcal{E}^i(M) \xrightarrow{d^i} \mathcal{E}^{i+1}(M) \to \cdots$$

is an isomorphism.

Proof. We will use the assumption that M is compact in choosing a finite positive density on M which determines the scalar product on sections. Let s be any section of $\mathcal{E}^i$. Then we have $\langle (d\partial + \partial d)s, s \rangle = \langle \partial s, \partial s \rangle + \langle ds, ds \rangle$. Hence we conclude that s is harmonic if and only if $ds = \partial s = 0$. In particular we have a natural linear map of the space of harmonic sections into $H^i(\mathcal{E}^\circ(M))$. If a harmonic section s represents the trivial cohomology class, it is of the form dt where $t \in \mathcal{E}^{i-1}(M)$. But then since s is harmonic, it follows that $\partial dt = \partial s = 0$. Consequently, $\langle \partial dt, t \rangle = \langle dt, dt \rangle = 0$, which implies that $s = dt = 0$. This shows that the map from $\mathcal{V}$ to H^i is injective.

We have finally to show that every cohomology class is represented by a harmonic section. We claim that any element x in $\mathcal{V}$ is orthogonal to Δy for all y. For, $\langle x, (d\partial + \partial d)y \rangle = \langle dx, dy \rangle + \langle \partial x, \partial y \rangle$. Since x is harmonic, both dx and ∂x are zero and our assertion is proved.

The image of $\Delta : \mathcal{H}_{2r} \to L^2(E)$ is closed by Theorem 1.10, and $\mathcal{V}$ is finite-dimensional by 1.9. We easily conclude from this that the space spanned by $\mathcal{V}$ and $\mathrm{Im}(\Delta)$ is also closed. We claim that it is the whole of $L^2(E)$. In fact, if $v \in L^2(E)$ is orthogonal to $\mathrm{Im}(\Delta)$, we have $\langle v, \Delta w \rangle = 0$ for all $w \in \mathcal{E}(M)$. But this is the same as saying that $\langle \Delta v, w \rangle = 0$. Hence $\Delta v = 0$, i.e. v belongs to $\mathcal{V}$. If it is also orthogonal to $\mathcal{V}$ then it is zero. Thus $\mathcal{V} \oplus \mathrm{Im}(\Delta) = L^2(E)$.

Given any element x of $\mathcal{E}(M)$ with $dx = 0$, note that x belongs to $L^2(E)$, and take the harmonic projection $h(x)$ of x in $\mathcal{V}$. The difference $x - h(x)$ belongs to $\mathrm{Im}(\Delta)$, say $x - h(x) = \Delta y$. Here y belongs, a priori, to $\mathcal{H}_{2r}$. But we started out with a differentiable section x, and hx, being harmonic, is also differentiable. Thus Δy is differentiable. From Theorem 1.6, it follows that y is also differentiable. Moreover, since dx is zero by assumption and $dh(x)$ is also zero, we have $d\Delta y = d\partial dy = 0$. Taking scalar product with dy we get $\partial dy = 0$. Again taking scalar product with y, we conclude that $dy = 0$. Hence we have $\Delta y = d\partial y$. Thus $x - h(x) = d\partial y$ with $y \in \mathcal{E}(M)$, proving that x and hx represent the same cohomology class.

In the course of the proof, we also showed the following decomposition.

2.7. Harmonic decomposition. *The L^2-space of sections of E is the orthogonal direct sum of S and the image of Δ. If $s \in \mathcal{E}(M)$, then it can be written as $t + \Delta u$ with $t, u \in \mathcal{E}(M)$, t harmonic.*

2.8. Corollary. *The cohomology spaces of an elliptic complex on a compact manifold are all finite-dimensional.*

Proof. We assumed that all the operators of the complex are of the same order. In that case, the above assertion is a consequence of Theorem 2.6 and Finiteness Theorem 1.9. In the general case, we may modify the sequence

$E^{i-1} \to E^i \to E^{i+1}$ by composing the differential $E^i \to E^{i+1}$ (or $E^{i-1} \to E^i$) with an elliptic operator $E^{i+1} \to E^{i+1}$ (or $E^{i-1} \to E^{i-1}$) whose order is the difference between the orders of d^i and d^{i-1} and which induces isomorphism on sections. In order to find one such operator, notice first that we can take any elliptic symbol and lift it to an operator to get an elliptic operator. Adding to this a large positive multiple of the identity, we get the required operator. This reduces the case to one in which all operators are elliptic of the same order.

In particular, applying the above to the standard complexes, we get the following corollaries.

1) *The cohomology spaces with values in $\mathbb{R}$ or $\mathbb{C}$ (or even in a local system of vector spaces) of a compact differentiable manifold are finite-dimensional.*

Let E be a holomorphic bundle over a compact complex manifold X. Consider the Dolbeault resolution [Ch. 7, 3.7–3.10]

$$0 \to \mathcal{E}_h \to \mathcal{E} \otimes \mathcal{A} \to \Lambda^{0,1}(T^*) \otimes \mathcal{E} \to \cdots \to \Lambda^{0,n}(T^*) \otimes \mathcal{E} \to 0.$$

This is a soft resolution [Ch. 7, Proposition 3.8] of the sheaf of holomorphic sections of E. Hence $H^i(\mathcal{E}_h)$ can be computed by taking global sections of the Dolbeault complex, which is elliptic. Thus we conclude

2) *The cohomology spaces $H^i(X, \mathcal{E}_h)$ of a holomorphic bundle E on a compact complex manifold are finite-dimensional.*

3) *The solution space of the Dirac operator of a compact $Spin_c$ manifold is finite-dimensional.*

We also have the following duality statement.

2.9. Theorem. *Let E° be an elliptic complex on a compact connected manifold. Then the ith cohomology of E° and the ith cohomology of the adjoint complex $\mathrm{adj}(E)_\circ$ are canonically dual to each other.*

Proof. If $s \in \Gamma(E)$, and $t \in \Gamma(\mathrm{adj}(E))$, then their pointwise pairing gives a density on M which, on integration, gives a complex number $\langle s, t \rangle$. If $u \in \Gamma(E^i)$ and $s \in \Gamma(\mathrm{adj}(E)_{i+1})$, this pairing satisfies $\langle d^i u, s \rangle = \langle u, \partial_i s \rangle$. Hence if $du = 0$, then u is orthogonal to the image of ∂_i. Restrict this pairing to cycles s, namely, s such that $\partial_i s = 0$. If $u = d^{i-1} v$, we have $\langle u, s \rangle = \langle d^{i-1} v, s \rangle = \langle v, \partial_i s \rangle = 0$. Thus the pairing $(u, s) \mapsto \langle u, s \rangle$ goes down to a canonical pairing between $H^i(E^\circ)$ and $H^i(\mathrm{adj}(E^\circ))$. It is this pairing that gives the duality asserted in the theorem. In order to prove that this is a perfect pairing, we may choose Hermitian metrics on E^i and a positive density on M and use the harmonic decomposition. We will now show that if x is any cohomology class such that $\langle x, y \rangle = 0$ for all cohomology classes $y \in H^i(\mathrm{adj}(E^\circ))$, then $x = 0$. Let hx be the harmonic representative of x. Then we have $\langle hx, s \rangle = 0$ for all sections s such that $ds = 0$. But hx is also orthogonal to the image of Δ, since $\langle hx, \Delta t \rangle = \langle \Delta(hx), t \rangle = 0$. Hence hx is orthogonal to the whole of $\Gamma(\mathrm{adj}(E^i))$ and is consequently 0. This shows that the induced map of $H^i(E^\circ)$ into the dual of $H^i(\mathrm{adj}(E^\circ))$ is injective. In particular, $\dim H^i(E^\circ) \leq \dim H^i(\mathrm{adj}(E^\circ))$. By symmetry the opposite inequality is also valid. Hence the two dimensions are the same and the above injective map is also an isomorphism.

2.10. Examples.

1) If $L : E \to F$ is any elliptic operator, then the cokernel of the map $\mathcal{E}(M) \to \mathcal{F}(M)$ (which is also finite-dimensional) is dual to the kernel of the adjoint operator.

2) We have seen in [Ch. 3, Example 3.8, 1)] that the adjoint of the de Rham complex can be identified with the de Rham complex tensored with the orientation local system $OR(M)$ with reindexing. Hence we have

2.11. Poincaré duality. *The ith de Rham cohomology group $H^i(M, \mathbb{R})$ of a compact differential manifold M is canonically dual to $H^{n-i}(M, OR)$. As a consequence, we see that if M is a compact connected differential manifold, then $H^n(M, OR) = \mathbb{R}$. In particular, if M is also oriented, then $H^n(M, \mathbb{R}) = \mathbb{R}$.*

Slightly more generally, we have a conjugate duality between $H^i(M, \mathbf{L})$ and $H^{n-i}(M, \mathrm{adj}(\mathbf{L}))$ where $\mathbf{L}$ is a local system of complex vector spaces, and $\mathrm{adj}(\mathbf{L})$ is the tensor product of OR and the complex conjugate of the dual of $\mathbf{L}$. If $\mathbf{L}$ admits a (constant) unitary metric along the fibres, then it gives an isomorphism of $(\overline{\mathbf{L}})^*$ and $\mathbf{L}$ and hence there is a Hermitian duality between $H^i(M, \mathbf{L})$ and $H^{n-i}(M, \mathbf{L} \otimes OR)$ in the case of unitary local systems.

3) Similarly we have the following theorem.

2.12. Serre duality. *The ith cohomology group $H^i(X, \mathcal{E}_h)$ of the sheaf of holomorphic sections of a holomorphic vector bundle E over a compact, connected, complex manifold X of complex dimension n, is canonically dual to $H^{n-i}(M, \mathcal{K}_h \otimes \mathcal{E}_h^*)$, where K is the canonical line bundle, namely the nth exterior power of the holomorphic cotangent bundle.*

Proof. Since the cohomology spaces of $\mathcal{E}_h$ are identified with the cohomology spaces of the Dolbeault complex of E, we only need to apply the duality theorem above to the Dolbeault complex. The constituents of the adjoint complex are $(\Lambda^{(0,i)}(T^*))^* \otimes \mathcal{S} \otimes \mathcal{E}^*$.

Since the manifold X is oriented, the sheaf S is simply $\Lambda^{2n}(T^*) = \Lambda^{(n,n)}(T^*)$.

On the other hand, the pairing $\Lambda^p(T^*) \otimes \Lambda^{2n-p}(T^*) \to \Lambda^{2n}(T^*)$ restricts to a perfect pairing $\Lambda^{(0,i)}(T^*)^* \otimes \Lambda^{(n,n-i)}(T^*) \to \Lambda^{n,n}(T^*)$. Hence we have the isomorphisms $\Lambda^{(0,i)}(T^*)^* \otimes S \otimes \mathcal{E}^* \simeq \Lambda^{(n,n-i)}(T^*) \otimes \mathcal{E}^* \simeq \Lambda^{(0,n-i)}(T^*) \otimes \mathcal{K} \otimes \mathcal{E}^*$. The differentials are easily identified with the Dolbeault differentials, again up to sign. This proves the duality claimed.

2.13. Index of an elliptic operator.

Given an elliptic operator on a compact manifold, one would like to know if it has any solution at all. In other words, one looks for criteria to conclude that the space $\mathcal{V}$ of harmonic sections is zero. More ambitiously, one might wish to compute the dimension of $\mathcal{V}$. By computing, we mean an integral formula, namely, that the integral of a suitable density on M (determined by the operator) gives the dimension of $\mathcal{V}$. Such a formula would imply that a (continuous) perturbation of the equation does not change the number. In other words, if we have a continuous family of elliptic operators, then the associated densities will also vary continuously and so the integral (being an integer) will be independent of the operator. This is actually false. We may take the Laplacian on a compact Riemannian manifold for which there are nonzero harmonic sections (for example, on functions, constants are harmonic), and perturb it by adding a scalar times the identity homomorphism. The latter does not have any solutions if the scalar is a large positive number. We may summarise this by saying that the dimension of the space of solutions is *not deformation invariant*. However, it is easy enough to show that the *difference* between the dimension of the solution space of an elliptic operator and that of the adjoint, is deformation invariant.

2.14. Definition.

If E° is a finite elliptic complex on a compact manifold, then the integer $\sum(-1)^i \dim H^i(E^\circ)$ is defined to be its *index*. In particular, the *index* of an elliptic operator $L : E \to F$ on a compact manifold is the integer given by $\dim \ker(L : \mathcal{E}(M) \to \mathcal{F}(M)) - \dim \operatorname{coker}(L : \mathcal{E}(M) \to \mathcal{F}(M))$.

One can indeed give an integral formula for computing this index as described above. It is called the *Atiyah-Singer index formula*. This formula involves the Pontrjagin classes of the manifold and also classes associated to the elliptic operator, or complex. Thus one defines a density on M which, when integrated, gives the index. In particular, this also gives a formula for the index of the Dolbeault complex of a holomorphic bundle on a compact complex manifold. This then computes $\chi(E) = \sum(-1)^i \dim H^i(E)$ by an integral formula which is known as the *Hirzebruch-Riemann-Roch theorem*.

We will not prove the index formula here, but content ourselves with stating the Riemann-Roch theorem.

Firstly, we need a piece of notation. Suppose M is a compact oriented differential manifold and E a vector bundle on it of rank r. Let P be a power series in r variables X_i, which is symmetric in the variables. Then we can express it as a power series in the elementary symmetric functions of the X_i's. In this, substitute for the ith elementary symmetric function, the ith Chern class of the bundle. Since the cohomologies of the manifold vanish beyond its dimension, this makes perfect sense as an element of the total cohomology of M. We will denote this by P_c. There are two ways in which these kinds of classes arise in the Riemann-Roch theorem. Firstly, start with a power series Q in *one* variable. Then we can take for P the power series $\sum Q(X_i)$ and we call the resulting cohomology class, the character class $\mathrm{ch}(Q)(E)$.

When $P = e^X$ this is called the *Chern character* and denoted $\mathrm{ch}(E)$.

On the other hand, we may take P to be the *product* $\Pi Q(X_i)$ and define the corresponding cohomology class to be the *Hirzebruch class* $h(Q)(E)$. When we take P to be the power series $\frac{X}{1-e^{-X}}$, we call the resulting Hirzebruch class the *Todd class* and denote it td. Finally if x is an element of the total cohomology class we denote by $\int x$ the element of $\mathbb{R}$ obtained by taking the top component and taking the corresponding element of $\mathbb{R}$ obtained by integrating the top form. Then the Riemann-Roch formula is the following.

2.15. Theorem. *For any holomorphic vector bundle E on a compact complex manifold, we have*

$$\chi(E) = \sum (-1)^i \dim H^i(E) = \int \mathrm{ch}(E).\,\mathrm{td}(T_M).$$

2.16. Remark. When E is a line bundle L we may write simply $e^{c(L)}$ for $\mathrm{ch}(E)$, so that the formula becomes $\int \sum \frac{c(L)^i}{i!}(\mathrm{td}(T_M))_{(n-i)}$. In particular, if M is of dimension 1, the integrand is $\mathrm{td}(T_M)_1 + c(L)$.

The original version of the proof of the index formula may be found in [15]. A later proof, using the heat kernel, was found by Atiyah, Bott and Patodi [1]. The Hirzebruch Riemann-Roch theorem for projective manifolds appeared in [8]. A relative version was given by Grothendieck in a far-reaching generalization. Other proofs are due to Fulton [4] and Nori [14].

2.17. Riemann-Roch theorem for a compact Riemann surface.

We wish to see the Riemann-Roch formula for a compact Riemann surface C in a user-friendly form. Looking at the formula, one realises that the first invariant to understand is the class $\mathrm{td}(T_C)$. Fortunately, for this we need only the first coefficient of the power series $\frac{x}{(1-e^{-x})}$, which is $x/2$.

Hence the Todd class is in this case $1 + c_1(T_C)/2$. Now there are several instructive ways of computing c_1 for a Riemann surface. For example, one notes that it is the negative of the *Euler class* of the tangent bundle, treating it as a real vector bundle [Ch. 5, Remark 6.15, 2)]. But the Euler class is an element of $H^2(C)$ and can be thought of as a number. We have remarked [Ch. 5, Remark 6.15, 2)] that it is the Euler-Poincaré characteristic. This number is $2 - 2g$ where $2g$ is the first Betti number of C. On the other hand if L is any line bundle, then $\mathrm{ch}(L) = 1 + c_1(L)$. Here again we can think of $c_1(L)$ as a number identifying it with $\int c_1(L)$. This number is called the *degree* of the line bundle. The formula reduces then to

2.18. $\qquad \dim H^0(L) - \dim H^1(L) = \int (c_1(L) + c_1(T_C)/2) = \deg(L) + 1 - g.$

Take for L the trivial bundle. Then $H^0(L)$ consists of global functions, but the only global holomorphic functions on a compact connected manifold are constants, so that its dimension is 1. Clearly $\deg(L) = 0$ and so the Riemann-Roch theorem implies that $\dim H^1(C, \mathcal{O}) = g$.

The Riemann-Roch theorem sets out to compute $\chi(L)$ in terms of the topological invariants. If we can compute $\chi(L)$ by other means, it can also serve to compute the topological invariants.

We will denote by ω the holomorphic line bundle T_h^* and take now $L = \omega$. Then we get $\dim H^0(C, \omega) - \dim H^1(C, \omega) = \deg(\omega) + 1 - g$. Noting that by Serre's duality theorem we have $\dim H^1(C, \omega) = \dim H^0(C, \mathcal{O}) = 1$ and $\dim H^0(C, \omega) = \dim H^1(C, \mathcal{O}) = g$, this computes the *degree* of ω:

2.19. $\qquad\qquad\qquad\qquad \deg(\omega) = 2g - 2.$

We will see another example of a similar sort. Let $x \in C$. We have seen that the ideal sheaf is locally free of rank 1. It is in fact defined by the following exact sequence:

$$0 \to \mathcal{I}_x \to \mathcal{O} \to \mathcal{O}_x \to 0.$$

Here $\mathcal{O}_x$ is the 'skyscraper sheaf' which is 0 outside x and has $\mathbb{C}$ for its stalk at x. It can be defined as the direct image of the sheaf $\mathcal{O}$ for the inclusion of the single point manifold $\{x\}$ in C. Now $H^i(C, \mathcal{O}_x)$ is easily seen to be $H^i(\{x\}, \mathcal{O}) = 0$ if $i \neq 0$ and $\mathbb{C}$ when $i = 0$. Hence we have $\chi(\mathcal{I}_x) = \chi(\mathcal{O}) - 1 = -g$. Appealing to the Riemann-Roch theorem, we conclude that

2.20. $\qquad\qquad\qquad\qquad \deg(\mathcal{I}_x) = -1.$

2.21. Remark. Equation 2.20 can of course be directly checked as follows. Note that $\mathcal{I}_x$ is trivial outside x. On the other hand it is also trivial in an open set containing x. We choose a coordinate system (U, z) at x and identify this with a disc with 0 as centre. Note that the ideal sheaf $\mathcal{I}_0$ on the disc is trivialised using the multiplication $\mathcal{O} \to \mathcal{I}_0$ by the function z. Hence if we take the trivial connection on the trivialised line bundle on $M \setminus \{x\}$, it also gives the trivial connection in $U \setminus \{0\}$. Under the transformation of multiplication by $1/z$ the connection becomes $f \mapsto df + f\,dz/z$. Choose a smaller disc V around 0 with $\overline{V} \subset U$. We now have a trivial bundle on $U \setminus V$ with the above connection on it. This can be extended to the whole of U by extending the differential form dz/z from $V \subset U$ to a (differentiable) form α on U. Now clearly the trivial connection on $M \setminus \{V\}$ and the connection $d + \alpha$ on U coincide on the intersection, and so we get a connection on $\mathcal{I}_x$. We now compute its curvature form R. Since the connection is flat on $M \setminus \overline{V}$, the curvature form is zero there. On the other hand it is $d\alpha$ inside U. We need to integrate this on M to get the Chern class. Now $\int_M R$ is $\int_V d\alpha = \int_S \alpha$ where S is any circle between the two discs. But $\alpha = dz/z$ in this annulus and so we need to compute $\int_S dz/z$. This is clearly $2\pi i$. By our definition of the topological Chern class, namely $(-1/2\pi i)$ times the above, we get the degree as claimed.

The dual of the line bundle $\mathcal{I}_x$ is usually denoted $\mathcal{O}(x)$. So the above equality is equivalent to saying that the degree of $\mathcal{O}(x)$ is 1.

2.22. Remark. If L is any line bundle on a compact Riemann surface, tensoring L with the exact sequence

$$0 \to \mathcal{I}_x \to \mathcal{O} \to \mathcal{O}_x \to 0$$

one deduces that $\chi(L) - \chi(L \otimes \mathcal{O}(-x)) = 1$. On the other hand, we just saw that $\deg(L) - \deg(L \otimes \mathcal{O}(-x)) = 1$. From this, one can easily prove the Riemann-Roch theorem for line bundles on a compact Riemann surface.

3. Composition Formula

3.1. Lifting of higher order symbols.

We have seen that if M is a differential manifold and N a closed submanifold of codimension 1, then its ideal sheaf is locally free of rank 1. Much the same is true for complex manifolds and for the same reason, namely the submanifold is locally defined by a single equation. If M is a complex manifold and D a closed submanifold of complex codimension 1, the ideal sheaf $\mathcal{I}_D$ of D is a holomorphic line bundle. Its dual is denoted $\mathcal{O}(D)$. Sections of $\mathcal{I}_D$ are holomorphic functions on M which vanish on D, while those of $\mathcal{O}(D)$ consist of meromorphic functions which have at most a pole on D. It is therefore clear that sections of holomorphic line bundles or of vector bundles are of great importance in the study of complex geometry. The first question that arises in this connection is how to compute $\dim H^0(E)$. If we know for example that $H^i(E)$ are zero for all positive i (where E is a holomorphic vector bundle over a compact complex manifold), then the Hirzebruch-Riemann-Roch formula mentioned above computes the dimension of $H^0(E)$.

This question can be asked in the wider context of elliptic complexes, the case of the complex manifold being covered by the Dolbeault complex. We will be interested in criteria for the vanishing of the cohomology spaces $H^i(E^\circ)$ of an elliptic complex E°. We will give a general sufficient condition for the vanishing of $H^i(E^\circ)$ when the differentials are given by lifting first order symbols using connections, according to the theory in Chapter 5.

We begin by discussing the question of lifting a higher order symbol $E \to S^k(T) \otimes F$ into a differential operator $E \to F$ of order $\leq k$. For first order symbols we have seen in Chapter 5 that this is accomplished by choosing a connection on E. We will therefore assume given a connection on E. One would hope to get on iteration, first, an operator from E to $\bigotimes^r T^* \otimes E$ and therefore, an operator $E \to S^k(T^*) \otimes E$ of order k whose kth order symbol is essentially the identity map. But, in order to iterate d_∇, we need a connection on $T^* \otimes E$, $T^* \otimes T^* \otimes E$ and so forth. So let us assume that a *linear* connection is also given. Actually it is practical to take a torsion free linear connection, as we will soon see.

We get on such iteration, an operator $\bigotimes^k(\nabla) : E \to \bigotimes^k T^* \otimes E$ of order k, and hence on composing with the natural map $\bigotimes^k T^* \to S^k(T^*)$, an operator $S^k(\nabla) : E \to S^k(T^*) \otimes E$ as well. Its symbol can be computed according to the recipe given in [Ch. 2, 7.23]. In fact, the symbol of $\bigotimes^k(\nabla)$ is the image in $S^k(T) \otimes (\bigotimes^k T^*) \otimes \mathrm{End}(E)$ of the identity element in $\bigotimes^k(T) \bigotimes^k(T^*) \otimes \mathrm{End}(E)$. Hence the symbol of $S^k(\nabla)$ is the map $E \to S^k(T) \otimes S^k(T^*) \otimes E$ given by $\eta \otimes \mathrm{Id}_E$ where η is the image in $S^k(T) \otimes S^k(T^*)$ of the identity section of $\bigotimes^k(T) \bigotimes^k(T^*)$. Given a symbol $s : E \to S^k(T) \otimes F$, it is clear now how to lift it to an operator. We compose $S^k(\nabla) : E \to S^k(T^*) \otimes E$ and $I_{S^k(T^*)} \otimes s : S^k(T^*) \otimes E \to S^k(T^*) \otimes S^k(T) \otimes F$ and contract $S^k(T^*) \otimes S^k(T)$ to $\mathcal{A}$ and get an operator $E \to F$.

3.2. Definition. Let $s : E \to S^k(T) \otimes F$ be a given kth order symbol. Assume given a linear connection on M and a connection on E. Then the composite of the differential operator $S^k(\nabla) : E \to S^k(T^*) \otimes E$, the $\mathcal{A}$-linear map $I_{S^k(T^*)} \otimes s : S^k(T^*) \otimes E \to S^k(T^*) \otimes S^k(T) \otimes F$ and the $\mathcal{A}$-linear map $S^k(T^*) \otimes S^k(T) \otimes F \to F$, which is the tensor product of the contraction map and Id_F, is called the *lift* of the symbol s.

Starting with the above data, namely a connection on E and a linear connection, we have given a procedure to lift all symbols $E \to S^k(T) \otimes F$ to differential operators. In other words, this yields splittings of the exact sequences (for all k)

$$0 \to D^{k-1}(E, F) \to D^k(E, F) \to \mathrm{Hom}(E, S^k(T) \otimes F) \to 0.$$

Operators obtained by lifting symbols are analogues of homogeneous differential operators, in the sense that any differential operator can be written uniquely as a linear combination of lifted differential operators. For, given any differential operator D of order k, we may lift its kth order symbol to an operator D_k using the above splitting. Then since D and D_k have the same kth order symbol, $D - D_k$ is an operator of order $\leq k - 1$ and so this proves our assertion inductively. Thus we have shown the following.

3.3. Proposition. *Given a connection on a vector bundle E and a linear connection on the manifold M, every differential operator $E \to F$ can be written uniquely as a sum of differential operators $E \to F$ obtained by lifting symbols.*

3.4. Composition of lifted operators.

Let $s : E \to S^k(T) \otimes F$ and $t : F \to S^l(T) \otimes G$ be two symbols. By composition and multiplication in the symmetric algebra of T, we get a *composite symbol* $t \circ s : E \to S^{k+l}(T) \otimes G$ of order $k + l$. Assume given a connection each on E, F and T. Using the connection on E and T we may lift s to a differential operator $D_s : E \to F$. Similarly, using the connections on F and T, we get an operator $D_t : F \to G$ with symbol t. On the other hand, using those on E and T we may also lift the symbol $t \circ s$ to a differential operator $D_{t \circ s} : E \to G$. Since $D_t \circ D_s$ and $D_{t \circ s}$ have the same symbol, namely $t \circ s$, their difference is an operator of order $\leq k + l - 1$.

If $l = 0$, then $D_t = t$ is an $\mathcal{A}$-homomorphism $F \to G$, and the connection on F plays no role in the definition of D_t nor of course in D_s and $D_{t \circ s}$. By definition of the lifts, we have obviously the following result.

3.5. Proposition. *If the order of the symbol t is zero, then we have $D_{t \circ s} = t \circ D_s$.*

On the other hand, if $k = 0$, again we have that $D_s = s$ is an $\mathcal{A}$-linear homomorphism $E \to F$, but the lift of t involves the connection on F while that of $t \circ s$ involves the connection in E. Therefore one cannot expect such a simple formula as above. Notice that we have $(1_{T^*} \otimes s \circ \nabla_E)$ and $\nabla_F \circ s$ are both first order operators from E to $T^* \otimes F$. Their difference is by definition ds where d is the absolute derivative of s with respect to the connection on $\mathrm{Hom}(E, F)$ given rise to by ∇_E and ∇_F. As such, ds can be regarded as a map $E \to T^* \otimes F$. Then we have $(1_{T^*} \otimes s) \circ \nabla_E - \nabla_F \circ s = ds$. If $l = 1$, we denote as usual by $\tilde{t}$ the map $T^* \otimes F \to G$ determined by t and conclude that

3.6.
$$D_{t \circ s} = D_t \circ s + (\tilde{t}) \circ ds.$$

The next interesting case is when $k = l = 1$. We will obtain a formula for the expansion in terms of what we may call ∇-homogeneous operators, of the composite of two ∇-homogeneous operators. We have seen that the top term in the expansion is the lift of the composite symbol. Thus we need to identify the first order ∇-homogeneous operator and the constant term in the expansion.

In order to make the computations easy to understand, we will use a notational artifice. Whenever we have a diagram **D** of the following type:

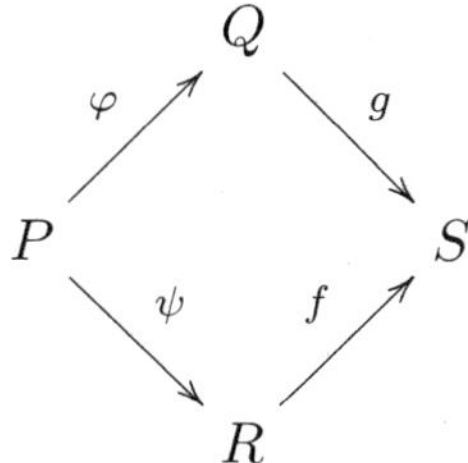

we denote the map $P \to S$ obtained as $f \circ \psi - g \circ \varphi$ by $\mathrm{OC}(\mathbf{D})$ (obstruction to commutativity of Diagram **D**). Consider now the following diagram:

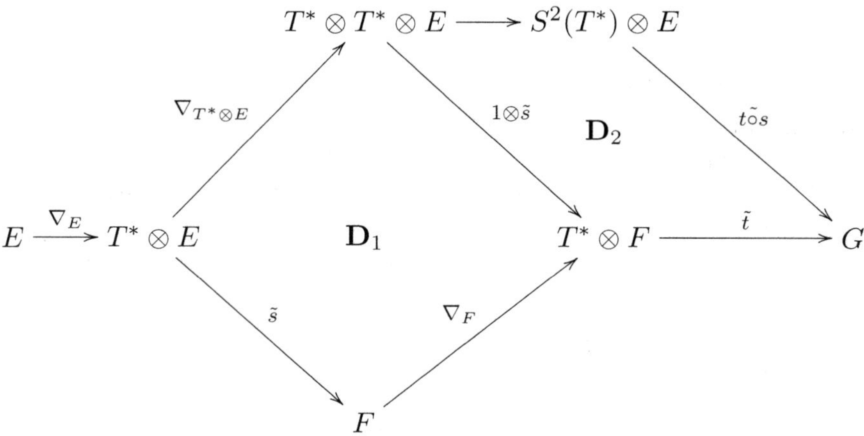

The two diamonds in the above diagram, denoted $\mathbf{D}_1$ and $\mathbf{D}_2$, are not claimed to be commutative. We will in fact compute the obstruction to the commutativity of the two diagrams.

3.7. Lemma *Let $\tilde{s} : T^* \otimes E \to F$ be the map given rise to by the symbol $s : E \to T \otimes F$. Then the absolute derivative $d_\nabla \tilde{s}$, interpreted as a homomorphism $T^* \otimes E \to T^* \otimes F$, is the obstruction to the commutativity of Diagram $\mathbf{D}_1$ above.*

Proof. In fact, the evaluation of $d_\nabla \tilde{s}$ on any vector field X is by definition given by $(\nabla_F)_X \circ \tilde{s} - (1 \otimes \tilde{s}) \circ (\nabla_{T^* \otimes E})_X$.

3.8. Lemma. *The obstruction to the commutativity of $\mathbf{D}_2$ is the bilinear map on T^* with values in $\mathrm{Hom}(E, G)$ obtained by alternatising $\tilde{t}(1_{T^*} \otimes \tilde{s})$.*

Proof. Note that the composite of the natural map $T^* \otimes T^* \otimes E \to S^2(T^*) \otimes E$ and $\widetilde{t \circ s} : S^2(T^*) \otimes E \to G$ is by definition $\tilde{t} \circ (1 \otimes \tilde{s}) \circ \mathrm{sym}$, where $\mathrm{sym} : T^* \otimes T^* \otimes E \to T^* \otimes T^* \otimes E$ is induced by the bilinear map $(X, Y) \mapsto (1/2)(X \otimes Y + Y \otimes X)$. Hence the term $\mathrm{OC}(\mathbf{D}_2)$ is $\tilde{t}(1 \otimes \tilde{s}) - \tilde{t}(1 \otimes \tilde{s}) \circ \mathrm{sym} = \tilde{t}(1 \otimes \tilde{s}) \circ \mathrm{alt}$.

Now in the above diagram, the map $E \to G$ obtained by skirting along the counterclockwise (resp. clockwise) path is by definition $D_t \circ D_s$ (resp. $D_{t \circ s}$). We introduce the via media $V : E \to G$ by setting $V = \tilde{t} \circ (1 \otimes \tilde{s}) \circ \nabla_{T^* \otimes E} \circ \nabla_E$. Then write $D_t \circ D_s - D_{t \circ s}$ as $(D_t \circ D_s - V) + (V - D_{t \circ s})$. The expression in the first parenthesis is $\tilde{t} \circ \mathrm{OC}(\mathbf{D}_1) \circ \nabla_E$, while that in the second parenthesis is $\mathrm{OC}(\mathbf{D}_2) \circ \nabla_{T^* \otimes E} \circ \nabla_E$.

If we plug in the values of OC from the above two lemmas, we get the formula for $D_t \circ D_s$ as follows:

3.9. $\qquad D_t \circ D_s = D_{t \circ s} + \tilde{t}(d_\nabla \tilde{s}) \circ \nabla_E + \tfrac{1}{2}\tilde{t}(1 \otimes \tilde{s}) \circ \mathrm{alt} \circ \nabla_{T^* \otimes E} \circ \nabla_E.$

3.10. Composition formula. *Let $s : E \to T \otimes F$ and $t : F \to T \otimes G$ be two first order symbols. Given connections on E and F we may lift them to differential operators $D_s : E \to F$ and $D_t : F \to G$ respectively. Let us also fix a linear connection without torsion and lift the composite second order symbol $t \circ s : E \to S^2(T) \otimes G$ (obtained by symmetrising the composite $(1 \otimes t) \circ s : E \to T \otimes T \otimes G$) into a differential operator $D_{t \circ s}$ from E to G. Then we have*

$$D_t \circ D_s = D_{t \circ s} + D_{\tilde{t} \circ d_\nabla \tilde{s}} + \tfrac{1}{2}(\tilde{t}(1_{T^*} \otimes \tilde{s})) \circ R_E.$$

Proof. In this formula, $d_\nabla \tilde{s}$ is the absolute derivative of $\tilde{s}$ with respect to the connection on $\mathrm{Hom}(T^* \otimes E, F)$ given rise to by the three connections, and is to be interpreted as a homomorphism $T^* \otimes E \to T^* \otimes F$ so that $\tilde{t} \circ d_\nabla \tilde{s}$ is a homomorphism $T^* \otimes E \to G$. This, being a first order symbol, can be lifted to an operator using the connection on E. Also, R_E is the curvature form and is a homomorphism $E \to T^* \otimes T^* \otimes E$. The homomorphism $(1 \otimes \tilde{s})\tilde{t}$ is a homomorphism $T^* \otimes T^* \otimes E \to G$, and the composite with R_E gives an $\mathcal{O}$-linear map $E \to G$. That is the explanation of the formula.

The composition formula is essentially a restatement of formula 3.9. We first note that the term $\tilde{t} \circ d_\nabla \tilde{s} \circ \nabla_E$ is by definition the lift of the symbol which we have explained above. Now let us evaluate $\nabla_{T^* \otimes E} \circ \nabla_E$ on vector fields X, Y. We obtain, for any section u of E, the expression $\nabla_X(Y \mapsto \nabla_Y u) = \nabla_X \nabla_Y u - \nabla_{\nabla_X Y} u$. Its composite with alt is the same as the alternate bilinear form $(X, Y) \mapsto \nabla_X \nabla_Y u - \nabla_Y \nabla_X u - \nabla_{\nabla_X Y} u + \nabla_{\nabla_Y X} u$. The last two terms add up to $-\nabla_{[X,Y]} u$ since the torsion of the linear connection is zero. Now the sum simplifies to $R_E(X, Y)u$ giving the constant term as claimed.

3.11. Remark. The constant term in the above formula gives an $\mathcal{A}$-homomorphism of E into G. It can therefore be computed fibrewise. Let $m \in M$. Choose a basis $\{e_i\}$ of T_m^*. We have homomorphisms $(s)_i : E \to F$ and $(t)_i : F \to G$ defined as evaluation of the respective symbols at e_i. Then consider the homomorphism $c_{ij} : t_i \circ s_j - t_j \circ s_i$ of E into G. Also write the curvature form of E as $\sum R_{ij}\, e_i \wedge e_j$ where R_{ij} are endomorphisms of E. Then the constant term is $\frac{1}{2} \sum_{i<j} c_{ij} \circ R_{ij}$.

Let E° be an elliptic complex. Assume given Hermitian structures on the bundles E^i and a positive density on M. Suppose also that all the E^i's are provided with Hermitian connections and that the differentials of the complex are given by lifting first order symbols. Assume given a linear connection on M which is torsion free and preserves the density. Finally also assume that the symbols of the operators d_i are invariant with respect to the connections on E^i, E^{i+1} and T. In the following we will also assume that the differentials are of order 1. Under these conditions, we wish to derive a formula for the Laplacian of the elliptic complex.

3.12. Definition. An elliptic complex satisfying all the above conditions is called a *geometric complex.*

Then we can trivialise $\mathcal{S}$ and identify E^i with $\mathrm{adj}(E^i)$ so that the adjoint ∂_i of d^i may be identified with a differential operator from E^{i+1} to E^i. Using the connection on E^{i+1}, we may lift the adjoint symbol $E^{i+1} \to T \otimes E^i$ to an operator $E^{i+1} \to E^i$. From [Ch. 6, Theorem 2.12] it follows that this lifted operator coincides with ∂_i, if the symbol of d^i is invariant with respect to the connections on E, F and T.

In the computation below, we will denote by $\mathrm{adj}(s)$ the adjoint symbol of s, which, we recall, associates to any covector v the negative of the adjoint (i.e. conjugate transpose) of $s(v)$. Let s_i be the symbol of d_i evaluated at a cotangent vector $v \in T_m^*$ at a point $m \in M$. The top term of the Laplacian is of course the lift of the composite symbol, namely the negative of the quadratic map $s_i^* s_i + s_{i-1} s_{i-1}^*$ which is positive definite for each $v \in$

$T_m^* \setminus \{0\}$. Since the symbols are invariant under the connections, there is no first order term. The constant term is the composite $A \circ R_{E^i}$ where $R_E : E \to T^* \otimes T^* \otimes E$, or equivalently a homomorphism $T \otimes T \otimes E \to E$, is the curvature form of the connection on E, and A is the homomorphism $E^i \to T \otimes T \otimes E^i$ associated to the alternate bilinear form $(v, w) \mapsto (s_i(v))^* s_i(w) - (s_i(w))^* s_i(v) + s_{i-1}(v)(s_{i-1})^*(w) - s_{i-1}(w)(s_{i-1})^*(v)$.

Thus we derive the following formula for the Laplacian of a geometric complex.

3.13. Formula for the Laplacian. *The Laplacian of a geometric complex is given by the expression $\Delta = D_{s^*s} + AR_E$, where A is the alternating 2-form on T^* taking (v, w) to $\frac{1}{2}(s_i^*(v) \circ s_i(w) - s_i^*(w) \circ s_i(v) - s_{i-1}(w) \circ s_{i-1}^*(v) + s_{i-1}(v) \circ s_{i-1}^*(w))$ with values in $\mathrm{End}(E)$, R_E is the curvature form as a section of $\Lambda^2(T^*) \otimes \mathrm{End}(E)$ and AR_E is the element of $\mathrm{End}(E^i)$ obtained by contraction on $\Lambda^2(T^*)$ and the composition in $\mathrm{End}(E)$.*

3.14. Remark. Choose, as in Remark 3.11, a basis e_i of T_m^* for any $m \in M$. Then the constant term is $\frac{1}{2} \sum_{k<l}((s_k^i)^* \circ s_l^i - s_l^{i-1} \circ (s_k^{i-1})^*) \circ R_{kl}$, where s^* is the conjugate transpose of s.

4. A Vanishing Theorem

Let W (resp. V) be a real (resp. complex) vector space. Assume given a Hermitian positive definite metric on V. We will denote the corresponding inner product by $\langle \, , \, \rangle$. If q is a quadratic map from W into $\mathrm{End}(V)$ such that $q(w)$ is a Hermitian positive definite endomorphism of V for all $w \in W \setminus \{0\}$, let b be the corresponding (polarised) bilinear map of W into the space of Hermitian endomorphisms of V. Consider the map $(W \times V) \times (W \times V) \to \mathbb{C}$ given by $(w_1, v_1, w_2, v_2) \mapsto \langle b(w_1, w_2)v_1, v_2 \rangle$. It is clearly $\mathbb{R}$-multilinear, $\mathbb{C}$-linear in v_1 and $\mathbb{C}$-antilinear in v_2. Thus we get a Hermitian form h on $W \otimes V$. Assume that this is positive definite as well.

We are interested in applying the above in the following situation. Let E be a Hermitian vector bundle and ∇ a Hermitian connection on it. We wish to compute the composite $\mathrm{adj}(\nabla) \circ \nabla$. For this to make sense we need a density and a torsion free linear connection leaving it invariant. Besides we also need a Hermitian metric on $T^* \otimes E$. It may be thought that one may take a Hermitian metric on T^* and use it. However it is more fruitful to use the above setup for introducing a Hermitian metric on $T^* \otimes E$. In other words we assume given a bilinear map b of T^* with values in $\mathrm{End}(E)$ such that the above condition is satisfied on each fibre with $W = T_m^*$ and $V = E_m$. We need also to assume that this bilinear map is invariant under the connections in question.

We will now compute the composite $\tilde{t} \circ (1 \otimes \tilde{s})$, where s and t are the symbols of the connection ∇ and its adjoint respectively. For any $v \in T_m^*$, the symbol of the connection ∇ takes any $x \in E_m$ to $v \otimes x \in T_m^* \otimes E_m$. The symbol t of its adjoint at v is determined, up to sign, by $\langle x, t(w \otimes y) \rangle = \langle v \otimes x, w \otimes y \rangle = \langle b(v, w)x, y \rangle$. The composite can be thought of as the bilinear form associating to (v, w) the endomorphism $t_v \circ s_w$ determined by the equation $\langle x, (t_v \circ s_w)(y) \rangle = \langle s_v x, s_w y \rangle = \langle v \otimes x, w \otimes y \rangle = \langle b(v, w)x, y \rangle = \langle x, b(v, w)y \rangle$. Hence $(t \circ (1 \otimes s))(y) = b(v, w)y$. Finally this computes the composite as $(v, w) \mapsto b(v, w).\,\mathrm{Id}$. This being symmetric, its composite with R_E is zero. Thus we have

4.1. Proposition. *Let E be a Hermitian bundle. Assume given a symmetric bilinear map b of T^* into the bundle of Hermitian forms on E, such that the associated Hermitian form on $T^* \otimes E$ is positive definite. We also assume given a torsion free linear connection which leaves a density invariant and also the above bilinear map. Then the composite of a Hermitian connection $E \to T^* \otimes E$ and its adjoint is a lifted operator with symbol $-b$.*

Proof. We are of course going to use the composition formula 3.10. The above computation gives the composite symbol to be $-b$. So the top term is as claimed. We have also assumed that b is invariant under the connections, so that the first order term in the expansion vanishes. We have already remarked that the constant term is zero as well. This completes the proof of the assertion.

4.2. Corollary. *Let E° be a geometric elliptic complex. If the negative symbol b of the Laplacian at E^i satisfies the condition above, namely, the Hermitian form on $T^* \otimes E_i$ given by $(v, x, w, y) \mapsto \langle b(v, w)x, y \rangle$ is positive definite, then the top homogeneous term in the decomposition of the Laplacian gives a nonnegative Hermitian operator on $L^2(E^i)$.*

Proof. The bilinear map associated to the negative symbol of the Laplacian is obtained by symmetrising $(v, w) \mapsto (s^i)_v^* \circ (s^i)_w + s_v^{i-1} \circ (s^{i-1})_w^*$. It is clear that for every v, w, the image is a Hermitian endomorphism of E_i (with respect to the chosen Hermitian structure on it). Moreover, if $v = w$, it is positive semidefinite, and by the ellipticity assumption, it is even positive definite for $v \neq 0$. From the definition of a geometric complex, it follows that the connections preserve the symbols of the operators involved, which implies that the above bilinear form on T^* with values in Hermitian endomorphisms of E_i is also ∇-invariant. From the above proposition, we see therefore that the top term in the expansion for the Laplacian is $\mathrm{adj}(\nabla) \circ \nabla$, where we use the above Hermitian form on $T^* \otimes E^i$. Hence it induces a nonnegative Hermitian operator on $L^2(E^i)$.

From this we get the following sufficient condition for the vanishing of the ith cohomology of a geometric elliptic complex.

4.3. Theorem. *Let E° be a geometric elliptic complex on a compact complex manifold as in 4.2. If the $\mathcal{A}$-linear homomorphism $E^i \to E^i$ given by*

$$\kappa(E^\circ) = \widetilde{(s^i)}^* \circ (1_{T^*} \otimes \widetilde{s}^i) + \widetilde{s}^{i-1} \circ (1_{T^*} \otimes \widetilde{(s^{i-1})}^*) R_E$$

is positive definite, then $H^i(E^\circ) = 0$.

Proof. The cohomology space is isomorphic to the space of harmonic sections by Theorem 2.6. Now use the formula for the Laplacian of a geometric complex given in 3.13. Finally use the fact, proved in Corollary 4.2, that the top term in this expansion is a nonnegative operator, to conclude that all harmonic sections are zero under the hypothesis of the theorem.

Notice that the de Rham complex of a Riemannian manifold is a geometric complex. The Riemannian density is invariant under the Levi-Civita connection, and since there is a natural induced metric on all the bundles $\Lambda^i(T^*)$, the Laplacian of the de Rham complex is defined. We will now apply the above principle in this case.

The symbol of the exterior derivative $d : \Lambda^i(T^*) \to \Lambda^{i+1}(T^*)$ associates to $v \in T^*$ the exterior multiplication $\lambda(v)$ by v. The symbol of its adjoint is $-i_v$ [Ch. 7, 7.15]. Hence the symmetrised composite may be computed by taking $v = w$. To this it associates $-(i_v\lambda_v + \lambda_v i_v) = -g(v)$. Id. Hence the Hermitian form on $T^* \otimes E^i$ is $(v, x, w, y) \mapsto \langle g(v, w)x, y \rangle = g(v, w)\langle x, y \rangle$. In other words, it is the tensor product of the metric on T^* and that on E^i. In particular it is positive definite, as we have checked even generally, and is also ∇-invariant. We will now apply the above vanishing theorem to this complex.

The alternatised composite of the symbols, associates to (v, w) the endomorphism $(1/2)(i_w\lambda_v - i_v\lambda_w + \lambda_w i_v - \lambda_v i_w)$. This is easily seen to be the same as $i_w\lambda_v - i_v\lambda_w$, and it is a derivation (of even type) of the exterior algebra of T^*. On T^*, this action is simply the natural isomorphism of Λ^2 with the subspace of skew-symmetric endomorphisms. Hence for any p, the above map $\Lambda^2(T^*) \to \mathrm{End}(\Lambda^p(T^*))$ is $A \mapsto D(A)$, where $D(A)$ is the derivational extension of the corresponding skew-symmetric endomorphism of T^* to the exterior algebra. The curvature form on $\Lambda^p(T^*)$ is given by $(X, Y) \mapsto D^p(R(X, Y))$. Hence the substitution of R in the above map yields the endomorphism of $\Lambda^p(T^*)$ obtained by identifying R with an element of $\mathrm{End}(T^*) \otimes \mathrm{End}(T^*)$ and applying $D \otimes D$ to this to get an element of $\mathrm{End}(\Lambda(T^*)) \otimes \mathrm{End}(\Lambda(T^*))$ and then composing. We denote this element by

$D(R)$. We thus have the following particular case of the vanishing theorem.

4.4. Theorem (Bochner). *The ith Betti number of a compact Riemannian manifold is zero if $D^i(R)$ is positive definite. In particular, if the Ricci curvature is positive definite, the first Betti number is zero.*

4.5. Remark. The formula $\Delta = \nabla^*\nabla + D^i(R)$ thus obtained is called the *Weitzenbock formula.*

5. Hodge Decomposition

We will soon look at the case of a compact Kähler manifold and the Dolbeault complex of a holomorphic bundle on it and derive a nice corollary of our vanishing theorem. For now, we will take up a quite different application of the formula for the Laplacian.

5.1. Proposition. *Let M be a Riemannian manifold and $A : \Lambda^i(T^*) \to \Lambda^j(T^*)$ any homomorphism of bundles which commutes with the connections on the two bundles induced by the Riemannian connection. Then it takes harmonic forms to harmonic forms.*

Proof. It is of course sufficient to show that the map A commutes with the Laplacian. To show this we may use the above formula. Clearly A commutes with the top term since it is the lift of the symbol by a connection which leaves A invariant. If we show that the constant term also commutes with A, we will be through. If we consider R to be an element of $\mathrm{End}(T_m^*) \otimes \mathrm{End}(T_m^*)$ for each $m \in M$, then we claim that it actually belongs to $W \otimes W$, where W is the subspace of $\mathrm{End}(T_m^*)$ generated by $R(v_1, v_2)$, $v_i \in T_m$. Since R is symmetric, this follows from the obvious fact that R belongs to $\mathrm{End}(T_m^*) \otimes W$ in any case.

5.2. Corollary. *If M is a Riemannian manifold, then the star operator maps harmonic forms to harmonic forms.*

5.3. Remark. In particular, there is an isomorphism between $H^i(M, \mathbb{R})$ and $H^{n-i}(M, OR)$. We already know by the duality theorem that for a compact, connected Riemannian manifold M, the spaces $H^i(M, \mathbb{R})$ and $H^{n-i}(M, OR)$ are *dual* to each other. Using both we get a nondegenerate bilinear form on all the cohomology spaces. To two forms α, β this form assigns $\int \alpha \wedge *\beta$.

Suppose M is a compact Kähler manifold. Then we get the following consequence of Proposition 5.1.

5.4. Corollary (Hodge decomposition). *Let M be a compact Kähler manifold. There is a canonical decomposition of $H^i(M, \mathbb{C})$ into a direct sum of $H^{p, i-p}(M)$ where $H^{p, i-p}$ is the Dolbeault cohomology $H^{i-p}(M, \Lambda^p(T^*))$.*

Proof. This is a direct application of Proposition 5.1 after we note that J commutes with the Riemannian connection in a Kähler manifold. Hence the projection to the $(p, i - p)$ component is also an endomorphism of $\Lambda^i(T^*)$ which commutes with the Riemannian connection. According to Proposition 5.1, this implies that it takes harmonic forms into harmonic forms. This gives the decomposition claimed. Now the cohomology classes in $H^{p, i-p}$ are represented by harmonic forms of type $(p, i - p)$. This means that both d and ∂ vanish on them. But obviously this implies that d'' and ∂'' also vanish on them. It follows that these are also Dolbeault harmonic forms. Since complex conjugation also takes harmonic forms to harmonic forms, we conclude that the Dolbeault harmonic forms which are d''-closed and ∂''-closed are also d'-closed and ∂'-closed. Hence these are Δ-harmonic as well. This identifies the space of harmonic forms of type $(p, i - p)$ with the corresponding Dolbeault cohomology spaces.

5.5. Definition. The above decomposition is called the *Hodge decomposition.*

One can also compute the Laplacian corresponding to the Dolbeault complex, using the composition formula. Again, at a (real) cotangent vector v at a point m, the sum of the composite of the symbols of d'' and ∂'' and that of the symbols of ∂'' and d'' can be computed as for the de Rham complex. It turns out that it takes the value $(1/4).g(v + iJv, v - iJv). \mathrm{Id} = (1/2)g(v). \mathrm{Id}$, where g is the complex bilinear extension of the Riemannian metric. This actually proves, in view of the formula for the Laplacian, the following fact.

5.6. Theorem. *Let M be a Kähler manifold. The Laplacian of the underlying Riemannian metric leaves all the sheaves $\Lambda^{p,q}(T^*)$ invariant and is twice the Laplacian of the Dolbeault complex.*

This proves again the Hodge decomposition of the de Rham cohomology spaces of a compact Kähler manifold.

5.7. Remark. While the Hodge decomposition was arrived at using the Kähler metric, the decomposition itself is *independent* of the metric. Consider the sheaf homomorphism given by the exterior derivative d of the sheaf Ω^{i-1} of holomorphic $(i - 1)$-forms into Ω^i. We claim that the image is the sheaf of closed i-forms. In other words, any closed holomorphic form can be written locally as the image by d of a holomorphic form. This is proved exactly as in [Ch. 2, Proposition 6.14]. The kernel of this sheaf homomorphism is the sheaf of closed $(i-1)$-forms. Hence we have the exact sequence of sheaves

$$0 \to \Omega_{cl}^{i-1} \to \Omega^{i-1} \to \Omega_{cl}^{i} \to 0.$$

The cohomology exact sequence gives

$$\cdots \to H^j(\Omega^{i-1}) \to H^j(\Omega_{cl}^{i}) \to H^{j+1}(\Omega_{cl}^{i-1}) \to H^{j+1}(\Omega^{i-1}) \to \cdots .$$

The induced map $H^j(\Omega^{i-1}) \to H^j(\Omega_{cl}^{i})$ is zero, since any cohomology class can be represented by a harmonic form for d'' but then it is also harmonic for d and consequently its image by d is zero. So the above cohomology sequence can be written as a short exact sequence

$$0 \to H^j(\Omega_{cl}^{i}) \to H^{j+1}(\Omega_{cl}^{i-1}) \to H^{j+1}(\Omega^{i-1}) \to 0.$$

Notice that this gives an increasing sequence of vector spaces finally ending up with $H^{i+j}(\Omega_{cl}^{0})$. But a closed holomorphic function is locally constant and so we have the isomorphism $\Omega_{cl}^{0} \approx \mathbb{C}$ of sheaves. Thus the space $H^r(X, \mathbb{C})$ is filtered by vector subspaces $H^i(M, \Omega_{cl}^{r-i})$ and the associated quotients are isomorphic canonically to $H^i(\Omega^{r-i})$. This filtration *does not depend* on the Kähler metric and is called the *Hodge filtration.* Indeed, by taking the complex conjugation and intersecting, we can actually get back the decomposition from the filtration.

Note that at the 0th stage, $H^0(M, \mathbb{C})$ as well as $H^0(M, \mathcal{O})$ are constants, since holomorphic functions on a compact connected manifold are constants. Next, we will illustrate the decomposition for the first cohomology. The above sequence gives (on taking $i=1$ and $j=0$)

$$0 \to H^0(\Omega_{cl}^{1}) \to H^1(\Omega_{cl}^{0}) \to H^1(\mathcal{O}) \to 0.$$

Since all holomorphic forms are d''-closed and are also obviously ∂''-closed, they are d''-harmonic. Hence they are also d-harmonic. This shows that $H^0(\Omega_{cl}^{1}) = H^0(\Omega^1)$. Thus the above sequence becomes

$$0 \to H^0(\Omega^1) \to H^1(\mathbb{C}) \to H^1(\mathcal{O}) \to 0.$$

This is the Hodge filtration in this case. Since there is a complex conjugation automorphism in $H^1(\mathbb{C})$, we may apply it to $H^0(\Omega^1)$ and check that it maps isomorphically to $H^1(\mathcal{O})$ giving the decomposition.

6. Lefschetz Decomposition

We will give another application of a similar nature to the study of the cohomology of a compact Kähler manifold, this time depending on the metric. A map A satisfying the hypotheses of Proposition 5.1 takes harmonic forms to harmonic forms. This means that although cohomology spaces are global, they are represented by harmonic forms and as such the induced maps on them can be understood from a fibrewise computation since A is by assumption $\mathcal{O}$-linear. We will now give an example of the power of this method. What follows is what happens at the fibre level, and after explaining it we will return to the application to Kähler manifolds.

6.1. Digression on $\mathfrak{sl}(2)$-modules.

Let V be a finite-dimensional vector space over $\mathbb{C}$. Assume that H, E and F are linear transformations of V which satisfy the relations $[H, E] = 2E$, $[H, F] = -2F$, and $[E, F] = H$. We will refer to such a triple of linear transformations as an $\mathfrak{sl}(2)$-module. Two such modules V and W are considered isomorphic if there is a linear isomorphism of V with W which transforms the two triples, one to the other.

6.2. Remark.
The Lie algebra of $SL(2, \mathbb{C})$ consists of $(2,2)$ matrices of trace zero. A basis for the underlying vector space is given by the matrices $h = \left(\begin{smallmatrix} 1 & 0 \\ 0 & -1 \end{smallmatrix}\right)$, $e = \left(\begin{smallmatrix} 0 & 1 \\ 0 & 0 \end{smallmatrix}\right)$, and $f = \left(\begin{smallmatrix} 0 & 0 \\ 1 & 0 \end{smallmatrix}\right)$. These have the above commutation relations, and so V as above is simply a representation of the Lie algebra of $SL(2)$.

6.3. Example.
The above realisation of the three elements with the given commutation relations makes $V = \mathbb{C}^2$ an $\mathfrak{sl}(2)$-module. From this we also get the symmetric powers of V as $\mathfrak{sl}(2)$-modules. These have the following description too. Consider the space of homogeneous polynomials P of degree m, in two variables x and y. Define $H(P) = x\frac{\partial P}{\partial x} - y\frac{\partial P}{\partial y}, E(P) = x\frac{\partial P}{\partial y}$ and $F(P) = y\frac{\partial P}{\partial x}$. Denote this module by V_m. Note that the binomials $v_p = x^p y^q, p + q = m$ are all eigenvectors for H with eigenvalues $p - q$. Also E takes v_m to 0 and for $p < m$, v_p to qv_{p+1}, and F takes v_0 to 0 and for $p > 0$, v_p to pv_{p-1}. Any nonzero $\mathfrak{sl}(2)$-invariant subspace contains an eigenvector for H, namely one of the v_p's and applying E and F successively we get all the basis vectors. This shows that this subspace is the whole of V_m. In other words, V_m is an *irreducible* $\mathfrak{sl}(2)$-module. Finally the fact that $L = x\frac{\partial}{\partial x} + y\frac{\partial}{\partial y}$ acts as the scalar $m.\mathrm{Id}$ (Euler's formula) can be expressed in terms of H, E and F. We have $L^2 - H^2 = 4xy\frac{\partial^2}{\partial x \partial y} = 2(EF + FE) - 2L$. Hence the transformation $H^2 + 2EF + 2FE$ acts as the scalar $m^2 + 2m$.

6.4. Classification of $\mathfrak{sl}(2)$-modules.

We wish to show that all $\mathfrak{sl}(2)$-modules are essentially made up of the examples V_m above. In the analysis that follows we take the cue from the above example.

Let V be an arbitrary $\mathfrak{sl}(2)$-module. We will first show that E is singular. Let v be an eigenvector for H with eigenvalue λ. Then Ev, if nonzero, is also an eigenvector of H, with eigenvalue $\lambda + 2$. For, $HEv = EHv + [H, E]v =$

$(\lambda + 2)Ev$. (In a similar manner, we also see that Fv, if nonzero, is an eigenvector for H with eigenvalue $\lambda - 2$.) Consider the sequence of vectors $v, Ev, E^2v, \ldots\,$. All these cannot be nonzero, since they would then be eigenvectors of H with distinct eigenvalues, but H has only finitely many eigenvalues. This proves our assertion that E has nonzero kernel.

Next note that the kernel of E is invariant under H. For if $Ew = 0$, then $EHw = HEw - [H, E]w = -2Ew = 0$. Thus there is an eigenvector w for H (say, with eigenvalue λ) such that $Ew = 0$. Consider the sequence of vectors $w, Fw, \ldots, F^i w$. We claim that E takes $F^i w$ to a multiple of $F^{i-1}w$. Assume inductively that $EF^{i-1}w = \mu_i F^{i-2}w$. We know that this is the case if $i = 1$ with $\mu_1 = 0$. Then we have $EF^i w = ([E, F] + FE)F^{i-1}w = HF^{i-1}w + FEF^{i-1}w = (\lambda - 2(i - 1))F^{i-1}w + FEF^{i-1}w = (\lambda - 2(i - 1) + \mu_i)F^{i-1}w$. We have therefore proved our claim and in fact the scalars μ_i are evaluated to be $(i-1)(\lambda - i + 2)$. Again as before all the members of the sequence $F^i w$ cannot be nonzero. Let m be the least positive integer such that $F^{m+1}w = 0$. The vector space W spanned by $w, Fw, \ldots, F^m w$, is invariant under H, E and F. Moreover, H takes the diagonal form $(\lambda, \lambda - 2, \ldots, \lambda - 2m)$. Since H is a commutator, it has trace 0. Hence we have $(m + 1)\lambda - m(m + 1) = 0$, i.e. λ is the nonnegative integer m. We have already computed $EF^i w$ to be $i(m - i + 1)F^{i-1}w$. The linear map of W into V_m which takes $F^p w$ to $(x^{m-p}y^p)/(m - p)!$ is an $\mathfrak{sl}(2)$-isomorphism of W with V_m. This proves that every $\mathfrak{sl}(2)$-module V contains as a submodule, an isomorphic copy of one of the V_m's.

6.5. Exercise. In any $\mathfrak{sl}(2)$-module, check that the element $C = H^2 + 2EF + 2FE$ commutes with H, E and F. Show that C has only eigenvalues of the type $m^2 + 2m$, with m a positive integer.

Now we have the following classification of all $\mathfrak{sl}(2)$-modules.

6.6. Theorem. *Let V be any finite-dimensional $\mathfrak{sl}(2)$-module. Then V is canonically a direct sum of subspaces $V(m)$ in each of which the element $C = H^2 + 2EF + 2FE$ is represented by the scalar endomorphism $(m^2 + 2m)$.Id. The subspace $V(m)$ is itself the direct sum of modules isomorphic to V_m, namely the space of homogeneous polynomials of degree m in two variables.*

Proof. We will decompose the given module into a direct sum of $\mathfrak{sl}(2)$-modules by taking the generalised eigenspaces of C. Since all the three elements commute with C, these subspaces are invariant under $\mathfrak{sl}(2)$. Call them $V(m)$. We will now restrict ourselves to one of these modules. Accordingly we will assume that the only eigenvalue of C in this module is $m^2 + 2m$ and show that it is a direct sum of submodules isomorphic to V_m.

As a first step, we will consider the case $V = V(m)$ with $m = 0$. In this case, any irreducible submodule of V is of the form V_m for some m. But it has to be trivial, i.e. $H = E = F = 0$, since C has the eigenvalue $m^2 + 2m$ on V_m which is assumed to be 0. Let W_1 be the direct sum of all trivial submodules of V. We may apply the same argument for the quotient of V by W_1, and conclude that there is a filtration $W_0 = 0 \subset W_1 \subset \cdots \subset W_{r-1} \subset W_r = V$ with the property that W_{i+1}/W_i is a direct sum of trivial modules for all i. Now H, E and F take W_i to W_{i-1}. We need to show that $r = 1$. If not, acting by $[H, E] = HE - EH$ on W_2 we conclude that $[H, E]$ annihilates it. Since $[H, E] = 2E$, this means E is zero on W_{r-2}. Similarly F also acts trivially on it. Since $H = [E, F]$, it follows that W_2 is the trivial module, contradicting our assumption.

We will finally consider the case when C has only one eigenvalue $m^2 + 2m$, $m \neq 0$. Let W be maximal among subspaces of V which are direct sums of modules isomorphic to V_m as $\mathfrak{sl}(2)$-modules. We will show that $W = V$. If not, the $\mathfrak{sl}(2)$-module V/W contains a submodule S isomorphic to V_m. Consider the space L of all linear maps of V_m into V such that their composites with the natural map $V \to V/W$ are $\mathfrak{sl}(2)$-module homomorphisms into S. The space L can be considered as an $\mathfrak{sl}(2)$-module by setting $xf = x_V \circ f - f \circ x_{V_m}$, for $x = H$, E, or F. The generalised eigenspace $L(0)$ of C in L for the eigenvalue 0 is, as we have seen above, a trivial representation. In other words, elements $f \in L(0)$ are $\mathfrak{sl}(2)$-homomorphisms. To any $f \in L(0)$ associate its composite with the natural map $V \to W$ to get a surjective map to the one-dimensional trivial representation, namely the space of $\mathfrak{sl}(2)$-homomorphisms of V_m into S. In particular, the isomorphism $V_m \to S$ can be lifted to an $\mathfrak{sl}(2)$-homomorphism $V_m \to V$. Its image is a module isomorphic to V_m and is supplementary to W, contradicting the maximality of W.

A posteriori, we see that C, H are all semisimple in any module since they are so in V_m's and therefore the generalised eigenspaces $V(m)$ of C are actually eigenspaces.

6.7. Remark. We proved in 6.4 that all irreducible modules are isomorphic to V_m for some m. In the proof of Theorem 6.6, we saw that every representation of $\mathfrak{sl}(2)$ is a direct sum of irreducible modules, using the element C which acts on any $\mathfrak{sl}(2)$-module.

But any $\mathfrak{sl}(2)$-module arises by a complex extension of a representation of the *real* Lie algebra $\mathfrak{su}(2)$ consisting of skew-Hermitian matrices. The latter is the Lie algebra of the special unitary group $SU(2)$ which is simply connected and compact. So the complete reducibility of $\mathfrak{sl}(2)$-modules is entirely equivalent to the complete reducibility of complex representations of the compact group $SU(2)$. To prove this, one can use global methods, such as integration.

6.8. Exercise. Let G be a compact Lie group and V a G-module. Show by proving the existence of a Hermitian metric invariant under the group that V is completely reducible.

We are interested in the $\mathfrak{sl}(2)$-module that arises in the following manner. Let V be a real vector space of dimension $2n$. Assume given a J-structure and a Hermitian metric with respect to it, that is to say, a symmetric positive definite $\mathbb{R}$-bilinear form satisfying $g(X, JY) + g(JX, Y) = 0$. Let ω be the corresponding nondegenerate alternating bilinear form, namely $(X, Y) \mapsto g(X, JY)$. We can diagonalise the Hermitian form so that there exists a linearly independent set $(e_i), i = 1, \ldots, n$, such that $e_i, f_i = Je_i$ form an orthonormal basis for g, and ω has the form $\sum e_i' \wedge f_i'$ with respect to the dual basis. In the following we will denote e_i', f_i' by e_i, f_i for notational simplicity.

Define E to be multiplication in the sense of the exterior algebra $\Lambda(V)$ by ω. Define H to be the transformation given by $H\alpha = (i-n)\alpha$ for $\alpha \in \Lambda^i(V)$. It follows that $HE(\alpha) = H(\alpha \wedge \omega) = (i + 2 - n)\alpha \wedge \omega$. On the other hand $EH(\alpha) = (i - n)\alpha \wedge \omega$. Thus $(HE - EH)(\alpha) = 2\alpha \wedge \omega = 2E(\alpha)$. In other words, $[H, E] = 2E$. We will now define a linear transformation F in the exterior algebra. Note that the element $v = e_1 \wedge f_1 \wedge \cdots \wedge e_n \wedge f_n$ gives an orientation on V and also the metric gives a star operator denoted $*$. Use this operator to define the conjugate $F = *E(*)^{-1}$. Then F maps $\Lambda^r(V)$ into $\Lambda^{r-2}(V)$. Now $2F = 2 * E(*)^{-1} = *[H, E](*)^{-1} = [*H(*)^{-1}, F]$. But $*H(*)^{-1}$ acts on $\Lambda^i(V)$ as multiplication by $2n-i-n$. Hence $*H(*)^{-1} = -H$. Thus we get $[H, F] = -2F$.

We will now show that $EF - FE = H$. We will do this by computing it on the elements of the form $e_S \wedge f_T$. Here S and T are subsets of the set $N = \{1, \ldots, n\}$, with some ordering. It is convenient to split up the subsets S and T as follows. We will denote by A, B, C, D the subsets $S \cap T, S \setminus T, T \setminus S, N \setminus (S \cup T)$ respectively. Then we will write the element as $e_{AB} \wedge f_{AC}$.

Then $*(e_{AB} \wedge f_{AC}) = \pm e_{DC} \wedge f_{DB}$ the sign being the same as in $e_{AB} \wedge f_{AC} \wedge e_{DC} \wedge f_{DB} = \pm v$. We will denote by the corresponding small letters the cardinality of the sets involved. In both EF and FE the term $*^2$ occurs twice so that we may take $\pm v$ for the volume element throughout without bothering to find out which, as long as we use the same volume element on both occasions. After choosing A, B, C, D as above, we will actually take for the volume element, the term $e_{ABCD} \wedge f_{ABCD}$, which is clearly ϵv and it is not relevant to determine ϵ and carry it along.

The operators EF and FE preserve each $\Lambda^r(V)$ and we will compute $(E*)^2$ on $\alpha = e_{AB} \wedge f_{AC}$, noting that it differs from EF only by the factor $(-1)^{b+c}$. We have $*(\alpha) = \epsilon_1(e_{DC} \wedge f_{DB})$, the sign being the signature of the permutation taking $AB\underline{ACDC}DB$ to $ABCD\underline{ABCD}$. This gives the parity of the sign to be $c(a+c+d) + d(a+c) + b(c+d)$. When we apply E to $*\alpha$, we therefore get $(E*)(e_{AB} \wedge f_{AC}) = \epsilon_2(\sum_{i\in A} e_{iDC} \wedge f_{iDB})$ where ϵ_2 has the parity of $c+d+\epsilon_1 = c+d+ca+c^2+cd+da+dc+bc+bd$ or, what is the same, that of $d + ac + ad + bc + bd$. Noting that each term in the summation is of the same form with A, B, C, D replaced by iD, C, B and $A \setminus \{i\}$ respectively, we may repeat the formula and get $(E*)^2(\alpha) = \epsilon_3(\sum_{j\in\{i\}\cup D, i\in A} e_{j(A\setminus\{i\})B} \wedge e_{j(A\setminus\{i\})C}) = \epsilon_3(\sum_{i\in A} e_{AB} \wedge e_{AC} + \sum_{j\in D, i\in A} e_{j\cup(A\setminus i)B} \wedge f_{j\cup(A\setminus i)C})$, where ϵ_3 has the parity of $d+ac+ad+bc+bd+(a-1)+(d+1)b+(d+1)(a+1)+cb+c(a-1)$. This simplifies to $b + c$ which means that $\epsilon_3 = (-1)^{b+c}$. In other words, $EF(\alpha) = (\#A)\alpha + \sum_{j\in D, i\in A} e_{(A\setminus\{i\})\cup jB} \wedge f_{(A\setminus\{i\})\cup jC}$. A similar computation gives $FE(\alpha) = (\#D)\alpha + \sum_{j\in D, i\in A} e_{(A\setminus\{i\})\cup jB} \wedge f_{j\in D, i\in A} f_{(A\setminus\{i\})\cup jC}$. Taking the difference we get $[E, F](\alpha) = (\#A - \#D)\alpha$. But it is clear that $\#A - \#D = r - n$, proving that $[E, F] = H$, as claimed.

6.9. Space of harmonic forms as an $SL(2)$-module.

Let M be a Kähler manifold. Then the exterior algebra of T_m^* at any point $m \in M$ is an $\mathfrak{sl}(2)$-module in a canonical way as shown above. Thanks to Proposition 5.1, we conclude that the operations E, F, and H commute with the Laplacian and hence make the space of harmonic forms an $\mathfrak{sl}(2)$-module.

6.10. Definition. A harmonic form as well as the cohomology class represented by it, is said to be *primitive* if it is in the kernel of F under the above action.

The module of harmonic forms can be decomposed as a direct sum of irreducible $\mathfrak{sl}(2)$-modules. But the action has the property that all the eigenvalues of H in the kernel of F are at most 0. In other words, all primitive forms are of degree $i \leq n$.

Moreover a harmonic form α of degree $i \leq n$ is primitive if and only if $E(*\alpha) = 0$. Since $*\alpha$ is of degree $2n - i$, we conclude that it belongs to the eigenspace of H with eigenvalue $2n - i - n = n - i$. Also $*\alpha$ is in the kernel of E so that the $SL(2)$-module generated by $*\alpha$ is actually V_{n-i}. Equally α generates the same $SL(2)$-module. Iteration of E on α leads to $*\alpha$, i.e. $E^{n-i}\alpha$ is a nonzero multiple of $*\alpha$. To summarise, we have

6.11. Proposition. *Let M be a compact Kähler manifold. The action of E^m on $H^i(M, \mathbb{C})$, $i \leq n$, is injective if $m \leq n - i$ and the space of primitive forms of degree i is the kernel of E^{n-i+1}. Moreover, every harmonic form*

of degree i is a sum of harmonic forms obtained as $E^r \omega_r$, where ω_r's are primitive harmonic classes of degree $i - 2r$.

As in the case of the Hodge decomposition, this implies many interesting constraints on the Betti numbers of a compact Kähler manifold. For example, for a start, the even Betti numbers b_i are never zero for $i \leq 2n$. Note that this implies that even-dimensional spheres of dimension greater than 2 can never be Kähler, even if they admit complex structures.

Of course we can derive many more conditions. For instance, if $i \leq n-1$, we have $b_i \leq b_{i+2}$.

7. Kodaira's Vanishing Theorem

7.1. Definition. Let M be a complex manifold and E a holomorphic principal bundle on M with a complex Lie group G as structure group. Then a (differentiable) connection on E is said to be of *type* $(1,0)$ if the connection form, extended as a complex linear map of $T_\xi \otimes \mathbb{C}$ with values in $\mathfrak{g}$, $\xi \in P$, is of type $(1,0)$.

Notice that the condition can also be stated purely in terms of the (nonextended) connection form γ on E by saying that $\gamma(JX) = i\gamma(X)$, where J is the (almost) complex structure on E and i is the multiplication in the complex vector space $\mathfrak{g}$. In the case when $G = \mathbb{C}^*$, that is to say, E is the principal bundle associated to a line bundle, the complex-valued form γ on E is of type $(1,0)$.

A connection on a vector bundle E is the same as an operator $E \rightarrow T^* \otimes E$ with identity symbol. This gives two operators from E, one into $(T^*)^{(1,0)} \otimes E$ and another into $(T^*)^{(0,1)} \otimes E$. It is easily verified from the definition that the connection is of type $(1,0)$ if and only if the latter of these operators is the Dolbeault differential d'' which is associated to the holomorphic bundle E.

7.2. Proposition. *Given a Hermitian metric on a holomorphic line bundle on a complex manifold, there is a unique connection form of type $(1,0)$ on the principal bundle, which preserves the Hermitian structure. Its curvature is a form of type $(1,1)$.*

Proof. We can use the Hermitian metric to reduce the structure group of the bundle to $U(1)$. If there are two Hermitian connections, then their difference goes down to a form α on M with values in the Lie algebra of $U(1)$, namely $i\mathbb{R}$. On the other hand, if the connections are of type $(1,0)$, then α satisfies $\alpha(JX) = i\alpha(X)$, which is possible only if $\alpha = 0$. This proves the uniqueness.

Let us treat the Hermitian metric as an $\mathbb{R}^+$-valued function h on the principal bundle P of E satisfying $h(\xi s) = |s|^2 h(\xi)$ for all $\xi \in P$ and $s \in \mathbb{C}^\times$. In other words, it is a section of the associated bundle for the action of $\mathbb{C}^\times$ on $\mathbb{R}^+$ given by $\lambda \mapsto$ multiplication by $|\lambda|^{-2}$. The invariance of the Hermitian metric under the connection therefore gives $\nabla_X h = Xh + \gamma(X)h = 0$. The value of $\gamma(X)h$ is, by the above computation, $-2\,\mathrm{Re}(\gamma X).h(\xi)$. Hence we have $\mathrm{Re}\,\gamma(X) = (1/2)Xh/h$, or what is the same, $\mathrm{Re}\,\gamma = (1/2)d(\log h)$. Since γ is in addition a form of type $(1,0)$, this implies that $\gamma = d'(\log h)$ and that its curvature form is given by $d''d'(\log\ h)$.

7.3. Definition. The unique connection associated to the Hermitian structure on a holomorphic line bundle is called the *Chern connection*. The corresponding curvature form is called the *Chern form of the Hermitian structure*.

7.4. Remark. Compare this with the computation of curvature in the case of the Hopf bundle in [Ch. 5, 6.2]. In retrospect, the connection which we defined there is just the Hermitian connection, using the Hermitian metric along the fibres induced from that of the ambient vector space.

Let L be a holomorphic line bundle provided with a Hermitian metric. The above computation shows that the Chern form is then a purely imaginary form of type $(1,1)$. In particular we deduce that the Chern class of any holomorphic line bundle is actually in $H^{1,1}(\mathbb{C})$. In particular, we have a class in the Dolbeault cohomology space $H^1(M, \Omega^1)$. Holomorphic line bundles are classified up to isomorphism by $H^1(X, \mathcal{O}^\times)$. If v is the cohomology class corresponding to L, then the above class is obtained as the image of v under the map induced by the sheaf map $\mathcal{O}^\times \to \Omega^1$ given by $f \mapsto d(\log f)$. Indeed this follows from the commutative diagram

$$
\begin{array}{ccccccccc}
0 & \to & 2\pi i\mathbb{Z} & \to & \mathcal{O} & \to & \mathcal{O}^\times & \to & 0 \\
 & & \downarrow & & \| & & \downarrow & & \\
0 & \to & \mathbb{C} & \to & \mathcal{O} & \to & \Omega^1_{cl} & \to & 0.
\end{array}
$$

Here the lower sequence has its right arrow given by $f \mapsto df/f$. It is exact because any closed holomorphic form is locally a boundary. The top sequence is exact since any everywhere nonzero function can be locally written as the exponential of a holomorphic function.

We will now address the question: what forms are Chern forms of a Hermitian connection? As we have seen above, these have to be closed forms of type $(1,1)$ representing cohomology classes which are ($2\pi i$ times) integral.

Firstly, any $(2\pi i$ times) integral cohomology class, represented by a closed form of type $(1,1)$, is the (geometric) Chern class of a suitable holomorphic line bundle L. Indeed, in the above diagram, by hypothesis, we are given an element c of $H^2(M, \mathbb{C})$ which comes from $H^2(M, 2\pi i\mathbb{Z})$, and whose image in $H^2(M, \mathcal{O})$ is zero. It follows that it comes from an element of $H^1(M, \mathcal{O}^\times)$. In other words, there is a holomorphic line bundle L whose Chern class is c.

Next we would like to know if any closed form α of the above type is actually the Chern form of a holomorphic line bundle with a Hermitian structure.

If we introduce an arbitrary Hermitian metric on a holomorphic line bundle L with this Chern class, its Chern form β may not coincide with α. But the difference $\alpha - \beta$ is an imaginary 2-form representing the trivial class and hence of the form $d\omega$ for some purely imaginary 1-form ω. Write ω as $\xi - \bar{\xi}$, with ξ of type $(0,1)$. Then we get the set of equations, $\alpha - \beta = d'\xi - d''\bar{\xi}$, and $d''\xi = 0$. Let η be the Δ''-harmonic representative of the Dolbeault class of ξ. Then $\xi - \eta$ is of the form $d''f$. Since η is harmonic for the Laplacian Δ as well, we have $d'\eta = 0$. From this we conclude that $d'\xi = d'd''f$ and therefore that $\alpha - \beta$ is of the form $d'd''(f + \bar{f})$. In other words, there is a real-valued function g such that $\alpha - \beta = d''d'g$. Now we can modify the Hermitian metric by multiplying by $\exp(g)$ and check that the Chern form of the new Hermitian connection is actually α.

Thus we have proved the following

7.5. Theorem. *1) Let X be a compact Kähler manifold. A cohomology class in $H^2(X, \mathbb{C})$ is the (geometric) Chern class of a holomorphic line bundle if and only if it is of type $(1,1)$ and represents $(2\pi i$ times) an integral class.*

2) Any form of type $(1,1)$ which represents the Chern class of a holomorphic line bundle, is the Chern form of a suitable Hermitian metric on the line bundle.

7.6. Theorem (Kodaira). *Let L be a holomorphic line bundle on a compact Kähler manifold M, whose topological Chern class can be represented by a closed $(1,1)$-form which is positive definite, and E any other holomorphic vector bundle. Then the cohomology groups $H^q(M, E \otimes L)$ vanish for large enough r whenever q is at least 1.*

Proof. We will apply the vanishing theorem (Theorem 4.3) to the Dolbeault complex of $E \otimes L$. We note first that there exists a Hermitian structure on L such that the Chern form R_L is negative definite. Extend the form R_L complex-bilinearly to $T^* \otimes \mathbb{C}$. Since it is of type $(1,1)$, one sees that its restriction to the spaces of type $(1,0)$ as well as type $(0,1)$ are zero. The Hermitian form $(v,w) \mapsto R_L(v,\overline{w})$ on $(T^*)^{0,1}$ is positive definite by our assumption.

The composite symbol in the formula for the Laplacian is given by $(v,w) \mapsto \iota_v \circ \lambda_w + \lambda_v \circ \iota_w$, where we identify the real cotangent space T^* with the complex vector space of $(0,1)$ differentials, and use the Hermitian metric h on the latter, in order to define ι. If α is a differential of type $(0,1)$, then the composite symbol corresponding to (v,w) takes α to $\iota_v(w \wedge \alpha) + (\iota_w\alpha)v = h(v,w)\alpha - h(v,\alpha)w + h(w,\alpha)v$. It is easily seen that under the above identification of T^* with $(T^*)^{0,1}$, the symmetrisation of h gives $(1/2)g$. Hence the Dolbeault complex satisfies the conditions for the formula.

We will now compute the constant term. The curvature of $\Lambda^{(0,p)}(T^*) \otimes E \otimes L$ is the sum of three terms, the first two being independent of L. The constant term in the formula for the Laplacian is correspondingly the sum of three terms, the first two being independent of L. The last term is obtained by substituting in the alternatised composite symbol the curvature form R_L. This composite is, as in the Riemannian case, a derivation of the exterior algebra of $(T^*)^{0,1}$. Its action on $(T^*)^{0,1}$ has been computed above. This is a sum of two terms. The first is a scalar endomorphism, obtained by substituting in the alternating form ω on T^*, the curvature form R_L, and the second is the derivation $D^p(R_L)$ given by the Hermitian endomorphism R_L of $(T^*)^{0,1}$. The first one is zero, since the curvature form preserves J.

Finally, if we replace L by L^r, the constant term in the expression of the corresponding Laplacian becomes a sum of terms all of which are the same as for L, except the last one, namely $D^p(R_L)$, which gets replaced by $D^p(R_{L^r}) = \mathrm{pr}\, D^p(R_L)$. Since R_L is positive definite by assumption, the constant term is positive definite for large enough r, proving the theorem.

We also obtain the following precise statement.

7.7. Corollary. *If L is a line bundle with a Hermitian structure on a compact complex manifold, such that the Chern form is negative definite, then $H^i(K \otimes L) = 0$ for all $i \geq 1$.*

Proof. Note first that the Chern form is a closed form of type $(1,1)$, and we are assuming that it is the imaginary part of a Hermitian positive definite form. In particular, the manifold is compact Kähler and so Theorem 7.6 is applicable to our situation. In the above computation we will take $p = n$. If we replace L by $K \otimes L$ in the above formula, we have $c_{K \otimes L} = -Ricci + c_L$ and the curvature term corresponding to $\det(T^*)^{0,1}$ is the Ricci form. Thus the term κ reduces to $D^q(c_L)$ which is positive definite for $q \geq 1$.

7.8. Remark. Using Serre's duality this can also be stated as the vanishing of $H^i(X, L)$ for all $i < n$, if the geometric Chern class of L can be represented by a positive definite form.

Let us illustrate how Kodaira's theorem works in the case of a compact Riemann surface C and look at the vanishing theorem in this case.

Since the degree of ω_C is $2g - 2$ by equation 2.19, the vanishing theorem reduces to the following.

7.9. Vanishing for a Riemann surface. *Let L be a line bundle of degree d on a compact Riemann surface C of genus g, with $d > 2g - 2$. Then $H^1(C, L) = 0$.*

7.10. Lichnerowicz vanishing theorem.

We will end this section by giving one more corollary of the vanishing theorem, by applying it to the case of the Dirac operator of a spin manifold. The result is the following.

7.11. Theorem (Lichnerowicz). *If the scalar curvature on an even-dimensional Spin manifold is positive at all points, then the space of harmonic spinors vanishes.*

Proof. Recall that the Dirac operator (8.6, 8.7) D is obtained by lifting the symbol $T^* \otimes Spin^+(M) \to Spin^-(M)$ induced by Clifford multiplication on the total Spin representation, with respect to the canonical lift of the Levi-Civita connection to the principal Spin bundle. We use here $-g$ as the quadratic form to construct the Clifford algebra. The Spin representation is unitary, and any element v such that $g(v) = 1$ is represented by a unitary matrix. Hence the conjugate transpose of the action of v taking $Spin^+$ to $Spin^-$ is given by $v^{-1} = -v$ taking $Spin^-$ to $Spin^+$. Thus the adjoint symbol is given by v again. So the composite symbol takes (v, w) to vw acting as an endomorphism of $Spin^+$. When symmetrised, this is the map of (v, w) to $\frac{1}{2}(vw + wv) = -\frac{1}{2}b(v, w)$. The symbol being invariant under the connection, the first order term in the composition formula is zero. Finally, we will compute the constant term. If (e_i) is an orthonormal basis for g in T_m^*, the Riemannian curvature tensor is represented by $\sum_{i<j} R_{ij} E_{ij}$ in the Lie algebra of the orthogonal group, where R_{ij} is a 2-form for each i, j. The element E_{ij} is the skew symmetric endomorphism of T^* given by the prescription $x \to b(e_i, x)e_j - b(e_j, x)e_i$. Its action on the Spin representation space is through the element $\frac{1}{2}e_i e_j$ of the Clifford algebra. (See [Ch. 7, 8.6].) Hence the constant term gives the element $\frac{1}{4}\sum_{i<j,k<l} R_{ijkl} e_i e_j e_k e_l$ in the even Clifford algebra (acting as an endomorphism of $Spin^+$ via the Clifford module structure). Using the Bianchi identity, we see that the summation

needs to be taken only over those (i, j, k, l) where the set $\{i, j, k, l\}$ consists of two elements. Since R is alternating in the first two and the last two indices, we may assume that $i = k$ and $j = l$, $i < j$. This gives the term $\frac{1}{4} \sum R_{ijij} e_i e_j e_i e_j$ in the Clifford algebra. Using the fact that $e_i^2 = e_j^2 = -1$ and $e_i e_j = -e_j e_i$, we conclude that the constant term reduces to multiplication by $1/4$ times the scalar curvature, proving the assertion.

There are many interesting generalisations of the Dirac operator, and similar vanishing theorems have been proved for them. One may consult [**12**] for a discussion of these questions. It will be instructive for the reader to derive these vanishing theorems also from the above principle.

8. The Imbedding Theorem

Kodaira's vanishing theorem has a beautiful application, namely a characterisation of those complex manifolds that can be imbedded as complex submanifolds of the complex projective space $\mathbb{CP}^N$ for some N.

8.1. Imbedding theorem (Kodaira). *If a compact complex manifold admits a closed form of type $(1, 1)$ which is positive definite and represents an integral class in H^2, then it is isomorphic to a submanifold of $\mathbb{P}^N$ for some N.*

8.2. Remark. By a positive definite form, we mean a form which is the imaginary part of a Hermitian positive definite form.

Note that the assumption ensures that the manifold with the Hermitian structure having the above $(1, 1)$-form as associated alternating form (which has been assumed to be closed) is Kähler. Secondly, by Theorem 7.5, there is a line bundle L whose (topological) Chern class is given by the closed form.

8.3. Definition. A line bundle whose (topological) Chern class can be represented by a $(1, 1)$-form which is positive definite is said to be *positive*.

Note that if L is a line bundle with a Hermitian structure such that the Chern form is negative definite, then its topological Chern class is positive.

8.4. Remark. In the case when M is a compact Riemann surface, a line bundle L is positive if and only if its degree is positive. This is only a question of checking that a real $(1, 1)$ form is positive in the sense that it is the imaginary part of a Hermitian positive definite form, if and only if its integral is positive.

Using the vanishing theorem one would like to construct an imbedding of M into some complex projective space.

8.5. Maps into the projective space.

We will first analyse what it means to give a holomorphic map of a complex manifold M into the complex projective space $\mathbb{CP}^N$. If $z_0, z_1, \ldots, z_N$ are the homogeneous coordinates in $\mathbb{P}^N$, then one might try to get functions $\varphi_0, \varphi_1, \ldots, \varphi_N$ on M such that the given map sends any $m \in M$ to the point in $\mathbb{CP}^N$ with homogeneous coordinates $\varphi_0(m), \varphi_1(m), \ldots, \varphi_N(m)$. But then the z_i are *not* functions on $\mathbb{P}^N$, nor are the φ_i. However, the z_i are holomorphic (in fact, linear) functions on $\mathbb{C}^{N+1} \setminus \{0\}$. With respect to the action of $\mathbb{C}^\times$ on $\mathbb{C}^{N+1} \setminus \{0\}$, these functions satisfy $f(v\lambda) = f(v).\lambda$. This means that these functions may be considered as sections of the line bundle associated to the Hopf bundle by the action of $\mathbb{C}^\times$ on $\mathbb{C}$ in which λ acts as multiplication by λ^{-1}. The $\mathcal{O}$-module associated to this line bundle is traditionally denoted by $\mathcal{O}(1)$. Thus z_i can be regarded as sections of $\mathcal{O}(1)$. Hence the functions φ_i we have in mind turn out to be sections of a line bundle L, namely the pull-back of $\mathcal{O}(1)$. Note that at no point of M can all these sections vanish, since they have to be homogeneous coordinates of the image point in $\mathbb{CP}^N$. To sum up, corresponding to any map $M \to \mathbb{CP}^N$, we obtain the data consisting of a line bundle L and an $(N+1)$-tuple of sections such that at every point of M at least one of the sections is nonzero.

Conversely, we can start with such data, namely a line bundle L and holomorphic sections $s_0, s_1, \ldots, s_N$. Then we can map any $m \in M$ to the point of the projective space whose homogeneous coordinates are $s_i(m)$. Note that the $s_i(m)$ belong to the fibre L_m and not to $\mathbb{C}$. But we can identify L_m with $\mathbb{C}$ as a one-dimensional vector space any way we like, in order to consider $s_0(m), s_1(m), \ldots, s_N(m)$ as homogeneous coordinates of a point. The point does not depend on this choice, since a different identification changes all the $s_i(m)$ by the *same* scalar factor, and therefore we can thus define a point in $\mathbb{P}^N$ unambiguously. Of course we have also to make the assumption here that not all the s_i vanish at any point of M. We thus have the following simple translation, into the line bundle language, of maps into $\mathbb{P}^N$.

8.6. Proposition. *There is a natural bijection between holomorphic maps of M into $\mathbb{P}^N$ and data consisting of a line bundle L and an $(N+1)$-tuple $(s_0, s_1, \ldots, s_N)$ of holomorphic sections of L with the property that for every $m \in M$ there exists some i with $s_i(m) \neq 0$. Here we identify two such data if there exists an isomorphism of the line bundles taking the sections of the one to the corresponding sections of the other.*

However, we do not just want a holomorphic map into $\mathbb{P}^N$ but an imbedding. So we investigate what we need to assume about our data in order to ensure that the associated map into $\mathbb{P}^N$ is injective. In other words, we need that if m, m' are distinct points, then the two vectors $(s_0(m), s_1(m), \ldots, s_N(m))$ and $(s_0(m'), s_1(m'), \ldots, s_N(m'))$ are linearly independent. This is equivalent to saying that there exists a linear form which vanishes on one vector but not on the other. Accordingly we have

8.7. Observation. *There is a bijection between the set of injective holomorphic maps into $\mathbb{P}^N$ and the data consisting of line bundles L and an $(N+1)$-tuple $(s_0, s_1, \ldots, s_N)$ of holomorphic sections of L such that for any distinct two points m, m', there is a linear combination of s_i's which is zero at m and nonzero at m'.*

Similarly, one can also check that the map has injective differential at m if for every nonzero $v \in T_m(M)$, there is a linear combination of s_i's which is zero at m but its differential at m is nonzero on v.

8.8. Definition. A line bundle with enough sections to imbed M into a projective space is said to be *very ample*. If a positive power of a line bundle is very ample, we say that the bundle is *ample*.

8.9. Remark. Using the vanishing theorem, we will show that if L is a positive line bundle, L^m has enough sections to imbed M in a projective space, for large m.

Firstly we note the following simple fact.

8.10. Proposition. *A line bundle is positive if it is ample.*

Proof. If a line bundle L is ample, then by definition some positive power L^k is very ample, and it is enough to show that very ample bundles are positive. In other words, we can imbed M in some $\mathbb{P}^N$ using (a basis of) sections of L and then show that the restriction of the Chern class of $\mathcal{O}(1)$ to M can be represented by a positive definite form. But then it is enough to show that the Chern class of $\mathcal{O}(1)$ itself can be represented by a positive definite form. For if such is the case, then so is the restriction of that form to M, and it represents the Chern class of the pull-back of $\mathcal{O}(1)$, namely L. Thus we only need to verify the statement for the line bundle $\mathcal{O}(1)$. This follows from the computation in [Ch. 5, 6.2].

8.11. Remark. If a compact Kähler manifold has second Betti number 1, then it admits a Kähler structure such that the Kähler class is integral. For we know that the Kähler class is nonzero, and hence the Hermitian metric can be altered by a scalar multiple so that the Kähler class is integral. Although this is a trivial statement, it has many important consequences. Any compact Riemann surface, the projective space and many varieties like Grassmannians, are indeed Kähler and have second Betti number 1. Hence they can all be imbedded in the complex projective space, thanks to the imbedding theorem.

8.12. Exercise. Let Γ be a lattice in a complex vector space V. Show that if there exists a Kähler structure on the quotient, representing an integral class, then there exists such a structure invariant under translations by elements of the quotient. Hence deduce that there exists a Hermitian positive definite form on V whose imaginary part takes integral values when restricted to the lattice.

8.13. Imbedding of Riemann surfaces.

We will first see how the imbedding theorem works in the case of a compact Riemann surface C. Let L be a line bundle of sufficiently large degree. Such a line bundle does exist. We can take any point $p \in C$ and consider the line bundle $\mathcal{O}(p)$. It has degree 1 by equation 2.20, and the kth power of this line bundle therefore has degree k. We will denote such a bundle by $\mathcal{O}(kx)$.

For any $x \in C$, consider the exact sequence

$$0 \to \mathcal{I}_x = \mathcal{O}(-x) \to \mathcal{O} \to \mathcal{O}_x \to 0.$$

Tensor this with L and assume that $d = \deg(L) > 2g - 1$. Then $H^1(C, L \otimes \mathcal{O}(-x)) = 0$ by 7.9, and hence the map $H^0(L) \to H^0(\{x\}, L) = L_x$ is surjective. In other words, for every $x \in C$ there exists a section of L which is not zero at x. If for the data of Proposition 8.6, we take $(L, s_0, \ldots, s_N)$ where the s_i form basis for H^0, we get a holomorphic map of C into $\mathbb{CP}^N$. But we can do better. Suppose x, y are two distinct points. Then consider the ideal sheaf of holomorphic functions which vanish at x and y. This sheaf is in fact $\mathcal{I}_x \otimes \mathcal{I}_y = \mathcal{O}(-x) \otimes \mathcal{O}(-y)$. We will denote this by $\mathcal{O}(-x - y)$. We then get the following exact sequence:

$$0 \to \mathcal{O}(-x - y) \to \mathcal{O} \to \mathcal{O}_{\{x,y\}} \to 0.$$

If L has degree $> 2g$, then $L \otimes \mathcal{O}(-x - y)$ has degree $> 2g - 2$ and so again $H^1(C, L \otimes \mathcal{O}(-x - y)) = 0$ for any two points $x, y \in C$. Hence the evaluation map $H^0(L) \to L_x \oplus L_y$ is surjective. This shows that we can take a basis of $H^0(L)$ and use Observation 8.7 to get an injective map of C into $\mathbb{P}^N$ where $\dim H^0(L) = N + 1$.

Finally we also need to ensure that the differential at each point is injective at all points. That is taken care of by using the exact sequence

$$0 \to \mathcal{O}(-2x) \to \mathcal{O} \to \mathcal{O}/\mathcal{I}_x^2 \to 0.$$

The same argument assures us that the map $H^0(C, L) \to \mathcal{L}_x/\mathcal{I}_x^2\mathcal{L}_x$ is surjective. (Here $\mathcal{L}_x$ denotes the stalk of the associated $\mathcal{O}$-sheaf $\mathcal{L}$ at x.) This shows that there is a section of L which vanishes at x, but not with multiplicity 2. It is easily checked that this is exactly equivalent to saying that the above holomorphic map into $\mathbb{C}\mathbb{P}^N$ has nonzero differential at x.

This proves the following

8.14. Theorem. *Any Riemann surface C can be imbedded as a submanifold of some projective space. More precisely, any line bundle L of degree $d > 2g$ is very ample, where the genus of C is g.*

8.15. Blow-up.

Let now L be a positive line bundle on a compact complex manifold M. One tries to use the vanishing theorem in order to prove the imbedding theorem on exactly the same lines as for Riemann surfaces. The first problem is that if we take any point of M, the corresponding ideal sheaf is not locally free. Only the ideal sheaves of submanifolds of codimension 1 are locally free. So we have to find a way of replacing points by submanifolds of codimension 1 (in perhaps some other manifold).

In order to execute this, we need a construction known as 'blow-up'.

8.16. Example. We will start with an example. Let V be a vector space. Consider the Hopf bundle on $\mathbb{P}(V)$. Then the associated line bundle for the action of $\mathbb{C}^\times$ on $\mathbb{C}$, given by multiplication, can be described as follows. Consider the product $V \setminus \{0\} \times \mathbb{C}$ and make $\mathbb{C}^\times$ act on it by $\lambda(v, a) = (\lambda v, \lambda^{-1}a)$. The quotient is the total space Q of this line bundle. We have a natural map of this complex manifold into $\mathbb{P}(V)$ since it is a bundle over it, but there is also a holomorphic map π of Q into V which maps (v, a) to av. Consider the open set W in Q given by $a \neq 0$. It is mapped isomorphically to the open set $V \setminus \{0\}$, the inverse being given by $v \to (v, 1)$. The entire submanifold $D = \mathbb{P}(V) \times \{0\}$, namely the zero section of this bundle, is mapped to 0 in V. We may therefore think of Q as replacing the origin in V by a whole projective space, while leaving other points undisturbed.

One can check that at any point $x = (v, 0)$ of D, the differential of the map π maps the tangent subspace $T_x(D)$ to zero of course, but it induces a map of the normal space $T_x(Q)/T_x(D)$ to the one-dimensional subspace of the tangent space $T_{\pi x}(V)$ given by v itself. This is obvious if we interpret π as the inclusion of the line bundle in the trivial vector bundle $\mathbb{P}(V) \times V$ in the obvious way, followed by the projection to V.

Now let X be any complex manifold of dimension n and x any point of X. Then we can construct a new complex manifold by removing x from X and replacing it with the complex projective space of dimension $n - 1$ as above. Take a coordinate open set U around x. Identify U with an open neighbourhood of 0 in a vector space V. Then we can make the above construction over V and take the open submanifold $U' = \pi^{-1}(U)$ of Q. As we have seen, $W = \pi^{-1}(U \setminus \{0\})$ is mapped by π isomorphically to $U \setminus \{0\}$. Hence we can take the two manifolds U' and $X \setminus \{x\}$, and use the above isomorphism $\pi : W \simeq U \setminus \{0\}$ to glue them together. The resulting complex manifold $Q_x(X)$ is obviously independent of the identifications made and defines what is called the *blow-up* of the manifold X at x.

8.17. Definition. The complex manifold $Q_x(X)$ defined above, together with the holomorphic map π of $Q_x(X)$ onto X, is called the *blow-up* of X at x. The submanifold D of $Q_x(X)$, namely $\pi^{-1}(x)$, is called the *exceptional divisor*.

The map π is an isomorphism of $Q_x(X) \setminus D$ onto $X \setminus \{x\}$. Moreover, $\pi^{-1}(\{x\})$ is actually D. The differential of the map at any point of the exceptional divisor maps the tangent space of D to zero and goes down to an injection of the normal space into the tangent space at x. This gives an isomorphism of D with the projective tangent space $PT_x(X)$. Hence one may think of the points of the exceptional divisor as tangent directions at x. Moreover, this analysis has shown that the normal bundle of the exceptional divisor is canonically isomorphic to the Hopf bundle $\mathcal{O}(-1)$ on $D = \mathbb{P}^{n-1}$. The ideal sheaf $\mathcal{I}_D$ restricts to D as the normal bundle. Hence we have

8.18. Lemma. *The ideal sheaf of the exceptional divisor in the blow-up of a point restricts to the exceptional divisor, as the line bundle $\mathcal{O}(1)$.*

Now let L be any line bundle on X and $Q(L) = \pi^*L$ its pull-back on $Q_x(X)$. Then we have

8.19. Proposition. *There exists a canonical isomorphism between $H^0(Q_x(X), \pi^*(L))$ and $H^0(X, L)$.*

Proof. In fact, we have a natural linear map of the space $H^0(X, L)$ into $H^0(Q_x(X), Q(L))$, namely pulling back by π. This map is injective since π is surjective. To prove that it is surjective, note that any section of $Q(L)$ gives a section of L over $X \setminus \{x\}$. Trivialising L in a neighbourhood U of x, we can identify its restriction to $U \setminus \{x\}$ with a holomorphic function.

The only thing to check is that this function can be extended to the whole of U. Since it came from an extension of the section to the pull-back, it is bounded in this punctured neighbourhood of x, and so it can be extended to the whole neighbourhood.

8.20. Remark. Actually we can extend *any* holomorphic function in the complement of a point in any complex manifold of dimension at least 2 to the whole manifold, thanks to a theorem of Hartogs.

Now our strategy is clear. Start with a positive line bundle L on a compact Kähler manifold. As in the example of the Riemann surface, given any point $m \in M$, we will show that there is a positive integer p and a section of L^p which does not vanish at m.

Notice that this already implies that we can arrange things so that the *same* power will do for *all* points. For if p is chosen as above depending on m, then it serves for all points in a neighbourhood of m as well. Covering the compact manifold by finitely many such open sets, we get sections s_i of L^{p_i} such that at any point $x \in M$ at least one of them does not vanish. Now take $p = \prod p_i$, and note that if s_i does not vanish at a point x, the section of L^p obtained by raising s_i to the power $\prod_{j \neq i} p_j$ does not vanish at x either.

In order to show that there is a section which is nonzero at m, we will consider the blow-up $Q_m(M)$ and the bundle $Q(L)$. Now this line bundle is not positive, since it is trivial on D. But it is well behaved otherwise. We will explain. Choose a suitable Hermitian structure on L so that (i times) the Chern form is positive definite. Then the pulled back form α is positive at all points of $Q_m(M) \setminus D$, zero on vectors tangent to D and positive definite on the normal vectors at points of D.

There is a Hermitian structure on the line bundle $\mathcal{I}_D | D = \mathcal{O}(1)$, such that its Chern form is a negative definite form β. Extending this Hermitian structure on $\mathcal{I}_D$ from D to the whole of $Q_m(M)$, we get a global form representing its Chern class. Now we can find a positive power k of $Q(L)$ such that the Chern form of its tensor product with $\mathcal{I}_D$, namely $\beta + k\alpha$, is negative definite at all points of D. Hence it is so also in a neighbourhood V of D. Let W be an open neighbourhood of D such that $\overline{W} \subset V$. Now the complement of W in $Q_m(M)$ is compact, and so we can find a power of L such that (i times) the Chern form of its tensor product with $\mathcal{I}_D$ is positive definite at all points of $Q(M) \setminus W$.

From the exact sequence

$$0 \to T(D) \to T(M)|D \to \mathrm{Nor}(D, M) \to 0$$

we deduce that K_M restricts to D as the bundle $K_D \otimes \mathcal{O}(1)$. But $K_{\mathbb{P}^{n-1}}$ is easily computed to be isomorphic to $\mathcal{O}(-n)$. Thus we see that there exists a positive power l of L such that $Q(L^l) \otimes \mathcal{I}_D \otimes K_M^{-1}$ is positive. Hence there exists $q > 0$ such that for all $p > q$, we have $H^i(Q_m(M), Q(L^p) \otimes \mathcal{I}_D) = 0$ for all $i > 0$. This implies that the map $H^0(Q(M), Q(L^p)) \to H^0(D, L^p)$ is surjective. In view of Proposition 8.19, this only means that the evaluation map of $H^0(M, L^p)$ on the fibre at m is surjective. This proves the following.

There exists $q > 0$ such that for every $p \geq q$, there is a holomorphic map of M into a suitable projective space such that the pull-back of $\mathcal{O}(1)$ is L^p.

It hardly needs to be said how to complete the rest of the proof. Let m, m' be any two points. Then use the blow-up $Q_{\{m,m'\}}(M)$ at both points. Let $D_m, D_{m'}$ be the two exceptional divisors. Exactly as before, we can find a positive integer q such that $H^i(Q_{m,m'}(M), Q(L^p) \otimes \mathcal{I}_D \otimes \mathcal{I}_{D'}) = 0$ for all $p \geq q$ and $i > 0$. This implies that the map $H^0(M, L^p) \to L_m^p \oplus L_{m'}^p$ is surjective and proves that the associated map into the projective space is injective.

Finally suppose given a nonzero vector v at m. The point in $PT_m(M)$ defined by v can be interpreted as a point $[v]$ of the exceptional divisor in $Q_m(M)$. Now blow $Q_m(M)$ up again at the point $[v]$ and argue exactly as above to complete the proof of the imbedding theorem. We leave this as an exercise to the reader, who may wish to consult [20] or [19] for a more detailed account of Kähler manifolds.

Finally, by a theorem of Chow, any submanifold of $\mathbb{P}^N$ can be defined as the set of common zeros of homogeneous *polynomials*. Thus the study of complex manifolds admitting positive line bundles, can actually be accomplished by purely algebraic machinery. This belongs to the domain of Algebraic Geometry.

Appendix

We gather here some of the facts and concepts used in the book.

1. Algebra

1.1. Tensor products.

All rings (except when explicitly stated) are assumed to have a unit element, denoted 1. All modules over the ring are supposed to be unitary.

Let A be a ring. A *balanced map* $b : L \times M \to N$ where L is a right A-module, M a left A-module and N an abelian group, is bilinear over $\mathbb{Z}$ (i.e. biadditive) and satisfies $b(la, m) = b(l, am)$ for all $a \in A, l \in L$ and $m \in M$. All balanced maps of $M \times N$ into an abelian group are classified by an abelian group $M \otimes N$ in the sense that there is a balanced map of $M \times N$ into the abelian group $M \otimes N$ such that any balanced map of $M \times N$ into any abelian group L factors to a unique homomorphism of $M \otimes N$ into L.

Let A be a commutative ring and L, M, N, modules over A. A *bilinear map* $b : L \times M \to N$ is a map whose restrictions to $\{x\} \times M$ and $L \times \{y\}$ are A-linear, for all $x \in L$ and $y \in M$. Bilinear maps from $M \times N$ are classified by the tensor product $M \otimes N$ in the sense that there is a natural bilinear map of $M \times N$ into the A-module $M \otimes N$ such that any bilinear map of $M \times N$ into any module L factors to a unique *linear map* of $M \otimes N$ into L. The r-fold tensor product $\bigotimes^r M$ is called the rth *tensor power*. The direct sum $T(M)$ of all the tensor powers (setting $\bigotimes^0 M = A$) is actually an A-algebra called the *tensor algebra* of M. The tensor algebra has also a universal property. It comes with a linear map of M into $T(M)$ such that any linear map of M into any A-algebra B extends to an algebra homomorphism of $T(M)$ into B.

1.2. Alternating and symmetric multilinear maps.

Let A be commutative. A multilinear map f of $M \times M \times \cdots \times M$ into an A-module L is said to be *alternating* if f takes the value 0 whenever two or more of the arguments are equal. A multilinear map is *symmetric* if $f(x_1, \ldots, x_r)$ remains unchanged when any two of the arguments are interchanged. Any alternating or symmetric map gives a linear map of $\bigotimes^r M \to L$ which annihilates respectively elements of the form $x_1 \otimes \cdots \otimes x \otimes \cdots \otimes x \otimes \cdots \otimes x_r$ and $x_1 \otimes \cdots \otimes x \otimes \cdots \otimes y \otimes \cdots \otimes x_r - x_1 \otimes \cdots \otimes y \otimes \cdots \otimes x \otimes \cdots \otimes x_r$.

The quotients of $\bigotimes^r M$ by the submodules generated respectively by elements of the above type are called the exterior power $\Lambda^r(M)$ and the symmetric power $S^r(M)$ respectively. The direct sum $\Lambda(M)$ of all the A-modules $\Lambda^r(M)$ (resp. $S^r(M)$) is an algebra called *the exterior algebra* (resp. *the symmetric algebra*) of M, denoted $\Lambda(M)$ (resp. $S(M)$). The symmetric algebra of M is universal in the sense that there is a natural linear inclusion of M into it such that any linear map of M into any commutative algebra B can be extended to an algebra homomorphism of $S(M)$ into B. Likewise, there is an inclusion of M in $\Lambda(M)$ such that any linear map l of M into an algebra B, satisfying $l(m)^2 = 0$ for all m, extends to a unique algebra homomorphism of $\Lambda(M)$ into B.

An alternating map of M into A is called an *alternating form*. The direct sum of all alternating forms on M forms an associate algebra. One defines the (wedge) product $\alpha \wedge \beta$ of an alternating p-form α and an alternating q-form β to be the alternating $(p+q)$-form

$$(m_1, \ldots, m_{p+q}) \mapsto \sum \epsilon_\sigma \alpha(m_{\sigma(1)}, \ldots, m_{\sigma(p)}).\beta(m_{\sigma(p+1)}, \ldots, m_{\sigma(p+q)}),$$

where the sum extends over the so-called shuffle permutations and $\epsilon(\sigma)$ is the signature of the permutation. A *shuffle* is a permutation which satisfies $\sigma(1) < \cdots < \sigma(p)$ and $\sigma(r+1) < \cdots < \sigma(p+q)$. Note that if α is a linear form, then $\alpha \wedge \alpha$ is by definition zero. Therefore the inclusion of M^* in the algebra of alternating forms gives rise to an algebra homomorphism of $\Lambda(M^*)$ into the algebra of alternating forms on M. This is easily seen to be an isomorphism.

A *polynomial function* on V of degree r is an element of $S^r(V^*)$ the direct sum of which is an algebra called *the algebra of polynomial functions* on M.

1.3. Linear and bilinear maps of vector spaces.

We will deal with a finite-dimensional vector space V over real or complex numbers.

There is a natural duality between $\Lambda^r(V)$ and $\Lambda^r(V^*)$ given by

$$(v_1, \ldots, v_r), (l_1, \ldots, l_r) \mapsto \det(l_i(v_j)).$$

There is a slightly less elegant duality between $S^r(V)$ and $S^r(V^*)$. This is given by $(v_1, \ldots, v_r), (l_1, \ldots, l_r) \mapsto \frac{1}{r!} \sum \prod l_i(v_{\sigma(j)})$, where the summation runs through all permutations of $[1, \ldots, r]$.

A linear endomorphism of a finite-dimensional vector space is said to be *semisimple* if it satisfies a separable polynomial. A semisimple endomorphism can be diagonalised over $\mathbb{C}$.

A symmetric bilinear form b on V is said to be *nondegenerate* if there is only one element $x \in V$ for which $b(x, y) = 0$ for all $y \in V$, namely 0. Given a symmetric bilinear form b on a vector space V, we have a decomposition $V = V_1 \oplus V_2$ such that $b(x, y) = 0$ for all $x \in V_1$ and $y \in V$, and the restriction of b to V_2 is nondegenerate. If b is a nondegenerate bilinear form, then there exists a basis (e_i) such that $b(e_i, e_j) = \pm \delta_{ij}$. If V is a real vector space, then the number of positive signs in this formula is independent of the basis. If there are p positive and q negative signs, we say it is of type (p, q). If b satisfies $b(x, x) > 0$ for all nonzero x, then we say it is *positive definite*.

If V is a complex vector space and $h : V \times V \to \mathbb{C}$ is an $\mathbb{R}$-bilinear form, then we say it is Hermitian if it is $\mathbb{C}$-linear in the first variable and antilinear in the second variable in the sense that $h(x, \lambda y) = \overline{\lambda} h(x, y)$ and $h(x, y) = \overline{h(y, x)}$ for all $x, y \in V$ and $\lambda \in \mathbb{C}$. If h is a complex Hermitian form which satisfies $h(x, x) > 0$ for all nonzero x, we say it is *positive definite*. Then there exists a basis (e_i) such that $h(e_i, e_j) = \delta_{ij}$.

1.4. Graded and filtered algebras.

An algebra A over a commutative ring k is $\mathbb{Z}$-*graded* (resp. $\mathbb{Z}/2$-*graded*) if it is a direct sum of k-submodules (M_i), $i \in \mathbb{Z}$ (resp. $i \in \mathbb{Z}/2$) such that $M_i.M_j \subset M_{i+j}$ for all $i, j \in \mathbb{Z}$ (resp. $i, j \in \mathbb{Z}/2$). The modules M_i are called the *graded components*. In the case of $\mathbb{Z}/2$-gradation, the two components are often written M^+ and M^-. Any $\mathbb{Z}$-gradation gives rise to a $\mathbb{Z}/2$-gradation on defining $M^+ = \sum_{i \text{ even}} M_i$ and $M^- = \sum_{i \text{ odd}} M_i$.

The exterior algebra and symmetric algebras are examples of $\mathbb{Z}$-graded algebras.

An algebra A is said to be a *filtered algebra* if it is provided with submodules $F^i A$, $i \in \mathbb{Z}$, with $F^i A \subset F^j A$ for all $i \leq j$ and $F^i A.F^j A \subset F^{i+j} A$ for all i, j. If A is filtered then the direct sum $Gr(A) = \sum F^i A / F^{i-1} A$ is a graded algebra with a naturally induced multiplication. It is called the *associated graded algebra*.

1.5. Exact sequences.

If $M' \to M$ and $M \to M''$ are two linear maps, then we say that $M' \to M \to M''$ is *exact* if the image of M' in M is the same as the kernel of the map $M \to M''$. Clearly $0 \to M' \to M$ is exact if and only if $M' \to M$ is injective. The sequence $M \to M'' \to 0$ is exact if and only if the map $M \to M''$ is surjective. A sequence of the form

1.6.
$$0 \to M' \to M \to M'' \to 0$$

is said to be *a short exact sequence* if it is exact at all the three points. This means that a) $M' \to M$ is injective, b) $M \to M''$ is surjective, and c) if we identify M' with its image in M, then M/M' is mapped isomorphically by the induced map, to M''.

An exact sequence of the type 1.6 is said to *split* if there exists a linear map $M'' \to M$ such that its composite with $M \to M''$ is the identity. This is equivalent to saying that there exists a linear map $M \to M'$ which is the identity on M'.

1.7. 5-Lemma. *Consider a commutative diagram*

$$
\begin{array}{ccccccccc}
0 & \to & M' & \to & M & \to & M'' & \to & 0 \\
 & & \downarrow & & \downarrow & & \downarrow & & \\
0 & \to & N' & \to & N & \to & N'' & \to & 0
\end{array}
$$

the horizontal sequences being short exact. If the first and the last downward arrows are isomorphisms, then so is the middle downward arrow.

1.8. Simple algebras.

In what follows all algebras are over a field k and are finite-dimensional as k-vector spaces.

An algebra A is said to be *central* if its centre, namely $\{a \in A : ab = ba$ for all $b \in A\}$, is k. A *simple algebra* is one which does not have any proper two-sided ideals. An algebra is central simple if and only if it is isomorphic as a k-algebra to the matrix algebra over a division algebra over k. It has a unique simple module over the algebraic closure $\overline{k}$ of k. The only finite-dimensional central division algebras over $\mathbb{R}$ are $\mathbb{R}$ itself and the quaternion algebra. The only division algebra over an algebraically closed field (in particular, over $\mathbb{C}$) is the field itself.

1.9. Groups.

Let G be a group. It is said to *act* on a set S if a map $G \times S \to S$ denoted $(g, s) \mapsto gs$ is given satisfying a) $g_1(g_2 s) = (g_1 g_2)s$ and b) $1.s = s$. One can check that the relation: $s \simeq t$ if there exists $g \in G$ such that $gs = t$

is an equivalence relation. The quotient is denoted S/G. The equivalence classes are called *orbits* under the action. If there is only one orbit, that is to say, for any two elements $s, t \in S$ there exists $g \in G$ such that $gs = t$, then we say that the action is *transitive*. For any $s \in S$, the subset of G defined by $\{g \in G : gs = s\}$ is a subgroup and is called the *isotropy group* at s.

We say that a group G acts on another group A if in addition the maps $a \mapsto ga$ are all automorphisms of A. If G acts on A, then we can define another group called the *semidirect product* of G by A. As a set it is simply $A \times G$, the group structure being given by

$$(a, g).(a', g') = (a.ga', gg').$$

There is a natural surjective homomorphism of the semidirect product to G, and the kernel is canonically isomorphic to A.

The *free product* $G_1 * G_2$ of two groups G_1 and G_2 is a group containing G_1 and G_2 as subgroups whose union generates it as a group and satisfies the following universal property. If G is any group and $G_1 \to G$ and $G_2 \to G$ are two homomorphisms, then there is a unique homomorphism of $G_1 * G_2$ into G, extending the given homomorphisms. Such a group exists and is unique up to isomorphism.

A subgroup of a finitely generated group of finite index, is itself finitely generated.

2. Topology

2.1. Coverings.

We use the term 'Hausdorff' for topological spaces in which any two points can be separated by disjoint open neighbourhoods. Generally speaking, when we refer to locally compact, compact or paracompact spaces, they are supposed to be Hausdorff. A *compact space* is a Hausdorff space in which every open covering admits a finite subcovering. A product of any family of compact spaces is itself compact. A *relatively compact* subset is one whose closure is compact in the induced topology. A *locally compact space* is a Hausdorff space in which every point has a relatively compact open neighbourhood. The product of any finite family of locally compact spaces is locally compact. A continuous map f of a topological space X into Y is said to be *proper* if for every topological space Z, the map $(f \times I_Z) : X \times Z \to Y \times Z$ is closed in the sense that the image of any closed set is closed. If the spaces are locally compact, it is equivalent to saying that the inverse image of any compact set in Y is closed in X.

A normal space is a Hausdorff space in which either of the following equivalent conditions is satisfied.

a) Any two disjoint closed sets C_1 and C_2 can be separated by open sets in the sense that there exist disjoint open sets U_1 and U_2 containing respectively C_1 and C_2.

b) Any two disjoint closed sets C_1 and C_2 can be separated by continuous functions in the sense that there exists a continuous function f with values in the closed interval $[0, 1]$ such that f restricts to the constant functions 0 and 1 on the two closed sets.

Let $(U_i)_{i \in I}$ be an open covering of a topological space X. Then a *refinement* consists of another open covering $(V_j)_{j \in J}$ and a map $\eta : J \to I$ such that $V_j \in U_{\eta(j)}$ for all $j \in J$. A covering (C_i) is said to be *locally finite* if every $x \in X$ has an open neighbourhood U such that $\{i \in I : U \cap C_i \neq \emptyset\}$ is finite.

A *paracompact space* is a Hausdorff space in which every open covering admits a locally finite refinement. A locally compact space which has a countable base for open sets is paracompact. Every paracompact space is normal. A *shrinking* of an open covering $(U_i), i \in I$, is another open covering $(V_i), i \in I$ such that $\overline{V}_i \subset U_i$ for all $i \in I$. Every open covering of a normal space admits a shrinking.

The *topological union* of spaces $(X_i), i \in I$, is the disjoint set union $\bigcup_{i \in I} X_i$ with the topology in which a subset U is open if and only if $U \cap X_i$ is open in X_i for all $i \in I$.

If X is a compact metric space, and (U_i) is an open covering, then there exists a positive real number l such that any set whose diameter is less than l is contained in U_i for some $i \in I$. This number is called the *Lebesgue number* of the covering.

2.2. Connectedness properties.

A *connected space* is a topological space that is not the union of two disjoint proper subsets which are both open.

A *path* γ is a continuous map of the closed interval I into X. Then $\gamma(0)$ is called its *origin* and $\gamma(1)$ its *extremity*. A *loop* is a path whose origin and extremity are the same.

A *pathwise connected space* is one in which any two points can be connected by a continuous path.

A *locally connected (resp. locally pathwise connected) space* is one in which every point admits a fundamental system of open neighbourhoods which are all connected (resp. pathwise connected) in the induced topology.

A locally connected (resp. locally pathwise connected) space is homeomorphic to a topological union of subspaces each of which is connected (resp.

pathwise connected). Such a decomposition is unique, and the subspaces are called *connected components* (resp. *pathwise connected components*).

A *homotopy* h between two continuous maps $f, g : X \to Y$ is a continuous map of $X \times [0, 1]$ into Y such that $h(x, 0) = f(x)$ and $h(x, 1) = g(x)$ for all $x \in X$.

By a *homotopy* between two paths γ_1 and γ_2 with the same origin we generally mean a continuous map $h : I \times I \to X$ such that all the paths $\gamma_t(x) = h(t, x)$ have the same origin.

A *simply connected space* is a pathwise connected space in which all loops with the same origin are homotopic. A *locally simply connected space* is a space in which every point has a fundamental system of neighbourhoods which are simply connected.

A *covering space* of X is a space Y and a continuous map $p : Y \to X$ such that every point of X admits an open neighbourhood U such that $p^{-1}(U)$ is the topological union of U_i each of which is mapped homeomorphically to U by p.

Any pathwise connected, locally simply connected space has a covering space $\varphi : Y \to X$ with Y simply connected. A group π acts continuously on Y and its action on the fibre over any point is simply transitive. The group π is called the *fundamental group* of X, and Y is called the *universal covering space* of X.

Any covering space of a locally simply connected space is obtained by taking the quotient of the universal covering space by a subgroup of π.

3. Analysis

3.1. Measures and measure spaces.

Let X be a locally compact space having a countable base for open sets. Let S be the σ-ring generated by the class of all open subsets of X. A real- or complex-valued function is *Borel measurable* if it is measurable with respect to S.

A *Borel measure* μ is a measure defined on S such that $\mu(C)$ is finite for all compact sets C.

A Borel measure μ gives rise, by means of integration, to a continuous linear functional, denoted $f \mapsto \int f d\mu$ or simply $f \mapsto \mu(f)$, on the space C_c of continuous functions with compact support. Here continuity is intended in the sense that if $\{f_n\}$ is a sequence of functions with support in a fixed compact set K and tends to zero uniformly, then $\{\mu(f_n)\}$ tends to zero. The measure μ is determined by this functional, and all continuous functionals are obtained in this way.

A measure ν is said to be *absolutely continuous* with respect to another measure μ if $\nu(E) = 0$ for every measurable set E for which $|\mu|(E) = 0$. In this case there exists a measurable function f such that $\nu = f\mu$.

If E is any Borel set of positive measure in $\mathbb{R}^n$, then there exists an open set U containing 0 such that $U \subset E - E$, where $E - E$ is the set consisting of all $x - y$, $x, y \in E$.

3.2. Hilbert spaces.

A *Hilbert space H* is a vector space provided with a positive definite Hermitian inner product, usually denoted $(x, y) \mapsto \langle x, y \rangle$, such that it is complete with respect to the metric space structure given by $d(x, y) = \|x - y\|^2 = \langle x - y, x - y \rangle$. The space of measurable functions on a measure space which are square summable (identifying functions which are equal almost everywhere) is a Hilbert space with respect to the inner product $\langle f, g \rangle = \int f\overline{g}d\mu$.

Any Hilbert space admits an *orthonormal basis*, namely a set (e_i) with $\langle e_i, e_j \rangle = \delta_{ij}$ such that any $v \in H$ is a countable linear combination of e_i. If the space admits a countable orthonormal basis it is said to be *separable*. We only deal with separable Hilbert spaces.

3.3. Equicontinuity.

A set S of functions on a metric space is said to be *equicontinuous* at a point x if for every $\epsilon > 0$, there exists $\delta > 0$ such that $|f(x) - f(y)| < \epsilon$ whenever $d(x, y) < \delta$ for all $f \in S$. The point here is that δ is independent of the function. The set is said to be *equicontinuous* if it is so, at all points.

3.4. Implicit and inverse function theorems.

Let U be a domain in $\mathbb{R}^k \times \mathbb{R}^l$ containing 0, and $f : U \to \mathbb{R}^l$ a differentiable function taking 0 to 0. If the matrix $(\frac{\partial f_i}{\partial y_j})$ is invertible at 0, then there exist an open neighbourhood V of 0 in $\mathbb{R}^k$ and a differentiable function $\varphi : V \to \mathbb{R}^l$ such that $V \times \varphi(V)$ is contained in U and $f(x, \varphi(x)) = 0$ for all $x \in V$. Any two such functions coincide in a neighbourhood of 0. This statement is also true with parameters, that is to say, if in addition f depends differentiably on a parameter in $\mathbb{R}^m$, then φ exists depending differentiably on the parameter.

If $f : U \to \mathbb{R}^n$ is a differentiable map taking 0 to 0 and such that the matrix $(\frac{\partial f_i}{\partial x_j})$ is invertible, then there exists $U' \subset U$ such that f maps U' bijectively onto an open neighbourhood of 0 and the inverse is differentiable.

3.5. Existence theorem for ordinary differential equations.

If U is a domain in $\mathbb{R}^k$ containing 0, and f is a differentiable function $I \times U \to \mathbb{R}^k$, where I is an open interval containing 0, then there exists a differentiable function $\varphi : I' \to U$, where I' is a neighbourhood of 0 contained in I, such that

$$\frac{d\varphi(x)}{dx} = f(x, \varphi(x))$$

for all $x \in I'$. The solution $\varphi(t)$ is uniquely determined in a neighbourhood of 0 by its initial value $\varphi(0)$. The same is true again with differentiable dependence on some parameter in $\mathbb{R}^l$.

Bibliography

[1] M. F. Atiyah, R. Bott, and V. Patodi, On the heat equation and the index theorem, Inv. Math. **19** (1973), 279–330.

[2] Jacques Chazarain and Alain Piriou, Introduction to the theory of partial differential equations, Studies in Mathematics and Applications, North-Holland Publishing Company, Amsterdam-New York.

[3] C. Chevalley and S. Eilenberg, Cohomology theory of Lie groups and Lie algebras, Trans. Amer. Math. Soc. **63** (1948), 85–124.

[4] W. Fulton, Intersection theory, Ergebnisse der Mathematik und ihrer Grenzgebiete, Springer, 1984.

[5] Peter B. Gilkey, Invariance theory, the heat equation and the Atiyah-Singer index theorem, Math. Lecture series, Publish or Perish Inc., Delaware.

[6] P. R. Halmos, Measure theory, D. Van Nostrand Company, New York.

[7] Noel J. Hicks, Notes on differential geometry, D. Van Nostrand Company, New York.

[8] F. Hirzebruch, Topological methods in algebraic geometry, Springer, 1966.

[9] Jürgen Jost, Riemannian geometry and geometric analysis, Universitext, Springer.

[10] Shoshichi Kobayashi and Katsumi Nomizu, Foundations of differential geometry, Vols. I and II, Interscience Publishers, New York.

[11] J.-L. Koszul, Lectures on fibre bundles and differential geometry, T.I.F.R. Lectures on Mathematics, **20**, Bombay.

[12] H. Blaine Lawson and Marie-Louise Michelsohn, Spin geometry, Princeton Mathematical Series, **38**, Princeton University Press.

[13] Raghavan Narasimhan, Lectures on topics in analysis, T.I.F.R. Lectures on Mathematics, **34**, Bombay.

[14] M. V. Nori, The Hirzebruch-Riemann-Roch theorem, Michigan Math. J. **48** (2000), 473–482.

[15] Richard Palais, Seminar on the Atiyah-Singer index theorem, Annals of Mathematics Studies, **57**, Princeton University Press, Princeton.

[16] L. Schwartz, Lectures on complex manifolds, T.I.F.R. Lectures on Mathematics, Bombay.

[17] R. W. Sharpe, Differential geometry, Graduate Texts in Mathematics, **166**, Springer.

[18] I. M. Singer, Differential geometry, Lectures at M.I.T., Cambridge, Massachusetts.

[19] A. Weil, Introduction à l'étude des variétés Kähleriennes, Publications de l'Institut de Mathématique de l'Université de Nancago, **1267**, Hermann, Paris.

[20] R. O. Wells, Differential analysis on complex manifolds, Graduate Texts in Mathematics, **65**, Springer, New York.

[21] T. J. Willmore, Riemannian geometry, Oxford Science Publications.

Index